T0189631

Progress in Mathematics
Volume 313

Series Editors
Hyman Bass, *University of Michigan, Ann Arbor, USA*
Jiang-Hua Lu, *The University of Hong Kong, Hong Kong SAR, China*
Joseph Oesterlé, *Université Pierre et Marie Curie, Paris, France*
Yuri Tschinkel, *Courant Institute of Mathematical Sciences, New York, USA*

More information about this series at http://www.springer.com/series/4848

Jaume Llibre • Rafael Ramírez

Inverse Problems in Ordinary Differential Equations and Applications

 Birkhäuser

Jaume Llibre
Departament de Matemàtiques
Universitat Autònoma de Barcelona
Barcelona, Spain

Rafael Ramírez
Departament d'Enginyeria Informàtica
Universitat Rovira i Virgili
Tarragona, Catalonia, Spain

ISSN 0743-1643   ISSN 2296-505X (electronic)
Progress in Mathematics
ISBN 978-3-319-79935-3   ISBN 978-3-319-26339-7   (eBook)
DOI 10.1007/978-3-319-26339-7

Mathematics Subject Classification (2010): 34A55, 34C07, 70F17

© Springer International Publishing Switzerland 2016
Softcover re rint of the hardcover 1st edition 2016
This work is subject to copyright. All rights are reserved by the Publisher, whether the whole or part of the material is concerned, specifically the rights of translation, reprinting, reuse of illustrations, recitation, broadcasting, reproduction on microfilms or in any other physical way, and transmission or information storage and retrieval, electronic adaptation, computer software, or by similar or dissimilar methodology now known or hereafter developed.
The use of general descriptive names, registered names, trademarks, service marks, etc. in this publication does not imply, even in the absence of a specific statement, that such names are exempt from the relevant protective laws and regulations and therefore free for general use.
The publisher, the authors and the editors are safe to assume that the advice and information in this book are believed to be true and accurate at the date of publication. Neither the publisher nor the authors or the editors give a warranty, express or implied, with respect to the material contained herein or for any errors or omissions that may have been made.

Printed on acid-free paper

This book is published under the trade name Birkhäuser.
The registered company is Springer International Publishing AG Switzerland (www.birkhauser-science.com)

*We dedicate this book
to the memory of
Professor A.S. Galiullin*

Contents

Preface

In the theory of ordinary differential equations we can distinguish two fundamental problems. The first, which we may call the *direct problem*, is, in a broad sense, to find all solutions of a given ordinary differential equation. The second, which we may call the *inverse problem* and which is the focus of this work, is to find the most general differential system that satisfies a given set of properties. For instance, we might wish to identify all differential systems in $\mathbb{R}^N$ that have a given set of invariant hypersurfaces or that admit a given set of first integrals.

Probably the first inverse problem to be explicitly formulated was the problem in celestial mechanics, stated and solved by Newton in Philosophiae Naturalis Principia Mathematica (1687), of determining the potential force field that yields planetary motions that conform to the motions that are actually observed, namely, to Kepler's laws.

In 1877 Bertrand [10] proved that the expression for Newton's force of attraction can be obtained directly from Kepler's first law. He also stated the more general problem of determining the positional force under which a particle describes a conic section for any initial conditions. Bertrand's ideas were developed in particular in the works [42, 51, 78, 149].

In the modern scientific literature the importance of this kind of inverse problem in celestial mechanics was already recognized by Szebehely, see [152].

In view of Newton's second law, that acceleration is proportional to the applied force, it is clear that the inverse problems just mentioned are equivalent to determining second-order differential equations based on prespecified properties of the right-hand side of the equations.

The first statement of an inverse problem as the problem of finding the most general first-order differential system satisfying a given set of properties was stated by Erugin [52] in dimension two and developed by Galiullin in [60, 61].

The new approach to inverse problems that we propose uses as an essential tool the Nambu bracket. We deduce new properties of this bracket which play a major role in the proof of all the results of this work and in their applications. We observe that the applications of the Nambu bracket that we will give in this book are original and represent a new direction in the development of the theory of the Nambu bracket.

In the first chapter we present results of two different kinds. First, under very general assumptions we characterize the ordinary differential equations in $\mathbb{R}^N$ that have a given set of M partial integrals, or a given set of $M < N$ first integrals, or a given set of $M \leq N$ partial and first integrals. Second, we provide necessary and sufficient conditions for a system of differential equations in $\mathbb{R}^N$ to be integrable, in the sense that the system admits $N - 1$ independent first integrals.

Because of the unknown functions that appear, the solutions of the inverse problems in ordinary differential equations that we give in the first chapter have a high degree of arbitrariness. To reduce this arbitrariness we must impose additional conditions. In the second chapter we are mainly interested in planar polynomial differential systems that have a given set of polynomial partial integrals. We discuss the problem of finding the planar polynomial differential equations whose phase portraits contain invariant algebraic curves that are either generic (in an appropriate sense), or contain invariant algebraic curves that are non-singular in $\mathbb{RP}^2$ or are nonsingular in $\mathbb{R}^2$, or that contain singular invariant algebraic curves. We study the particular case of quadratic polynomial differential systems with one singular algebraic curve of arbitrary degree.

In the third chapter we state Hilbert's 16th problem restricted to algebraic limit cycles. Consider Σ'_n, the set of all real polynomial vector fields $\mathcal{X} = (P, Q)$ of degree n having real irreducible invariant algebraic curves (where irreducibility is with respect to $\mathbb{R}[x, y]$). A simpler version of the second part of Hilbert's 16th problem restricted to algebraic limit cycles can be stated as follows: *Is there an upper bound on the number of algebraic limit cycles of any polynomial vector field of Σ'_n?* By applying the results given in the second chapter we solve this simpler version of Hilbert's 16th problem for two cases: (a) when the given invariant algebraic curves are generic (in a suitable sense), and (b) when the given invariant algebraic curves are non-singular in $\mathbb{CP}^2$. We state the following conjecture: *The maximum number of algebraic limit cycles for polynomial planar vector fields of degree n is $1 + ((n - 1)(n - 2)/2)$.* We prove this conjecture for the case where n is even and the algebraic curves are generic M-curves, and for the case that all the curves are non-singular in $\mathbb{R}^2$ and the sum of their degrees is less than $n + 1$.

We observe that Hilbert formulated his 16th problem by dividing it into two parts. The first part asks for the mutual disposition of the maximal number of ovals of an algebraic curve; the second asks for the maximal number and relative positions of the limit cycles of all planar polynomial vector fields $\mathcal{X} = (P, Q)$ of a given degree. Traditionally the first part of Hilbert's 16th problem has been studied by specialists in real algebraic geometry, while the second has been investigated by mathematicians working in ordinary differential equations. Hilbert also pointed out that connections are possible between these two parts. In the third chapter we exhibit such a connection through the Hilbert problem restricted to algebraic limit cycles.

In the fourth chapter, applying results of the first chapter *we state and solve the inverse problem for constrained Lagrangian mechanics*: for a given natural mechanical system with N degrees of freedom, determine the most general force field that depends only on the position of the system and that satisfies a given set of constraints linear in the velocity. One of the main objectives in this inverse problem is to study the behavior of constrained Lagrangian systems with constraints linear in the velocity in a way that is different from the classical approach deduced from the d'Alembert–Lagrange principle. As a consequence of the solution of the inverse problem for the constrained Lagrangian systems studied here we obtain the general solution for the inverse problem in dynamics for mechanical systems with N degrees of freedom. We also provide the answer to the generalized Dainelli inverse problem, which before was solved only for $N = 2$ by Dainelli. We give a simpler solution to Suslov's inverse problem than the one obtained by Suslov. Finally, we provide the answer to the generalized Dainelli–Joukovsky problem solved by Joukovsky in the particular case of mechanical systems with two or three degrees of freedom.

Chapter 5 is devoted to *the inverse problem for constrained Hamiltonian systems*. That is, for a given submanifold $\mathcal{M}$ of a symplectic manifold $\mathbb{M}$ we determine the differential systems whose local flow leaves the submanifold $\mathcal{M}$ invariant. We study two cases: (a) $\mathcal{M}$ is determined by l first integrals with $l \in [\dim \mathbb{M}/2, \dim \mathbb{M})$, and (b) $\mathcal{M}$ is determined by $l < \dim \mathbb{M}/2$ partial integrals. The solutions are obtained using the basic results of the first chapter. In general, the given set of first integrals is not necessarily in involution. The solution of the inverse problem in constrained Hamiltonian systems shows that in this case the differential equations having the invariant submanifold $\mathcal{M}$ are not in general Hamiltonian. The origin of the theory of noncommutative integration, dealing with Hamiltonian systems with first integrals that are not in involution, started with Nekhoroshev's Theorem.

Chapter 6 deals with the problem of *the integrability of a constrained rigid body*. We apply the results given in Chapter 4 to analyze the integrability of the motion of a rigid body around a fixed point. If the absence of constraints the integrability of this system is well known. But the integration of the equations of motion of this mechanical system with constraints is incomplete. We study two classical problems of constrained rigid bodies, the Suslov and the Veselova problems. We present new cases of integrability for these two problems which contain as particular cases the previously known results.

We also study the equations of motion of a constrained rigid body when the constraint is linear in the velocity with excluding the Lagrange multiplier. By using these equations we provide a simple proof of the well-known *theorem of Veselova* and improve *Kozlov's result* on the existence of an invariant measure. We give a new approach to solving the Suslov problem in the absence of a force field and of an invariant measure.

In Chapter 7 we give three main results:

(i) *A new point of view on transpositional relations.* In nonholonomic mechanics two points of view on transpositional relations have been maintained, one supported by Volterra, Hammel, and Hölder, and the other supported by Suslov, Voronets, and Levi-Civita. The second point of view has acquired general acceptance, while the first has been considered erroneous. We propose a third point of view, which is a generalization of the second one.

(ii) *A new generalization of the Hamiltonian principle.* There are two well-known generalizations of the Hamiltonian principle: the Hölder–Hamiltonian principle and the Suslov–Hamiltonian principle. We propose another generalization of the Hamiltonian principle, one that plays an important role in the solution of the inverse problem that we state in the next item.

(iii) *Statement and solution of the inverse problem in vakonomic mechanics.* We construct the variational equations of motion describing the behavior of constrained Lagrangian systems. Using the solution of the inverse problem in vakonomic mechanics, we present a modification of vakonomic mechanics (MVM). This modification is valid for holonomic and nonholonomic constrained Lagrangian systems. We deduce the equations of motion for nonholonomic systems with constraints that in general are nonlinear in the velocity. These equations coincide, except perhaps on a set of Lebesgue measure zero, with the classical differential equations deduced from the d'Alembert–Lagrange principle.

We observe that the solution of the inverse problem in vakonomic mechanics plays a fundamental role in the new point of view on transpositional relations and the new generalization of the Hamiltonian principle that we present.

Several aspects of our work support the following conjecture: *The existence of mechanical systems with constraints that are not linear in the velocity must be sought outside Newtonian Mechanics.*

Finally we remark that the inverse approach in ordinary differential equations which we propose in this book, based on the development of properties of the Nambu bracket, yields a unifed approach to the study of such diverse problems as finding all differential systems with given partial and first integrals, Hilbert's 16th problem, constrained Lagrangian and Hamiltonian systems, integrability of constrained rigid bodies, and vakonomic mechanics.

Chapter 1

Differential Equations with Given Partial and First Integrals

1.1 Introduction

In this chapter we present two different kind of results. First, under very general assumptions *we characterize the ordinary differential equations in $\mathbb{R}^N$ which have a given set of either $M \leq N$, or $M > N$ partial integrals, or $M < N$ first integrals, or $M \leq N$ partial and first integrals.* Second, in $\mathbb{R}^N$ we provide some results on integrability, in the sense that the characterized differential equations admit $N - 1$ independent first integrals.

The main results of this chapter are proved by using the Nambu bracket. We establish new properties of this bracket.

For simplicity, we shall assume that all the functions which appear in this book are of class $\mathcal{C}^\infty$, although most of the results remain valid under weaker hypotheses.

The results obtained in this chapter are illustrated with concrete examples.

1.2 The Nambu bracket. New properties

In the 1970s, Nambu in [119] proposed a new approach to classical dynamics based on an N-dimensional Nambu–Poisson manifold replacing the even-dimensional Poisson manifold and on $N - 1$ Hamiltonians $H_1, \ldots, H_{N-1}$ instead of a single Hamiltonian H. In the canonical Hamiltonian formulation, the equations of motion (Hamilton equations) are defined via the Poisson bracket. In Nambu's formulation, the Poisson bracket is replaced by the Nambu bracket. Nambu had originally considered the case $N = 3$.

Although the *Nambu formalism* is a generalization of the *Hamiltonian formalism*, its significant applications are not as rich as the applications of the latter.

1

Let D be an open subset of $\mathbb{R}^N$. Let $h_j = h_j(\mathbf{x})$ for $j = 1, 2, \ldots, M$ with $M \leq N$ be functions $h_j : D \to \mathbb{R}$. We define the matrix

$$S_{M,N} = \begin{pmatrix} dh_1(\partial_1) & \cdots & dh_1(\partial_N) \\ \vdots & & \vdots \\ dh_M(\partial_1) & \cdots & dh_M(\partial_N) \end{pmatrix} = \begin{pmatrix} \partial_1 h_1 & \cdots & \partial_N h_1 \\ \vdots & & \vdots \\ \partial_1 h_M & \cdots & \partial_N h_M \end{pmatrix},$$

where $\partial_j h = \dfrac{\partial h}{\partial x_j}$ and $dh = \displaystyle\sum_{j=1}^{N} \partial_j h \, dx_j$. The matrix $S_{M,N}$ is also denoted by $\dfrac{\partial(h_1, \ldots, h_M)}{\partial(x_1, \ldots, x_N)}$.

We say that the functions h_j for $j = 1, \ldots, M \leq N$ are *independent* if the rank of the matrix $S_{M,N}$ is M for all $\mathbf{x} \in D$, except perhaps in a subset of D of zero Lebesgue measure.

If $M = N$, we denote the matrix $S = S_{N,N}$. We note that S is the *Jacobian matrix* of the map $(h_1, \ldots, h_N)$. The *Jacobian* of S, i.e., the determinant of S, is denoted by

$$|S| = \left| \frac{\partial(h_1, \ldots, h_N)}{\partial(x_1, \ldots, x_N)} \right| = \begin{vmatrix} dh_1(\partial_1) & \cdots & dh_1(\partial_N) \\ \vdots & & \vdots \\ dh_N(\partial_1) & \cdots & dh_N(\partial_N) \end{vmatrix} := \{h_1, \ldots, h_N\}.$$

The last bracket thus defined is known in the literature as the *Nambu bracket* [7, 96, 119, 153].

The objective of this section is to establish a number of properties of the *Nambu bracket*, some of them new. These new properties will play an important role in some of the proofs of the main results of this book.

The Nambu bracket $\{h_1, \ldots, h_N\}$ has the following known properties.

(i) It is a *skew-symmetric* bracket, i.e.,

$$\{h_{\sigma(1)}, \ldots, h_{\sigma(N)}\} = (-1)^{|\sigma|} \{h_1, \ldots, h_N\}$$

for arbitrary functions $h_1, \ldots, h_N$ and arbitrary permutations σ of $(1, \ldots, N)$. Here $|\sigma|$ is the order of σ.

(ii) It is a derivation, i.e., satisfies the *Leibniz rule*

$$\{h_1, \ldots, fg\} = \{h_1, \ldots, f\}g + f\{h_1, \ldots, g\}.$$

(iii) It satisfies the *fundamental identity* (Filippov Identity)

$$\begin{aligned}
F\,(f_1 &\ldots, f_{N-1}, g_1, \ldots, g_N) \\
&:= \{f_1 \ldots, f_{N-1}, \{g_1 \ldots, g_N\}\} \\
&- \sum_{n=1}^{N} \{g_1, \ldots, g_{n-1}, \{f_1 \ldots, f_{N-1}, g_n\}, g_{n+1}, \ldots, g_N\} = 0,
\end{aligned} \tag{1.1}$$

where $f_1, \ldots, f_{N-1}, g_1, \ldots, g_N$ are arbitrary functions. For more details see [96, 57, 119, 153]. (i) follows directly from the properties of determinants. (ii) is obtained using the properties of the derivative plus the properties of the determinants. (iii) will be the property (ix) with $\lambda = 1$, and we shall prove it in Proposition 1.2.2.

The properties listed above of the Nambu bracket are not sufficient for solving some of the problems which will arise in this book. The new properties that we give below will play a fundamental role in the proofs of some of the theorems and in the applications of the results in this book.

We emphasize that the applications of the Nambu bracket that we will give are original and represent a new direction developing Nambu's ideas.

We shall need the following results.

Proposition 1.2.1. *The following four identities hold.*

(iv) $\displaystyle\sum_{j=1}^{N} \frac{\partial f}{\partial x_j} \{g_1, \ldots, g_{n-1}, x_j, g_{n+1}, \ldots, g_N\}$
$$= \{g_1, \ldots, g_{n-1}, f, g_{n+1} \ldots, g_N\}.$$

(v) $\displaystyle\frac{\partial f}{\partial x_n} = \{x_1, \ldots, x_{n-1}, f, x_{n+1}, \ldots, x_N\}.$

(vi) $\displaystyle K_n^N := \sum_{j=1}^{N} \frac{\partial}{\partial x_j} \{g_1, \ldots, g_{n-1}, x_j, g_{n+1} \ldots, g_N\} = 0$, *for* $n = 1, 2, \ldots, N$.

(vii) $\displaystyle\frac{\partial f_1}{\partial x_N} \left| \frac{\partial(G, f_2, \ldots, f_N)}{\partial(y_1, \ldots, y_N)} \right| + \cdots + \frac{\partial f_N}{\partial x_N} \left| \frac{\partial(f_1, \ldots, f_{N-1}, G)}{\partial(y_1, \ldots, y_N)} \right|$
$$= \frac{\partial G}{\partial y_1} \left| \frac{\partial(f_1, \ldots, f_N)}{\partial(x_N, y_2, \ldots, y_N)} \right| + \cdots + \frac{\partial G}{\partial y_N} \left| \frac{\partial(f_1, \ldots, f_N)}{\partial(y_1, \ldots, y_{N-1}, x_N)} \right|.$$

Here the functions $g_1, \ldots, g_N, f_1, \ldots, f_N, G,$ *and* f *are arbitrary.*

Proof. The proof of (iv) reads

$$\{g_1, \ldots, g_{n-1}, f, g_{n+1} \ldots, g_N\} = \begin{vmatrix} \partial_1 g_1 & \cdots & \partial_N g_1 \\ \vdots & & \vdots \\ \partial_1 g_{n-1} & \cdots & \partial_N g_{n-1} \\ \partial_1 f & \cdots & \partial_N f \\ \partial_1 g_{n+1} & \cdots & \partial_N g_{n+1} \\ \vdots & & \vdots \\ \partial_1 g_N & \cdots & \partial_N g_N \end{vmatrix}$$

$$= \partial_1 f \begin{vmatrix} \partial_1 g_1 & \partial_2 g_1 & \cdots & \partial_N g_1 \\ \vdots & \vdots & & \vdots \\ \partial_1 g_{n-1} & \partial_2 g_{n-1} & \cdots & \partial_N g_{n-1} \\ 1 & 0 & \cdots & 0 \\ \partial_1 g_{n+1} & \partial_2 g_{n+1} & \cdots & \partial_N g_{n+1} \\ \vdots & \vdots & & \vdots \\ \partial_1 g_N & \partial_2 g_N & \cdots & \partial_N g_N \end{vmatrix} + \cdots$$

$$+ \partial_N f \begin{vmatrix} \partial_1 g_1 & \cdots & \partial_{N-1} g_1 & \partial_N g_1 \\ \vdots & \vdots & \vdots & \vdots \\ \partial_1 g_{n-1} & \cdots & \partial_{N-1} g_{n-1} & \partial_N g_{n-1} \\ 0 & \cdots & 0 & 1 \\ \partial_1 g_{n+1} & \cdots & \partial_{N-1} g_{n+1} & \partial_N g_{n+1} \\ \vdots & \vdots & \vdots & \vdots \\ \partial_1 g_N & \cdots & \partial_{N-1} g_N & \partial_N g_N \end{vmatrix}$$

$$= \{g_1, \ldots, g_{n-1}, x_1, g_{n+1}, \ldots, g_N\} \partial_1 f + \cdots$$
$$+ \{g_1, \ldots, g_{n-1}, x_N, g_{n+1}, \ldots, g_N\} \partial_N f.$$

The proof of (v) follows easily from the definition of the Nambu bracket.

The proof of (vi) is done by induction. Without loss of generality we shall prove that

$$K_1^N = \sum_{j=1}^{N} \frac{\partial}{\partial x_j} \{x_j, g_2, \ldots, g_N\} = 0. \tag{1.2}$$

For $N = 2$ we have

$$K_1^2 = \sum_{j=1}^{2} \frac{\partial}{\partial x_j} \{x_j, g_2\} = \frac{\partial}{\partial x_1} \left(\frac{\partial g_2}{\partial x_2} \right) - \frac{\partial}{\partial x_2} \left(\frac{\partial g_2}{\partial x_1} \right) = 0.$$

Now suppose that

$$K_1^{N-1} = \sum_{j=1}^{N-1} \frac{\partial}{\partial x_j} \{x_j, g_2, \ldots, g_{N-1}\} = 0.$$

We shall prove (1.2). Indeed, since

$$\{x_j, g_2, \ldots, g_N\} = \sum_{k=2}^{N} (-1)^{N+k+1} \frac{\partial g_k}{\partial x_N} \{x_j, g_2, \ldots, g_{k-1}, g_{k+1}, \ldots, g_N\}$$
$$\text{for} \quad j = 1, \ldots, N-1,$$
$$\{x_N, g_2, \ldots, g_N\} = (-1)^{N+1} \{g_2, \ldots, g_N\},$$

we deduce that

$$K_1^N = \sum_{j=1}^{N} \frac{\partial}{\partial x_j} \{x_j, g_2, \ldots, g_N\}$$

$$= \sum_{j=1}^{N-1} \frac{\partial}{\partial x_j} \left(\sum_{k=2}^{N} (-1)^{N+k+1} \frac{\partial g_k}{\partial x_N} \{x_j, g_2, \ldots, g_{k-1}, g_{k+1}, \ldots, g_N\} \right)$$

$$+ \frac{\partial}{\partial x_N} \{x_N, g_2, \ldots, g_N\}$$

$$= \sum_{k=2}^{N} (-1)^{N+k+1} \sum_{j=1}^{N-1} \frac{\partial}{\partial x_j} \left(\frac{\partial g_k}{\partial x_N} \right) \{x_j, g_2, \ldots, g_{k-1}, g_{k+1}, \ldots, g_N\}$$

$$+ \sum_{k=2}^{N} (-1)^{N+k+1} \frac{\partial g_k}{\partial x_N} \sum_{j=1}^{N-1} \frac{\partial}{\partial x_j} (\{x_j, g_2, \ldots, g_{k-1}, g_{k+1}, \ldots, g_N\})$$

$$+ (-1)^{N+1} \frac{\partial}{\partial x_N} \{g_2, \ldots, g_N\}$$

$$= \sum_{k=2}^{N} (-1)^{N+k+1} \sum_{j=1}^{N-1} \frac{\partial}{\partial x_j} \left(\frac{\partial g_k}{\partial x_N} \right) \{x_j, g_2, \ldots, g_{k-1}, g_{k+1}, \ldots, g_N\}$$

$$+ (-1)^{N+1} \frac{\partial}{\partial x_N} \{g_2, \ldots, g_N\}.$$

Here we used the inductive assumption that $K_1^{N-1} = 0$ with the functions $g_2, \ldots,$ $g_{k-1}, g_{k+1}, \ldots, g_N$ instead of $g_2, \ldots, g_{k-1}, g_k, g_{k+1}, \ldots, g_{N-1}$.

In view of the property (iv) we obtain that

$$\sum_{k=2}^{N} (-1)^{N+k+1} \sum_{j=1}^{N-1} \frac{\partial}{\partial x_j} \left(\frac{\partial g_k}{\partial x_N} \right) \{x_j, g_2, \ldots, g_{k-1}, g_{k+1}, \ldots, g_N\}$$

$$= \sum_{k=2}^{N} (-1)^{N} \left\{ (-1)^{k+1} \frac{\partial g_k}{\partial x_N}, g_2, \ldots, g_{k-1}, g_{k+1}, \ldots, g_N \right\}$$

$$= \sum_{k=2}^{N} (-1)^{N} \left\{ g_2, \ldots, g_{k-1}, \frac{\partial g_k}{\partial x_N}, g_{k+1}, \ldots, g_N \right\}$$

$$= (-1)^{N} \frac{\partial}{\partial x_N} \{g_2, \ldots, g_N\}.$$

Hence,

$$K_1^N = \sum_{j=1}^{N} \frac{\partial}{\partial x_j} \{x_j, g_2, \ldots, g_N\}$$

$$= (-1)^{N} \frac{\partial}{\partial x_N} \{g_2, \ldots, g_N\} + (-1)^{N+1} \frac{\partial}{\partial x_N} \{g_2, \ldots, g_N\} = 0,$$

and consequently the property (vi) is proved.

The proof of (vii) is easy to obtain by observing that the value of determinant

$$
\begin{vmatrix}
\dfrac{\partial f_1}{\partial y_1} & \cdots & \dfrac{\partial f_1}{\partial y_N} & \dfrac{\partial f_1}{\partial x_N} \\[2mm]
\vdots & \cdots & \vdots & \vdots \\[2mm]
\dfrac{\partial f_N}{\partial y_1} & \cdots & \dfrac{\partial f_N}{\partial y_N} & \dfrac{\partial f_N}{\partial x_N} \\[2mm]
\dfrac{\partial G}{\partial y_1} & \cdots & \dfrac{\partial G}{\partial y_N} & 0
\end{vmatrix}
$$

can be obtained by expanding by the last row and by the last column.   □

Proposition 1.2.2. *We define*

$$
\begin{aligned}
\Omega\,&(f_1 \ldots, f_{N-1}, g_1 \ldots, g_N, G) \\
&:= -\{f_1 \ldots, f_{N-1}, G\}\{g_1 \ldots, g_N\} \\
&\quad + \sum_{n=1}^{N} \{f_1, \ldots, f_{N-1}, g_n\}\{g_1, \ldots, g_{n-1}, G, g_{n+1}, \ldots, g_N\},
\end{aligned}
$$

and

$$
\begin{aligned}
F_\lambda\,&(f_1 \ldots, f_{N-1}, g_1, \ldots, g_N) \\
&:= -\{f_1 \ldots, f_{N-1}, \lambda\{g_1 \ldots, g_N\}\} \\
&\quad + \sum_{n=1}^{N} \{g_1, \ldots, g_{n-1}, \lambda\{f_1 \ldots, f_{N-1}, g_n\}, g_{n+1}, \ldots, g_N\},
\end{aligned}
$$

for arbitrary functions $f_1, \ldots, f_{N-1}, G, g_1, \ldots, g_N, \lambda$.
  Then the Nambu bracket satisfies the identities:

(viii) $\Omega\,(f_1 \ldots, f_{N-1}, g_1 \ldots, g_N, G) = 0$, *and*

(ix) $F_\lambda\,(f_1 \ldots, f_{N-1}, g_1, \ldots, g_N) = 0$. *Note this identity is a generalization of the fundamental identity (1.1), which is obtained when $\lambda = 1$.*

Proof. Indeed, $\Omega := \Omega\,(f_1 \ldots, f_{N-1}, g_1 \ldots, g_N, G)$ can be written as

$$
\Omega = -
\begin{vmatrix}
dg_1(\partial_1) & \cdots & dg_1(\partial_N) & \{f_1, \ldots, f_{N-1}, g_1\} \\
\vdots & \cdots & \vdots & \vdots \\
dg_N(\partial_1) & \cdots & dg_N(\partial_N) & \{f_1, \ldots, f_{N-1}, g_N\} \\
dG(\partial_1) & \cdots & dG(\partial_N) & \{f_1, \ldots, f_{N-1}, G\}
\end{vmatrix},
$$

and using (iv) we obtain

$$
\Omega = -
\begin{vmatrix}
dg_1(\partial_1) & \cdots & dg_1(\partial_N) & \sum_{j=1}^{N}\{f_1,\ldots,f_{N-1},x_j\}dg_1(\partial_j) \\[2mm]
\vdots & \cdots & \vdots & \vdots \\[2mm]
dg_N(\partial_1) & \cdots & dg_N(\partial_N) & \sum_{j=1}^{N}\{f_1,\ldots,f_{N-1},x_j\}dg_N(\partial_j) \\[2mm]
dG(\partial_1) & \cdots & dG(\partial_N) & \sum_{j=1}^{N}\{f_1,\ldots,f_{N-1},x_j\}dG(\partial_j)
\end{vmatrix}
$$

$$
= -\sum_{j=1}^{N}\{f_1,\ldots,f_{N-1},x_j\}
\begin{vmatrix}
dg_1(\partial_1) & \cdots & dg_1(\partial_N) & dg_1(\partial_j) \\
\vdots & \cdots & \vdots & \vdots \\
dg_N(\partial_1) & \cdots & dg_N(\partial_N) & dg_N(\partial_j) \\
dG(\partial_1) & \cdots & dG(\partial_N) & dG(\partial_j)
\end{vmatrix}
= 0.
$$

This proves (viii).

The proof of (ix) is as follows. Taking $G = x_j$ in the identity of (viii) and multiplying it by λ we obtain

$$
\begin{aligned}
\lambda\Omega\,&(f_1,\ldots,f_{N-1},g_1,\ldots,g_N,x_j)\\
&:= \lambda\{f_1,\ldots,f_{N-1},g_1\}\{x_j,g_2,\ldots,g_N\}\\
&\quad+\cdots+\lambda\{f_1,\ldots,f_{N-1},g_N\}\{g_1,\ldots,g_{N-1},x_j\}\\
&\quad+\cdots-\lambda\{f_1,\ldots,f_{N-1},x_j\}\{g_1,\ldots,g_N\} = 0.
\end{aligned}
$$

Using (vi), the last expression yields

$$
\begin{aligned}
0 = &\sum_{j=1}^{N}\frac{\partial}{\partial x_j}\left(\lambda\Omega(f_1,\ldots,f_{N-1},g_1,\ldots,g_N,x_j)\right)\\
= &\sum_{j=1}^{N}\{x_j,g_2,\ldots,g_N\}\frac{\partial}{\partial x_j}\left(\lambda\{f_1,\ldots,f_{N-1},g_1\}\right)+\cdots\\
&+\sum_{j=1}^{N}\{g_1,\ldots,g_{N-1},x_j\}\frac{\partial}{\partial x_j}\left(\lambda\{f_1,\ldots,f_{N-1},g_N\}\right)\\
&-\sum_{j=1}^{N}\{f_1,\ldots,f_{N-1},x_j\}\frac{\partial}{\partial x_j}\left(\lambda\{g_1,\ldots,g_N\}\right).
\end{aligned}
$$

Now using (iv) the previous expression becomes

$$\{\lambda\{f_1, \ldots, f_{N-1}, g_1\}, g_2, \ldots, g_N\}$$
$$+ \cdots + \{g_1, g_2, \ldots, g_{N-1}, \lambda\{f_1, \ldots, f_{N-1}, g_N\}\}$$
$$+ \cdots - \{f_1, \ldots, f_{N-1}, \lambda\{g_1, \ldots, g_N\}\}$$
$$= F_\lambda\,(f_1, \ldots, f_{N-1}, g_1, \ldots, g_N) = 0.$$

This complete the proof of (ix). $\qquad\qquad\qquad\qquad\qquad\qquad\qquad\qquad\square$

The identity (viii) was proved for the first time in $[142]$.

Remark 1.2.3. We note that (ix) was obtained from (viii), so in some sense (viii) is more basic. In fact, from the proof of (ix) we obtain

$$F_\lambda\,(f_1, \ldots, f_{N-1}, g_1, \ldots, g_N) = \sum_{j=1}^{N} \frac{\partial}{\partial x_j}\,(\lambda\Omega\,(f_1, \ldots, f_{N-1}, g_1, \ldots, g_N, x_j))\,.$$

Next we establish the relationship between the Nambu bracket and the classical Poisson bracket. We suppose that $N = 2n$, and $x_j = x_j$ and $x_{j+n} = y_j$ for $j = 1, \ldots, n$. The Poisson bracket $\{H, F\}^*$ of the functions H and F is defined as

$$\{H, F\}^* := \sum_{j=1}^{n} \left(\frac{\partial H}{\partial x_j} \frac{\partial F}{\partial y_j} - \frac{\partial F}{\partial x_j} \frac{\partial H}{\partial y_j} \right).$$

Proposition 1.2.4. *The Poisson bracket and the Nambu bracket satisfy the following two equalities for arbitrary functions* H, f, G, f_1, $\ldots$, f_{2n}

(x) $\displaystyle\sum_{j=1}^{n}\{x_1 \ldots, x_{j-1}, H, x_{j+1}, \ldots, x_{n+j-1}, f, x_{n+j+1}, \ldots, x_{2n}\} = \{H, f\}^*,$

(xi) $\displaystyle\sum_{j=1}^{2n}\{H, f_j\}^*\{f_1, \ldots, f_{j-1}, G, f_{j+1}, \ldots, f_{2n}\} = \{H, G\}^*\{f_1, \ldots, f_{2n}\}.$

Proof. The identity (x) is obtained by using the property (iv). Indeed,

$$\sum_{j=1}^{n}\{x_1 \ldots, x_{j-1}, H, x_{j+1} \ldots, x_{n+j-1}, f, x_{n+j+1}, \ldots, x_{2n}\}$$
$$= \sum_{j=1}^{n}\sum_{k=1}^{2n} \frac{\partial H}{\partial x_k}\{x_1 \ldots, x_{j-1}, x_k, x_{j+1} \ldots, x_{n+j-1}, f, x_{n+j+1}, \ldots, x_{2n}\}$$
$$= \sum_{j=1}^{n}\sum_{k,m=1}^{2n} \frac{\partial H}{\partial x_k} \frac{\partial f}{\partial x_m}\{x_1 \ldots, x_{j-1}, x_k, x_{j+1}, \ldots, x_{n+j-1}, x_m, x_{n+j+1}, \ldots, x_{2n}\};$$

here we consider that $x_0 = x_1$. Since

$$\{x_1, \ldots, x_{j-1}, x_k, x_{j+1}, \ldots$$
$$\ldots, x_{n+j-1}, x_m, x_{n+j}, \ldots, x_{2n}\} = \begin{cases} 1, & \text{if } k = j, \ m = n + j, \\ -1, & \text{if } k = n + j, \ m = j, \\ 0, & \text{otherwise,} \end{cases}$$

we obtain

$$\sum_{j=1}^{n} \{x_1 \ldots, x_{j-1}, H, x_{j+1}, \ldots, x_n, x_{n+1}, \ldots, x_{n+j-1}, f, x_{n+j+1}, \ldots, x_{2n}\}$$

$$= \sum_{j=1}^{n} \left(\frac{\partial H}{\partial x_j} \frac{\partial f}{\partial y_j} - \frac{\partial f}{\partial x_j} \frac{\partial H}{\partial y_j} \right) = \{H, f\}^*,$$

which proves (x).

Now we prove (xi). We have

$$\sum_{k=1}^{2n} \{H, f_k\}^* \{f_1, \ldots, f_{k-1}, G, f_{k+1}, \ldots, f_{2n}\}$$

$$= \sum_{k=1}^{2n} \left(\sum_{j=1}^{n} \{x_1 \ldots, x_{j-1}, H, x_{j+1}, \ldots, x_n, x_{n+1}, \ldots, x_{n+j-1}, f_k, x_{n+j+1}, \ldots, x_{2n}\} \right.$$

$$\left. \cdot \{f_1, \ldots, f_{k-1}, G, f_{k+1}, \ldots, f_{2n}\} \right),$$

where we have used (x). Therefore,

$$\sum_{j=1}^{n} \left(\sum_{k=1}^{2n} \{x_1 \ldots, x_{j-1}, H, x_{j+1}, \ldots, x_{n+j-1}, f_k, x_{n+j+1}, \ldots, x_{2n}\} \right.$$

$$\left. \cdot \{f_1, \ldots, f_{k-1}, G, f_{k+1}, \ldots, f_{2n}\} \right)$$

$$= \sum_{j=1}^{n} \{x_1 \ldots, x_{j-1}, H, x_{j+1}, \ldots, x_{n+j-1}, G, x_{n+j+1}, \ldots, x_{2n}\} \{f_1, \ldots, f_{2n}\}$$

$$= \{H, G\}^* \{f_1, \ldots, f_{2n}\}.$$

where in the first equality we have used (viii), and in the second equality we have used (x). $\qquad \square$

The properties established above will play an important role in the proofs of the main results. We note that the equalities (viii)–(xi) are new.

1.3 Ordinary differential equations in $\mathbb{R}^N$ with $M \leq N$ partial integrals

Let D be an open subset of $\mathbb{R}^N$. By definition, an autonomous differential system is a system of the form

$$\dot{\mathbf{x}} = \mathbf{X}(\mathbf{x}), \quad \mathbf{x} \in D, \tag{1.3}$$

where the dependent variables $\mathbf{x} = (x_1, \ldots, x_N)$ are real, the independent variable (the 'time' t) is real, and the $\mathbb{R}^n$-valued $\mathcal{C}^1$ function $\mathbf{X}(\mathbf{x}) = (X_1(\mathbf{x}), dots, X_N(\mathbf{x}))$ is defined in D.

The $\mathcal{C}^1$ function $g : D \to \mathbb{R}$ and the set $\{\mathbf{x} \in D : g = g(\mathbf{x}) = 0\}$ are called a *partial integral* and an *invariant hypersurface* of the vector field $\mathbf{X}$, respectively, if $\mathbf{X}(g)|_{g=0} = 0$.

In this section we construct the most general autonomous differential system in $D \subset \mathbb{R}^N$ having a given set of *partial integrals* g_j, $j = 1, 2, \ldots, M$, with $M \leq N$.

Our first result characterizes the differential systems (1.3) having a given set of M partial integrals with $M \leq N$.

Theorem 1.3.1. *Let $g_j = g_j(\mathbf{x})$ for $j = 1, 2, \ldots, M$ with $M \leq N$ be a given set of independent functions defined in an open set $D \subset \mathbb{R}^N$. Then any differential system defined in D which admits the set of partial integrals g_j for $j = 1, 2, \ldots, M$ can be written as*

$$
\begin{aligned}
\dot{x}_j = \sum_{k=1}^{M} \Phi_k \frac{\{g_1, \ldots, g_{k-1}, x_j, g_{k+1}, \ldots, g_N\}}{\{g_1, g_2, \ldots, g_N\}} \\
+ \sum_{k=M+1}^{N} \lambda_k \frac{\{g_1, \ldots, g_{k-1}, x_j, g_{k+1}, \ldots, g_N\}}{\{g_1, g_2, \ldots, g_N\}} = \mathbf{X}(x_j),
\end{aligned}
\tag{1.4}
$$

where $g_{M+j} = g_{M+j}(\mathbf{x})$ for $j = 1, \ldots, N - M$, are arbitrary functions defined in D which we choose in such a way that the Jacobian

$$|S| = \{g_1, \ldots, g_N\} \not\equiv 0, \tag{1.5}$$

in the set D and $\Phi_j = \Phi_j(\mathbf{x})$, for $j = 1, 2, \ldots, M$ and $\lambda_{M+k} = \lambda_{M+k}(\mathbf{x})$ for $k = 1, 2, \ldots N - M$ are arbitrary functions such that

$$\mu_j|_{g_j=0} = 0 \quad for \quad j = 1, \ldots, M_1. \tag{1.6}$$

Remark 1.3.2. The vector field $\mathbf{X}$ generated by the differential system (1.4) by choosing

$$
\begin{aligned}
\Phi_j &= \{g_1, \ldots, g_N\}\mu_j, \quad \text{with} \quad \mu_j|_{g_j=0} = 0 \quad for \quad j = 1, \ldots, M, \\
\lambda_{M+k} &= \{g_1, \ldots, g_N\}\nu_{M+k} \quad\quad\quad\quad\quad for \quad k = 1, \ldots, N - M,
\end{aligned}
$$

can be rewritten as

$$
\dot{x}_j = \sum_{k=1}^{M} \mu_k \{g_1, \ldots, g_{k-1}, x_j, g_{k+1}, \ldots, g_N\}
$$

$$
+ \sum_{k=M+1}^{N} \nu_k \{g_1, \ldots, g_{k-1}, x_j, g_{k+1}, \ldots, g_N\},
$$

(1.7)

where $\mu_1, \ldots \mu_M$ and $\nu_1, \ldots, \nu_M$ are arbitrary functions defined in D such that

$$
\mu_j|_{g_j=0} = 0 \quad \text{for} \quad j = 1, \ldots, M \leq N.
$$

Proof of Theorem 1.3.1. We consider the vector field

$$
\mathbf{X} = -\frac{1}{\{g_1, \ldots, g_N\}}
\begin{vmatrix}
dg_1(\partial_1) & \ldots & dg_1(\partial_N) & \Phi_1 \\
dg_2(\partial_1) & \ldots & dg_2(\partial_N) & \Phi_2 \\
\vdots & \vdots & \vdots & \vdots \\
dg_M(\partial_1) & \ldots & dg_M(\partial_N) & \Phi_M \\
dg_{M+1}(\partial_1) & \ldots & dg_{M+1}(\partial_N) & \lambda_{M+1} \\
\vdots & \vdots & \vdots & \vdots \\
dg_N(\partial_1) & \ldots & dg_N(\partial_N) & \lambda_N \\
\partial_1 & \ldots & \partial_N & 0
\end{vmatrix}
$$

(1.8)

$$
= \sum_{k,j=1}^{N} \frac{S_{jk} P_j}{|S|} \partial_k = \langle S^{-1}\mathbf{P}, \partial_{\mathbf{x}} \rangle,
$$

where $S \not\equiv 0$, S_{jk} for $k, j = 1, \ldots, N$ is the determinant of the adjoint of the matrix S after removing the row j and the column k (see (1.5)), S^{-1} is the inverse matrix of S, and $\mathbf{P} = (P_1, \ldots, P_N)^T = (\Phi_1, \ldots, \Phi_M, \lambda_{M+1}, \ldots, \lambda_N)^T$.

From (1.8) by expanding the determinant with respect to the last column and denoting by $\{g_1, \ldots, g_{k-1}, *, g_{k+1}, \ldots, g_N\}$ the vector field

$$
\begin{vmatrix}
\partial_1 g_1 & \ldots & \partial_N g_1 \\
\vdots & & \vdots \\
\partial_1 g_{k-1} & \ldots & \partial_N g_{k-1} \\
\partial_1 * & \ldots & \partial_N * \\
\partial_1 g_{k+1} & \ldots & \partial_N g_{k+1} \\
\vdots & & \vdots \\
\partial_1 g_N & \ldots & \partial_N g_N
\end{vmatrix},
$$

we obtain for the vector field $\mathbf{X}$ the equivalent representation

$$
\begin{aligned}
\mathbf{X}(*) = {} & \Phi_1 \frac{\{*, g_2, \ldots, g_N\}}{\{g_1, g_2, \ldots, g_N\}} + \cdots \\
& + \Phi_M \frac{\{g_1, \ldots, g_{M-1}, *, g_{M+1}, \ldots, g_N\}}{\{g_1, \ldots, g_{M-1}, g_M, g_{M+1}, \ldots, g_N\}} \\
& + \lambda_{M+1} \frac{\{g_1, \ldots, g_M, *, g_{M+2}, \ldots, g_N\}}{\{g_1, \ldots, g_M, g_{M+1}, g_{M+2}, \ldots, g_N\}} + \cdots \\
& + \lambda_N \frac{\{g_1, \ldots, g_{N-1}, *\}}{\{g_1, \ldots, g_{N-1}, g_N\}}.
\end{aligned}
\tag{1.9}
$$

By using this representation it is easy to obtain the relationship

$$
\begin{aligned}
\mathbf{X}(g_j) = {} & \Phi_1 \frac{\{g_j, g_2, \ldots, g_N\}}{\{g_1, g_2, \ldots, g_N\}} + \cdots + \Phi_M \frac{\{g_1, \ldots, g_{M-1}, g_j, g_{M+1}, \ldots, g_N\}}{\{g_1, \ldots, g_{M-1}, g_M, g_{M+1}, \ldots, g_N\}} \\
& + \lambda_{M+1} \frac{\{g_1, \ldots, g_M, g_j, g_{M+2}, \ldots, g_N\}}{\{g_1, \ldots, g_M, g_{M+1}, g_{M+2}, \ldots, g_N\}} + \cdots + \lambda_N \frac{\{g_1, \ldots, g_{N-1}, g_j\}}{\{g_1, \ldots, g_{N-1}, g_N\}} \\
= {} & \begin{cases} \Phi_j & \text{for} \quad 1 \le j \le M \\ \lambda_j & \text{for} \quad M+1 \le j \le N. \end{cases}
\end{aligned}
$$

Thus

$$
\mathbf{X}(g_j) = \Phi_j, \quad \mathbf{X}(g_{M+k}) = \lambda_{M+k},
\tag{1.10}
$$

for $j = 1, 2, \ldots, M$ and $k = 1, \ldots, N - M$. In view of the assumption $\Phi_j|_{g_j=0} = 0$ we obtain that the sets $\{g_j = 0\}$ for $j = 1, 2, \ldots, M$ are *invariant hypersurfaces* of the vector field $\mathbf{X}$. The vector field $\mathbf{X}$ was already used in [134, 139].

Now we shall prove that system (1.4) is the most general differential system that admits the given set of independent partial integrals. Indeed, let

$$
\dot{\mathbf{x}} = \tilde{\mathbf{X}}(\mathbf{x}) = (\tilde{X}_1(\mathbf{x}), \ldots, \tilde{X}_N(\mathbf{x}))
$$

be another differential system having $g_1, g_2, \ldots, g_M$ as partial integrals, that is $\tilde{\mathbf{X}}(g_j)|_{g_j=0} = 0$ for $j = 1, 2, \ldots, M$. Then taking

$$
\Phi_j = \tilde{\mathbf{X}}(g_j) = \sum_{l=1}^{N} \tilde{X}_l \partial_l g_j = \sum_{l=1}^{N} \tilde{X}_l \{x_1, \ldots, x_{l-1}, g_j, x_{l+1}, \ldots, x_N\},
$$

for $j = 1, 2, \ldots, M$, and

$$
\lambda_{M+k} = \tilde{\mathbf{X}}(g_{M+k}) = \sum_{l=1}^{N} \tilde{X}_l \partial_l g_{M+k} = \sum_{l=1}^{N} \tilde{X}_l \{x_1, \ldots, x_{l-1}, g_{M+k}, x_{l+1}, \ldots, x_N\}
$$

for $k = 1, \ldots, N - M$ (here we use the identity (v)) and substituting Φ_j and λ_{M+k} into formula (1.12) we get for arbitrary functions F that

$$\mathbf{X}(F) = \sum_{l=1}^{N} \Phi_j \frac{\{g_1, \ldots, g_{j-1}, F, g_{j+1} \ldots, g_M, \ldots, g_N\}}{\{g_1, g_2, \ldots, g_N\}}$$

$$+ \sum_{j=M+1}^{N} \lambda_{M+j} \frac{\{g_1, \ldots, g_M, g_{M+1}, \ldots, g_{j-1}, F, g_{j+1} \ldots, g_N\}}{\{g_1, g_2, \ldots, g_N\}}$$

$$= \sum_{j=1}^{N} \sum_{l=1}^{N} \tilde{X}_l \{x_1, \ldots, x_{l-1}, g_j, x_{l+1}, \ldots, x_N\} \frac{\{g_1, \ldots, g_{j-1}, F, g_{j+1}, \ldots, g_N\}}{\{g_1, g_2, \ldots, g_N\}}$$

$$= \sum_{l=1}^{N} \tilde{X}_l \sum_{j=1}^{N} \{x_1, \ldots, x_{l-1}, g_j, x_{l+1}, \ldots, x_N\} \frac{\{g_1, \ldots, g_{j-1}, F, g_{j+1}, \ldots, g_N\}}{\{g_1, g_2, \ldots, g_N\}}$$

$$= \sum_{l=1}^{N} \tilde{X}_l \{x_1, \ldots, x_{l-1}, F, x_{l+1}, \ldots, x_N\} = \tilde{\mathbf{X}}(F).$$

Here we have used the identities (iv) and (ix). Hence, in view of arbitrariness of F the theorem has been proved. $\qquad\square$

An immediate consequence of Theorem 1.3.1 is

Corollary 1.3.3. *Under the assumptions of Theorem 1.3.1, if $M = 1$ and $N = 2$, then the system (1.4), (1.7) takes the form*

$$\dot{x}_1 = \Phi_1 \frac{\{x_1, g_2\}}{\{g_1, g_2\}} + \lambda_2 \frac{\{g_1, x_1\}}{\{g_1, g_2\}}, \quad \dot{x}_2 = \Phi_1 \frac{\{x_2, g_2\}}{\{g_1, g_2\}} + \lambda_2 \frac{\{g_1, x_2\}}{\{g_1, g_2\}}, \quad (1.11)$$

where g_2, Φ_1, and λ_2 are arbitrary functions such that

$$\{g_1, g_2\} \not\equiv 0 \quad in \quad D, \quad \Phi_1|_{g_1=0} = 0.$$

Moreover, if we take $\Phi_1 = \mu_1 \{g_1, g_2\}$ and $\lambda_2 = \lambda \{g_1, g_2\}$, then

$$\dot{x}_1 = \mu_1 \{x_1, g_2\} + \lambda \{g_1, x_1\}, \quad \dot{x}_2 = \mu_1 \{x_2, g_2\} + \lambda \{g_1, x_2\},$$

respectively, where g_2, λ, and μ_1 are arbitrary functions such $\mu_1|_{g_1=0} = 0$.

Corollary 1.3.4. *Under the assumptions of Theorem 1.3.1, if $M = N$, then the system (1.4), (1.7) takes the form*

$$\dot{x}_j = \Phi_1 \frac{\{x_j, g_2, \ldots, g_{N-1}, g_N\}}{\{g_1, g_2, \ldots, g_{N-1}, g_N\}} + \cdots + \Phi_N \frac{\{g_1, g_2, \ldots, g_{N-1}, x_j\}}{\{g_1, g_2, \ldots, g_{N-1}, g_N\}}, \quad (1.12)$$

for $j = 1, 2, \ldots, N$, where $\Phi_1, \ldots, \Phi_N$ are arbitrary functions such that if

$$\{g_1, \ldots, g_N\} \not\equiv 0 \quad in \quad D, \quad then \quad \Phi_j|_{g_j=0} = 0 \quad for \quad j = 1, \ldots, N.$$

Moreover, if we choose

$$\Phi_j = \{g_1, \ldots, g_N\} \mu_j, \quad \text{with} \quad \mu_j|_{g_j=0} = 0 \quad \text{for} \quad j = 1, \ldots, N,$$

then from (1.12) *we get the differential system*

$$\dot{x}_j = \mu_1\{x_j, g_2, \ldots, g_{N-1}, g_N\} + \cdots + \mu_N\{g_1, g_2, \ldots, g_{N-1}, x_j\}, \tag{1.13}$$

$j = 1, 2, \ldots, N.$

In particular, if $M = N = 2$, then the differential system (1.12), (1.13) takes the form

$$\dot{x}_j = \Phi_1 \frac{\{x_j, g_2\}}{\{g_1, g_2\}} + \Phi_2 \frac{\{g_1, x_j\}}{\{g_1, g_2\}}, \quad j = 1, 2, \tag{1.14}$$

where Φ_1, Φ_2 *are arbitrary functions such that if*

$$\{g_1, g_2\} \not\equiv 0 \quad \text{in} \quad D, \quad \text{then} \quad \Phi_j|_{g_j=0} = 0 \quad \text{for } j = 1, 2,$$

and if $\Phi_j = \{g_1, g_2\} \mu_j$, *with* $\{g_1, g_2\} \mu_j|_{g_j=0} = 0$ *for* $j = 1, 2$, *then from* (1.14) *we get the differential system*

$$\dot{x}_j = \mu_1\{x_j, g_2\} + \mu_2\{g_1, x_j\}, \quad j = 1, 2. \tag{1.15}$$

Example 1.3.5. Consider the case when the given partial integrals are

$$g_j = x_j, \quad j = 1, \ldots, M.$$

We choose the arbitrary functions $g_k = x_k$ for $k = M + 1, \ldots, N$. Under this election we have that $\{g_1, \ldots, g_N\} = \{x_1, \ldots, x_N\} = 1$. It follows that the vector field $\mathbf{X}$ is well defined in $\mathbb{R}^N$ and the system (1.4) reads

$$\dot{x}_j = \Phi_j, \quad j = 1, \ldots, M,$$
$$\dot{x}_k = \lambda_k, \quad k = M + 1, \ldots, N,$$

where $\Phi_1, \ldots, \Phi_N$ are arbitrary functions such that $\Phi_j|_{x_j=0} = 0$.

Example 1.3.6. Suppose the given partial integrals are

$$g_1 = \left(\prod_{j=1}^{N} x_j \right) - 1, \quad g_j = x_j, \quad \text{for} \quad j = 2, \ldots, N.$$

Since $\{g_1, \ldots, g_N\} = \prod_{j=2}^{N} x_j$, the determinant $|S|$ vanishes in the set $\{x_j = 0$ for $j = 2, \ldots, N\}$. By choosing for the arbitrary functions $\Phi_1, \ldots, \Phi_N$

$$\Phi_j = \mu_j \prod_{k=2}^{N} x_k, \quad \text{with} \quad \mu_j|_{g_j=0} = 0,$$

we see that the differential system (1.13) takes the form

$$\dot{x}_1 = \mu_1 + \sum_{j=2}^{N}(-1)^{j+1}\mu_j \prod_{k \neq j}^{N} x_k,$$

$$\dot{x}_j = \mu_j \prod_{k=2}^{N} x_k \quad \text{for} \quad j = 2, \ldots, N.$$

Example 1.3.7. We consider the case when $N = M = 2$ and the given invariant curves are the circles

$$g_1 = x^2 + y^2 - R^2 = 0, \quad g_2 = (x - a)^2 + y^2 - r^2 = 0,$$

where μ_1 and μ_2 are arbitrary functions such that $\mu_j|_{g_j=0} = 0$, $j = 1, 2$. In this case

$$|S| = \{g_1, g_2\} = 4ay.$$

The differential system (1.15) for the given partial integrals is

$$\dot{x} = (2\mu_1 - 2\mu_2)y, \quad \dot{y} = -(2\mu_1 - 2\mu_2)x + 2a\mu_1, \tag{1.16}$$

where μ_1 and μ_2 are arbitrary functions such that $\mu_j|_{g_j=0} = 0$. This is the most general planar system with two given invariant circles. We study the case when these arbitrary functions are given by

$$2\mu_2 = (x + y + \gamma_1)((x - a)^2 + y^2 - r^2), \quad 2\mu_1 = (x + y + \gamma_2)(x^2 + y^2 - R^2),$$

where $\gamma_1 \neq \gamma_2$ are constants.

We consider two cases: $R = r$ and $R > |a| + r$.

For the first case if we choose

$$2\mu_1 = (x + y)(x^2 + y^2 - r^2), \quad 2\mu_2 = (x + y - a)((x - a)^2 + y^2 - r^2),$$

then the system (1.16) takes the form

$$\dot{x} = ay(a^2 - r^2 - 3ax + 3x^2 - ay + 2xy + y^2),$$
$$\dot{y} = -a^3x + 3a^2x^2 - 2ax^3 - ar^2y + a^2xy - ax^2y + ay^3.$$

This is a cubic differential system with two invariant circles

$$g_1 = x^2 + y^2 - r^2 = 0, \quad g_2 = (x - a)^2 + y^2 - r^2 = 0.$$

In particular, for $a = 3$ and $r = 1$, the phase portrait is shown in Figure 1.1. For more details and the definition on the Poincaré disc, see Chapter 5 of [49, 9].

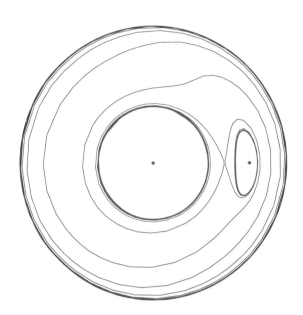

Figure 1.1: Phase portrait with two non-nested invariant circles which are limit cycles.

For the second case, if we choose

$$a = -1,\ r = 1,\ \gamma_1 = -\gamma_2 = 5, \quad \text{and} \quad R = 3,$$

then

$$2\mu_2 = (x + y + 5)((x - 1)^2 + y^2 - 1),$$
$$2\mu_1 = (x + y - 5)(x^2 + y^2 - 9).$$

Then the differential system (1.16) takes the form

$$
\begin{aligned}
\dot{x} &= y(12x^2 + 10y^2 + 2xy + 19x + 9y - 45), \\
\dot{y} &= -11x^3 + y^3 - 9xy^2 - x^2y - 24x^2 - 9xy - 5y^2 + 36x - 9y + 45,
\end{aligned}
\tag{1.17}
$$

and is a cubic differential system with two invariant circles

$$
\begin{aligned}
g_1 &= x^2 + y^2 - 9 = 0, \\
g_2 &= (x + 1)^2 + y^2 - 1 = 0.
\end{aligned}
$$

(see, for instance, [139]). System (1.17) has the phase portrait shown in Figure 1.2 in the Poincaré disc.

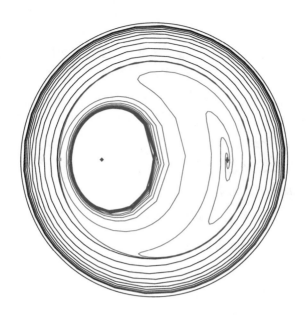

Figure 1.2: Phase portrait with two nested invariant
circles which are limit cycles.

1.4 Differential equations with partial and first integrals

In this section we construct the most general autonomous differential system in
$D \subset \mathbb{R}^N$ having a given set of partial integrals g_j for $j = 1, 2, \ldots, M_1$, and a given
set of first integrals f_k for $k = 1, 2, \ldots, M_2$, with $M = M_1 + M_2 \leq N$.

A non-locally constant function $H = H(\mathbf{x})$ defined in an open subset D_1 of
D such that its closure coincides with D is called a *first integral*, if it is constant
on the solutions of system (1.3) contained in D_1, i.e., $\mathbf{X}(H)|_{D_1} = 0$.

Our second main result characterizes the differential systems (1.3) having a
given set of M_1 partial integrals and M_2 first integrals with $1 \leq M_2 < N$ and
$M_1 + M_2 \leq N$.

Theorem 1.4.1. *Let $g_l = g_l(\mathbf{x})$ for $l = 1, 2, \ldots, M_1$ and $f_k = f_k(\mathbf{x})$ for $k = 1, 2, \ldots, M_2 < N$ with $M_1 + M_2 = M \leq N$, be independent functions defined in the open set $D \subset \mathbb{R}^N$. Then the most general differential systems in D which admit the partial integrals g_l for $j = 1, \ldots, M_1$ and the first integrals f_k for $k = 1, \ldots, M_2$*

are

$$\dot{x}_j = \sum_{k=1}^{M_1} \Phi_k \frac{\{g_1, \ldots, g_{k-1}, x_j, g_{k+1}, \ldots, g_{M_1}, f_1, \ldots, f_{M_2}, g_{M+1} \cdots g_N\}}{\{g_1, \ldots g_{M_1}, f_1, \ldots, f_{M_2}, g_{M+1}, \ldots, g_N\}} \tag{1.18}$$

$$+ \sum_{k=M+1}^{N} \lambda_k \frac{\{g_1, \ldots, g_{M_1}, f_1, \ldots, f_{M_2}, g_{M+1}, \ldots, g_{k-1}, x_j, g_{k+1}, \ldots, g_N\}}{\{g_1, \ldots, g_{M_1}, f_1, \ldots, f_{M_2}, g_{M+1}, \ldots, g_N\}},$$

$j = 1, 2, \ldots, N$, *where* g_{M+j} *for* $j = 1, \ldots, N - M$ *are arbitrary functions satisfying that* $|S| = \{g_1, \ldots g_{M_1}, f_1, \ldots, f_{M_2}, g_{M+1}, \ldots, g_N\} \not\equiv 0$ *in the set* D; *and* $\Phi_j = \Phi_j(\mathbf{x})$, *for* $j = 1, 2, \ldots, M_1$ *and* $\lambda_{M+k} = \lambda_k(\mathbf{x})$ *for* $k = M + 1, 2, \ldots, N$ *are arbitrary functions such that*

$$\Phi_j|_{g_j=0} = 0, \quad \text{for} \quad j = 1, \ldots, M, \tag{1.19}$$

Proof. Let $\mathbf{X}$ be the vector field

$$\mathbf{X} = -\frac{1}{|S|} \begin{vmatrix} dg_1(\partial_1) & \cdots & dg_1(\partial_N) & \Phi_1 \\ \vdots & \vdots & \vdots & \\ dg_{M_1}(\partial_1) & \cdots & dg_{M_1}(\partial_N) & \Phi_{M_1} \\ df_1(\partial_1) & \cdots & df_1(\partial_N) & 0 \\ \vdots & \vdots & \vdots & \vdots \\ df_{M_2}(\partial_1) & \cdots & df_{M_2}(\partial_N) & 0 \\ dg_{M+1}(\partial_1) & \cdots & dg_{M+1}(\partial_N) & \lambda_{M+1} \\ \vdots & \vdots & \vdots & \vdots \\ dg_N(\partial_1) & \cdots & dg_N(\partial_N) & \lambda_N \\ \partial_1 & \cdots & \partial_N & 0 \end{vmatrix} = \langle S^{-1}\mathbf{P}, \partial_\mathbf{x} \rangle,$$

where $\mathbf{P} = (P_1, \ldots, P_N)^T = (\Phi_1, \ldots, \Phi_{M_1}, 0, \ldots, 0, \lambda_{M+1}, \ldots, \lambda_N)^T$, which is the vector field associated to the differential system (1.18). Clearly this vector field is well defined in view of the assumptions.

Since $\mathbf{X}(g_j) = \Phi_j$, $\Phi|_{g_j=0} = 0$, for $j = 1, \ldots, M_1$, g_j are partial integrals of $\mathbf{X}$, and since $\mathbf{X}(f_j) = 0$ for $j = 1, \ldots, M_2$, f_j are first integrals of $\mathbf{X}$.

Now we prove that system (1.18) is the most general differential system admitting the partial integrals g_j and the first integrals f_k. Indeed, let $\dot{x} = \tilde{\mathbf{X}}(\mathbf{x})$ be another differential system which admits g_j for $j = 1, \ldots, M_1$ as partial integrals and f_k for $k = 1, \ldots, M_2$ as first integrals with $M_1 + M_2 \leq N$, i.e., $\tilde{\mathbf{X}}(g_j)|_{g_j=0} = 0$ for $j = 1, \ldots, M_1$ and $\tilde{\mathbf{X}}(f_k) = 0$ for $k = 1, \ldots, M_2$. Then taking $\Phi_j = \tilde{\mathbf{X}}(g_j)$ for $j = 1, \ldots, M_1$ and $\lambda_{M+k} = \tilde{\mathbf{X}}(g_k)$ for $k = M + 1, \ldots, N$ and arguing as in the proof of Theorem 1.3.1 we deduce that the vector field $\tilde{\mathbf{X}}$ is a particular case of the vector field $\mathbf{X}$. Thus the theorem is proved. $\qquad\square$

The next two results which follow easily from the proof of Theorem 1.4.1.

Corollary 1.4.2. *Under the assumptions of Theorem 1.4.1 but without partial integrals, i.e., if $M_1 = 0$, and $M_2 = M < N$, then the most general differential system in D which admit the first integrals f_k for $k = 1, \ldots, M_2$ is*

$$\dot{x}_j = \sum_{k=M+1}^{N} \lambda_k \frac{\{f_1, \ldots, f_M, g_{M+1}, \ldots, g_{k-1}, x_j, g_{k+1}, \ldots, g_N\}}{\{f_1, \ldots, f_{M_2}, g_{M+1}, g_{M+2}, \ldots, g_N\}}, \qquad (1.20)$$

$j = 1, 2, \ldots, N$.

Corollary 1.4.3. *Under the assumptions of Theorem 1.4.1 and if $M_1 + M_2 = M = N$, then the differential system (1.18) takes the form*

$$\dot{x}_j = \sum_{k=1}^{M_1} \Phi_k \frac{\{g_1, \ldots, g_{k-1}, x_j, g_{k+1}, \ldots, g_{M_1}, f_1, \ldots, f_{M_2}\}}{\{g_1, g_2, \ldots g_{M_1}, f_1, \ldots, f_{M_2}\}}, \qquad (1.21)$$

$j = 1, 2, \ldots, N$.

Corollary 1.4.4. *Under the assumptions of Theorem 1.4.1 the following statements hold.*

(a) *If $M_2 = N - 1$ and $M_1 = 1$, then the differential system (1.21) takes the form*

$$\dot{x}_j = \Phi_N \frac{\{f_1, \ldots, f_{N-1}, x_j\}}{\{f_1, \ldots, f_{N-1}, g_N\}}, \quad with \quad \Phi_N|_{g_N=0} = 0.$$

(b) *If $M_2 = N - 1$ and $M_1 = 0$, then the differential system (1.20) takes the form*

$$\dot{x}_j = \lambda_N \frac{\{f_1, \ldots, f_{N-1}, x_j\}}{\{f_1, \ldots, f_{N-1}, g_N\}},$$

where λ_N and g_N are arbitrary functions such that $\{f_1, \ldots, f_{N-1}, g_N\} \neq 0$.

Consequently, the vector field $\mathbf{X}$ in both cases admits the representation

$$\mathbf{X}(*) = \mu\{f_1, \ldots, f_{N-1}, *\}, \qquad (1.22)$$

where μ is an arbitrary function.

In order to reduce the arbitrariness in the equations constructed, one can introduce additional conditions. We illustrate this remark in the following particular case.

Now we provide a new proof of the classical result which states that a differential system in an open subset of $\mathbb{R}^N$ having $N - 2$ first integrals and with zero divergence is *integrable by quadratures*. This result goes back to Jacobi and Whittaker; for more details on this result, see the book [68].

Theorem 1.4.5. *Under the assumptions of Corollary 1.4.2, for $M_2 = N - 2$ the differential equations* (1.20) *take the form*

$$\dot{x}_j = X_j(\mathbf{x})$$
$$= \lambda_{N-2}\frac{\{f_1,\ldots,f_{N-2},x_j,g_N\}}{\{f_1,\ldots,f_{N-2},g_{N-1},g_N\}} + \lambda_N \frac{\{f_1,\ldots,f_{N-2},g_{N-1},x_j\}}{\{f_1,\ldots,f_{N-2},g_{N-1},g_N\}}, \quad (1.23)$$

$j = 1, 2, \ldots, N$. *Moreover, if the divergence of the system* (1.23) *is zero, then its solutions can be computed by quadratures.*

Remark 1.4.6. We observe that the divergence of the vector field (1.23) is

$$\sum_{j=1}^{N} \frac{\partial}{\partial x_j} \left(\lambda_{N-2}\frac{\{f_1,\ldots,f_{N-2},x_j,g_N\}}{\{f_1,\ldots,f_{N-2},g_{N-1},g_N\}} + \lambda_N \frac{\{f_1,\ldots,f_{N-2},g_{N-1},x_j\}}{\{f_1,\ldots,f_{N-2},g_{N-1},g_N\}} \right),$$

which in view of the identities (iv) and (v) for the Nambu bracket finally yields the relation

$$\mathrm{div}\,(h(\mathbf{x})X(\mathbf{x})) = \{f_1,\ldots,f_{N-2},\tau\lambda_{N-1},g_N\} + \{f_1,\ldots,f_{N-2},g_{N-1},\tau\lambda_N\},$$

for an arbitrary function h and $\tau = \dfrac{h}{\{f_1,\ldots,f_{N-2},g_{N-1},g_N\}}$. Consequently, the condition on the divergence allows to reduce the arbitrariness of the functions λ_{N-1}, λ_N, τ, and g_N. Indeed, in this case these functions must be such that

$$\{f_1,\ldots,f_{N-2},\tau\lambda_{N-1},g_N\} + \{f_1,\ldots,f_{N-2},g_{N-1},\tau\lambda_N\} = 0 \quad (1.24)$$

Proof of Theorem 1.4.5. In view of Corollary 1.4.2 the system must have the form (1.23) or, equivalently,

$$\dot{\mathbf{x}} = S^{-1}\mathbf{P}, \quad (1.25)$$

where $\mathbf{P} = (0,\ldots,0,\lambda_{N-1},\lambda_N)^T$ (see for more details the proof of Theorem 1.3.1). Hence Corollary 1.4.2 gives the most general differential system which admits first integrals f_j for $j = 1,\ldots,N-2$. After the change of variables $(x_1,\ldots,x_N) \rightarrow (y_1,\ldots,y_N)$, where $y_j = f_j$ for $j = 1,\ldots,N-2$, and $y_{N-1} = x_{N-1}$, $y_N = x_N$, we obtain that the differential system (1.25) on the set

$$E_c = \{(y_1,\ldots y_N) \in \mathbb{R}^N : y_1 = c_1,\ldots,y_{N-2} = c_{N-2}\}$$

becomes $\dot{\mathbf{x}} = B^{-1}\dot{\mathbf{y}} = B^{-1}\hat{S}^{-1}\tilde{\mathbf{P}}$, where $\hat{S}$ and B are defined by

$$S = \frac{\partial\,(f_1,\ldots,f_{N-2},\,g_{N-1},\,g_N)}{\partial\,(x_1,\ldots,x_N)}$$
$$= \frac{\partial\,(f_1,\ldots,f_{N-2},\,g_{N-1},\,g_N)}{\partial(y_1,\ldots,y_N)}\frac{\partial\,(y_1,\ldots,y_N)}{\partial\,(x_1,\ldots,x_N)} = \hat{S}B,$$
$$\dot{x}_j = \sum_{k=1}^{N} \left(\frac{\partial x_j}{\partial y_k} \right) \dot{y}_k,$$

for a function z, $\tilde{z}$ denotes the function $z(x_1, \ldots, x_N)$ expressed in the variables $\mathbf{y} = (y_1, \ldots, y_N)$.

Clearly we have that $\hat{S}$ is equal to

$$
\begin{pmatrix}
1 & 0 & \cdots & 0 & 0 & 0 \\
0 & 1 & \cdots & 0 & 0 & 0 \\
\vdots & \vdots & \vdots & \vdots & \vdots & \vdots \\
0 & 0 & \cdots & 1 & 0 & 0 \\
dg_{N-1}(\partial_1) & dg_{N-1}(\partial_2) & \cdots & dg_{N-1}(\partial_{N-2}) & dg_{N-1}(\partial_{N-1}) & dg_{N-1}(\partial_N) \\
dg_N(\partial_1) & dg_N(\partial_2) & \cdots & dg_N(\partial_{N-2}) & dg_N(\partial_{N-1}) & dg_N(\partial_N)
\end{pmatrix},
$$

where $\partial_j = \dfrac{\partial}{\partial y_j}$, and consequently

$$
|\hat{S}| = \begin{vmatrix} dg_{N-1}(\partial_{N-1}) & dg_{N-1}(\partial_N) \\ dg_N(\partial_{N-1}) & dg_N(\partial_N) \end{vmatrix} = \frac{\partial g_{N-1}}{\partial y_{N-1}} \frac{\partial g_N}{\partial y_N} - \frac{\partial g_{N-1}}{\partial y_N} \frac{\partial g_N}{\partial y_{N-1}} = \{g_{N-1}, g_N\}.
$$

After a change of variables $x_j = x_j(y_1, \ldots, y_N)$, $j = 1, \ldots, N$, the system $\dot{x}_j = X_j(\mathbf{x})$ can be rewritten as $\dot{\mathbf{y}} = \mathbf{Y}(\mathbf{y})$. A computation shows that

$$
\mathbf{Y} = \left\langle \hat{S}^{-1} \tilde{\mathbf{P}}, \partial_{\mathbf{y}} \right\rangle
$$

$$
= \frac{-1}{|\hat{\Upsilon}|} \begin{vmatrix}
1 & \cdots & 0 & 0 & 0 & 0 \\
0 & \cdots & 0 & 0 & 0 & 0 \\
\vdots & \cdots & \vdots & \vdots & \vdots & \vdots \\
0 & \cdots & 1 & 0 & 0 & 0 \\
dg_{N-1}(\partial_1) & \cdots & dg_{N-1}(\partial_{N-2}) & dg_{N-1}(\partial_{N-1}) & dg_{N-1}(\partial_N) & \lambda_{N-1} \\
dg_N(\partial_1) & \cdots & dg_N(\partial_{N-2}) & dg_N(\partial_{N-1}) & dg_N(\partial_N) & \lambda_N \\
\partial_1 & \cdots & \partial_{N-2} & \partial_{N-1} & \partial_N & 0
\end{vmatrix}
$$

$$
= \frac{-1}{|\hat{S}|} \begin{vmatrix}
dg_{N-1}(\partial_{N-1}) & dg_{N-1}(\partial_N) & \lambda_{N-1} \\
dg_N(\partial_{N-1}) & dg_N(\partial_N) & \lambda_N \\
\partial_{N-1} & \partial_N & 0
\end{vmatrix}.
$$

Thus

$$
\dot{y}_{N-1} = \mathbf{Y}(y_{N-1}) = \tilde{\lambda}_{N-1} \frac{\{y_{N-1}, \tilde{g}_N, \}}{\{\tilde{g}_{N-1}, \tilde{g}_N\}} + \tilde{\lambda}_N \frac{\{\tilde{g}_{N-1}, y_{N-1}, \}}{\{\tilde{g}_{N-1}, \tilde{g}_N\}}
$$

$$
= Y_{N-1}(\mathbf{y}),
$$

$$
\dot{y}_N = \mathbf{Y}(y_N) = \tilde{\lambda}_{N-1} \frac{\{y_N, \tilde{g}_N\}}{\{\tilde{g}_{N-1}, \tilde{g}_N\}} + \tilde{\lambda}_N \frac{\{\tilde{g}_{N-1}, y_N\}}{\{\tilde{g}_{N-1}, \tilde{g}_N\}} \tag{1.26}
$$

$$
= Y_N(\mathbf{y}),
$$

$$
\dot{y}_j = \mathbf{Y}(y_j) = 0, \quad j = 1, \ldots, N - 2.
$$

On the other hand, from (1.24) and the relation

$$\sum_{j=1}^{N} \frac{\partial X_j}{\partial x_j} = \frac{1}{K} \sum_{m=1}^{N} \frac{\partial \left(K Y_m\right)}{\partial y_m},$$

where $K = |S|$ (see (1.5)) is the Jacobian, we obtain the well-known relation

$$\sum_{j=1}^{N} \frac{\partial (h\, X_j)}{\partial x_j} = \frac{1}{K} \sum_{j=1}^{N} \frac{\partial \left(\tilde{h} K Y_j\right)}{\partial y_j} = \frac{1}{K} \left(\frac{\partial \left(\tilde{h} K Y_{N-1}\right)}{\partial y_{N-1}} + \frac{\partial \left(\tilde{h} K Y_N\right)}{\partial y_N} \right) = 0.$$

Consequently the function $K\tilde{h}$ is an integrating factor of (1.26). Thus the theorem is proved. □

1.5   Differential equations and integrability

In what follows we present new results on the integrability of systems (1.4), (1.18), (1.20), and (1.21).

We say that system (1.3) is *integrable* if it admits $N-1$ independent first integrals.

Theorem 1.5.1. *Under the assumptions of Theorem 1.3.1, the differential system (1.4) is integrable if and only if*

$$\Phi_l = \mu\{F_1, \ldots, F_{N-1}, g_l\}, \qquad \lambda_k = \mu\{F_1, \ldots, F_{N-1}, g_k\}$$

for $l = 1, \ldots, M$ and $k = M+1, \ldots, N$, where μ, $F_1, \ldots, F_{N-1}$ are suitable functions such that $F_1, \ldots, F_{N-1}$ are independent in D *and*

$$\mu\{F_1, \ldots, F_{N-1}, g_l\}|_{g_l=0} = 0.$$

Proof. Assume that the vector field $\mathbf{X}$ associated to the differential system (1.4) is integrable, i.e., admits $N-1$ independent first integrals $F_1, \ldots, F_{N-1}$. Without loss of the generality we suppose that

$$\begin{vmatrix} \partial_1 F_1 & \cdots & \partial_{N-1} F_1 \\ \vdots & & \vdots \\ \vdots & & \vdots \\ \partial_1 F_{N-1} & \cdots & \partial_{N-1} F_{N-1} \end{vmatrix} = \{F_1, \ldots, F_{N-1}, x_N\} \not\equiv 0.$$

Thus from the equations $\mathbf{X}(F_k) = \displaystyle\sum_{j=1}^{N} \frac{\partial F_k}{\partial x_j} X_j(\mathbf{x}) = 0$ for $k = 1, \ldots, N-1$ or, equivalently,

$$\sum_{j=1}^{N-1} \frac{\partial F_k}{\partial x_j} X_j(\mathbf{x}) = -X_N(\mathbf{x}) \frac{\partial F_k}{\partial x_N},$$

we obtain

$$X_j(\mathbf{x}) = X_N(\mathbf{x})\frac{\{F_1, \ldots, F_{j-1}, x_j, F_{j+1}, \ldots, F_{N-1}\}}{\{F_1, \ldots, F_{N-1}, x_N\}}.$$

Taking $X_N(\mathbf{x}) = \mu\{F_1, \ldots, F_{N-1}, x_N\}$, we obtain the representation

$$\mathbf{X}(*) = \mu\{F_1, \ldots, F_{N-1}, *\},$$

where μ is an arbitrary function. Thus

$$\mathbf{X}(g_l) = \Phi_l = \mu\{F_1, \ldots, F_{N-1}, g_l\}, \quad \mathbf{X}(g_k) = \lambda_k = \mu\{F_1, \ldots, F_{N-1}, g_k\},$$

for $l = 1, \ldots, M$ and $k = M + 1, \ldots, N$. So the "only if" part of the theorem follows. Now let us prove the "if" part.

We suppose that $\Phi_l = \mu\{F_1, \ldots, F_{N-1}, g_l\}$ and $\lambda_k = \mu\{F_1, \ldots, F_{N-1}, g_k\}$. Thus the vector field associated to the differential system (1.4) takes the form

$$\begin{aligned}
\mathbf{X}(x_j) &= \sum_{n=1}^{M} \Phi_n \frac{\{g_1, \ldots, g_{n-1}, x_j, g_{n+1}, \ldots, g_M, \ldots, g_N\}}{\{g_1, \ldots, g_N\}} \\
&\quad + \sum_{n=M+1}^{N} \lambda_n \frac{\{g_1, \ldots, g_M, g_{M+1}, \ldots, g_{n-1}, x_j, g_{n+1}, \ldots, g_N\}}{\{g_1, \ldots, g_N\}} \\
&= \mu \sum_{n=1}^{N} \{F_1, \ldots, F_{N-1}, g_n\} \frac{\{g_1, \ldots, g_{n-1}, x_j, g_{n+1}, \ldots, g_N\}}{\{g_1, \ldots, g_N\}}.
\end{aligned}$$

In view of the identity (ix) we have that

$$\mathbf{X}(x_j) = \mu\{F_1, \ldots, F_{N-1}, x_j\}\frac{\{g_1, \ldots, g_N\}}{\{g_1, \ldots, g_N\}} = \mu\{F_1, \ldots, F_{N-1}, x_j\}.$$

Thus the functions $F_1, \ldots, F_{N-1}$ are first integrals of $\mathbf{X}$. Hence the vector field is integrable. $\qquad\square$

Example 1.5.2. Assume that the given invariant hypersurfaces are the hyperplanes

$$g_{j,k} = x_j - a_{j,k} = 0, \quad j = 1, \ldots, N, \quad k = 1, \ldots, N_j.$$

Since $|S| = \{x_1, \ldots, x_N\} = 1$, the differential system (1.4) takes the form $\dot{x}_j = \Phi_j(\mathbf{x})$, where $\Phi_j|_{g_{j,k}=0} = 0$ for $j = 1, \ldots, N$, $k = 1, \ldots, N_j$. We shall study the particular case

$$\Phi_j = \Psi_j(x_j)\frac{\{\varphi_1, \ldots, \varphi_{N-1}, x_j\}}{\{\varphi_1, \ldots, \varphi_N\}},$$

where $\Psi_1, \ldots, \Psi_N, \varphi_1, \ldots, \varphi_N$ are arbitrary functions such that

$$\Psi_1(a_{1,k}) = \cdots = \Psi_N(a_{N,k}) = 0, \quad k = 1, \ldots, N_j,$$
$$\{\varphi_1, \ldots, \varphi_N\} \not\equiv 0.$$

Corollary 1.5.3. *Let φ_j be the functions*

$$\varphi_j = \sum_{k=1}^{N} \int \varphi_{kj}(x_k)dx_k, \quad j = 1, \ldots, N,$$

where $\varphi_{kj}(x_k)$ are arbitrary functions. Then the differential system

$$\dot{x}_j = \Psi_j(x_j)\frac{\{\varphi_1,\ldots,\varphi_{N-1},x_j\}}{\{\varphi_1,\ldots,\varphi_N\}}, \quad j = 1,\ldots,N, \qquad (1.27)$$

is integrable.

Proof. Indeed, we consider

$$\sum_{k=1}^{N} \frac{\partial\varphi_\alpha}{\partial x_k}\frac{\dot{x}_k}{\Psi_j(x_j)} = \sum_{\alpha=1}^{N} \frac{\partial\varphi_\alpha}{\partial x_k}\frac{\{\varphi_1,\ldots,\varphi_{N-1},x_k\}}{\{\varphi_1,\ldots,\varphi_N\}}.$$

Then in view of identity (iv) for the Nambu bracket we obtain that

$$\sum_{k=1}^{N} \frac{1}{\Psi_k(x_k)}\frac{\partial\varphi_\alpha}{\partial x_k}\dot{x}_k = \frac{\{\varphi_1,\ldots,\varphi_{N-1},\varphi_\alpha\}}{\{\varphi_1,\ldots,\varphi_N\}} = \begin{cases} 1, & \text{if } \alpha = N, \\ 0, & \text{otherwise.} \end{cases}$$

Thus the functions $F_1,\ldots,F_{N-1}$ given by the formulas

$$F_j = \sum_{k=1}^{N} \int \frac{\varphi_{kj}(x_k)}{\Psi_k(x_k)}dx_k \quad \text{for} \quad j = 1,\ldots,N-1 \qquad (1.28)$$

are independent first integrals. Consequently, the system (1.27) is integrable. $\square$

We observe that these first integral (1.28) appear in the study of the inverse Stäckel problem, see for instance [139, 166].

The following results are proved in a similar way to the proof of Theorem 1.5.1.

Theorem 1.5.4. *Under the assumptions of Theorem 1.4.1, the differential system (1.18) is integrable if and only if*

$$\Phi_l = \mu\{f_1,\ldots,f_{M_2},F_{M_2+1},\ldots,F_{N-1},g_l\},$$
$$\lambda_k = \mu\{f_1,\ldots,f_{M_2},F_{M_2+1},\ldots,F_{N-1},g_k\}$$

for $l = 1,\ldots,M_1$ and $k = M+1,\ldots,N$, where μ, $F_{M_2+1},\ldots,F_{N-1}$ are arbitrary functions such that $f_1,\ldots,f_{M_2},F_{M_2+1},\ldots,F_{N-1}$ are independent in D and

$$\mu\{f_1,\ldots,f_{M_2},F_{M_2+1},\ldots,F_{N-1},g_l\}|_{g_l=0} = 0 \quad \text{for} \quad l = 1,\ldots,M_1.$$

Corollary 1.5.5. *Under the assumptions of Corollary 1.4.2, the differential system* (1.20) *is integrable if and only if*

$$\lambda_k = \mu\{f_1, \ldots, f_M, F_{M+1}, \ldots, F_{N-1}, g_k\},$$

for $k = M+1, \ldots, N$ *where* μ, $F_{m+1}, \ldots, F_{N-1}$ *are arbitrary functions such that* $f_1, \ldots, f_M, F_{M+1}, \ldots, F_{N-1}$ *are independent in* D.

Corollary 1.5.6. *Under the assumptions of Corollary 1.4.3, the differential system* (1.21) *is integrable if and only if* $\Phi_l = \mu\{f_1, \ldots, f_{M_2}, F_{M_2+1}, \ldots, F_{N-1}, g_l\}$, *where* μ, $F_{M_2+1}, \ldots, F_{N-1}$ *are arbitrary functions,* $f_1, \ldots, f_{M_2}, F_{M_2+1}, \ldots, F_{N-1}$ *are independent in* D *and*

$$\mu\{f_1, \ldots, f_{M_2}, F_{M_2+1}, \ldots, F_{N-1}, g_l\}|_{g_l=0} = 0 \quad for \quad l = 1, \ldots, M_1.$$

In the next result we provide sufficient conditions for the existence of a first integral of the differential system (1.4).

Theorem 1.5.7. *Suppose that we are under the assumptions of Theorem 1.3.1, and that in the differential system* (1.4) *we choose*

$$\begin{aligned}
\lambda_{M+1} &= & L_0 g_{M+1} + L_1 g, \\
\lambda_{M+2} &= & L_0 g_{M+2} + L_1 g_{M+1} + L_2 g, \\
&\vdots & \vdots \\
\lambda_N &= L_0 g_N + L_1 g_{N-1} + \cdots + L_{N-M} g,
\end{aligned} \tag{1.29}$$

with $L_0 = \displaystyle\sum_{j=1}^{M} \frac{\Phi_j \tau_j}{g_j}$, *where* $g = \displaystyle\prod_{j=1}^{M} |g_j|^{\tau_j}$, *and* τ_j *for* $j = 1, 2, \ldots M$ *are constants, and* $L_1, \ldots L_{N-M}$ *are functions satisfying*

$$\sum_{j=0}^{N-M} \nu_j L_j = 0, \tag{1.30}$$

for suitable constants ν_j *for* $j = 0, 1, \ldots, M$.

Define $\mathbf{G} = (G_1, G_2, \ldots, G_{N-M})^T = \left(\dfrac{g_{M+1}}{g}, \dfrac{g_{M+2}}{g}, \ldots, \dfrac{g_N}{g} \right)^T$ *and*

$$B = \begin{pmatrix}
1 & 0 & 0 & 0 & \cdots & 0 & 0 \\
G_1 & 1 & 0 & 0 & \cdots & 0 & 0 \\
G_2 & G_1 & 1 & 0 & \cdots & 0 & 0 \\
G_3 & G_2 & G_1 & 1 & \cdots & 0 & 0 \\
\vdots & \vdots & \vdots & \vdots & \cdots & \vdots & \vdots \\
G_{N-M-2} & G_{N-M-3} & G_{N-M-4} & \vdots & \cdots & 1 & 0 \\
G_{N-M-1} & G_{N-M-2} & G_{N-M-3} & \vdots & \cdots & G_1 & 1
\end{pmatrix}.$$

Define a function $\mathbf{R} = (R_1, \ldots, R_{N-M})^T$ *satisfying*

$$\mathbf{R} = \int B^{-1} d\mathbf{G}, \quad with \quad d\mathbf{G} = (dG_1, dG_2, \ldots, dG_{N-M})^T,$$

where dG_k *denotes the differential of* G_k *for* $k = 1, 2, \ldots N$*. Then*

$$F = |g|^{\nu_0} \exp\left(\sum_{j=1}^{N-M} \nu_j R_j \right)$$

is a first integral of system (1.4)*.*

Proof. Let $\mathbf{X} = (X_1, \ldots, X_N)$ be the vector field associated to system (1.4).

Since $g_{M+j} = gG_j$ for $j = 1, 2, \ldots N - M$, using (1.29) and (1.10) we obtain

$$\mathbf{X}(g) = \sum_{j=1}^{M} \frac{g\tau_j}{g_j} \mathbf{X}(g_j) = g \sum_{j=1}^{M} \frac{\tau_j \Phi_j}{g_j} = L_0 g,$$

$$\mathbf{X}(g_{M+1}) = \lambda_{M+1}$$

or, equivalently, $\mathbf{X}(g)G_1 + g\mathbf{X}(G_1) = L_0\, g\, G_1 + L_1 g$, and similarly it follows that

$$L_0\, g\, G_2 + g\mathbf{X}(G_2) = L_0\, g\, G_2 + L_1 g G_1 + L_2 g,$$

$$\vdots$$

$$L_0\, g\, G_{N-M} + g\mathbf{X}(G_{N-M}) = L_0\, g\, G_{M-N} + \ldots + L_{N-M} g.$$

Thus

$$\mathbf{X}(g) = L_0 g,$$
$$\mathbf{X}(G_1) = L_1,$$
$$\mathbf{X}(G_2) = L_1\, G_1 + L_2,$$
$$\vdots$$
$$\mathbf{X}(G_{N-M}) = L_1\, G_{N-M-1} + L_2\, G_{N-M-2} \ldots + L_{N-M},$$

or, in matrix form,

$$\mathbf{X}(\mathbf{G}) = B\mathbf{L}, \quad where \quad \mathbf{L} = (L_1, L_2, \ldots, L_{N-M})^T. \tag{1.31}$$

We introduce the 1-forms $\omega_1, \omega_2, \ldots, \omega_{N-M}$ by

$$dG_1 = \omega_1$$
$$dG_2 = G_1\omega_1 + \omega_2$$
$$\vdots$$
$$dG_{N-M} = G_{M-M-1}\omega_1 + \cdots + G_1\omega_{N-M-1} + \omega_{N-M}$$

or, equivalently,

$$dG = BW, \qquad (1.32)$$

where $\mathbf{W} = (\omega_1, \ldots, \omega_{N-M})^T$. Hence, (1.31), (1.32), and the relation $\mathbf{X(G)} = d\mathbf{G(X)}$ yield

$$\mathbf{W(X)} = \mathbf{L}. \qquad (1.33)$$

A 1-form is *closed* if its exterior derivative is everywhere equal to zero. Denoting by $\wedge$ the *wedge product* on the *differential 1-forms*, we obtain

$$0 = d^2 G_1 = d\omega_1,$$

$$0 = d^2 G_2 = dG_1 \wedge \omega_1 + G_1 d\omega_1 + d\omega_2 = \omega_1 \wedge \omega_1 + G_1 d\omega_1 + d\omega_2 = d\omega_2,$$

$$0 = d^2 G_3 = dG_2 \wedge \omega_1 + G_2 d\omega_1 + dG_1 \wedge \omega_2 + G_2 d\omega_2 + d\omega_3,$$

$$= G_1 \left(\omega_1 \wedge \omega_1 \right) + \omega_2 \wedge \omega_1 + \omega_1 \wedge \omega_2 + G_2 d\omega_1 + G_2 d\omega_2 + d\omega_3 = d\omega_3.$$

Analogously, we deduce that $d\omega_j = 0$ for $j = 4, \ldots, N - M$, thus the 1-forms ω_j are closed. Therefore, $\omega_j = dR_j$, where R_j is a suitable function. Hence, by (1.33), we get

$$\omega_j(\mathbf{X}) = L_j, \quad j = 1, 2, \ldots, N - M.$$

Let $\mathbf{R} = (R_1, \ldots, R_{N-M})^T$ be the vector defined by

$$d\mathbf{R} = (\omega_1, \ldots, \omega_{N-M})^T = \mathbf{W} = B^{-1} d\mathbf{G},$$

obtained from (1.32).

After the integration of the system $d\mathbf{R} = B^{-1} d\mathbf{G}$ we obtain $\mathbf{R} = \int B^{-1} d\mathbf{G}$. Hence

$$R_1 = G_1,$$

$$R_2 = G_2 - \frac{G_1^2}{2!},$$

$$R_3 = G_3 - G_1 G_2 + \frac{G_1^3}{3!},$$

$$R_4 = G_4 - G_1 G_3 + G_1^2 G_2 - \frac{G_1^4}{4!} - \frac{G_2^2}{2!},$$

$$R_5 = G_5 - G_1 G_4 + G_1^2 G_3 - G_1^3 G_2 + \frac{G_1^5}{5!} + \frac{G_2^3}{3!},$$

$$\vdots$$

Therefore, since $G_j = \dfrac{g_{M+j}}{g}$, we deduce the representations

$$R_1 = \frac{g_{M+1}}{g} = \frac{A_1}{g},$$

$$R_2 = \frac{g_{M+2}}{g} - \frac{1}{2!}\left(\frac{g_{M+1}}{g}\right)^2 = \frac{A_2}{g^2},$$

$$R_3 = \frac{g_{M+3}}{g} - \frac{g_{M+1}g_{M+2}}{g^2} + \frac{1}{3!}\left(\frac{g_{M+1}}{g}\right)^3 = \frac{A_3}{g^3},$$

$$\vdots$$

So we have $R_j = A_j/g^j$ for $j = 1, 2, \ldots, N - M$, where A_j are the functions defined above.

From the equalities $\mathbf{X}(\ln|g|) = L_0$, $\mathbf{X}(R_j) = dR_j(X) = \omega_j(X) = L_j$ for $j = 1, \ldots, N - M$, and (1.30) we have that

$$\sum_{j=0}^{N-M} \nu_j L_j = \nu_0 \mathbf{X}(\ln|g|) + \sum_{j=1}^{N-M} \nu_j \mathbf{X}(R_j) = \mathbf{X}\left(\ln\left[|g|^{\nu_0}\exp\left(\sum_{j=1}^{N-M}\nu_j R_j\right)\right]\right) = 0.$$

Thus

$$F = |g|^{\nu_0}\exp\left(\sum_{j=1}^{N-M}\nu_j R_j\right) = |g|^{\nu_0}\exp\left(\sum_{j=1}^{N-M}\nu_j\frac{A_j}{g^j}\right),$$

is a first integral of the differential system (1.4). We observe that in general the functions g_j for $j = 1, \ldots, M$ are not algebraic. $\square$

Such a first integral was already obtained in [134]. We observe that these kinds of first integrals appear also in the study of the invariant algebraic hypersurfaces with multiplicity of a polynomial vector field, see [39, 98].

1.6 Ordinary differential equations in $\mathbb{R}^N$ with $M > N$ partial integrals

In this section we determine the differential systems (1.3) having a given set of M partial integrals with $M > N$.

Theorem 1.6.1. *Let $g_j = g_j(\mathbf{x})$ for $j = 1, 2, \ldots, M$, with $M > N$, be a set of functions defined in the open set $\mathrm{D} \subset \mathbb{R}^N$ such that at least one group of N of them are independent at the points of the set D, i.e., without loss of generality we can assume that $\{g_1, \ldots, g_N\} \not\equiv 0$ in D. Then the most general differential system* in D *which admit the partial integrals* g_j *for* $j = 1, 2, \ldots, M$ *is*

$$\dot{x}_j = \sum_{j_1,\ldots,j_{N-1}=1}^{M+N} G_{j_1,\ldots,j_{N-1}}\{g_{j_1},\ldots,g_{j_{N-1}},x_j\}, \quad j = 1, 2, \ldots, N, \qquad (1.34)$$

where $G_{j_1,\ldots,j_{N-1}} = G_{j_1,\ldots,j_{N-1}}(\mathbf{x})$ *are arbitrary functions satisfying*

$$\dot{g}_j\big|_{g_j=0} = \left(\sum_{j_1,\ldots,j_{N-1}=1}^{M+N} G_{j_1,\ldots,j_{N-1}}\{g_{j_1},\ldots,g_{j_{N-1}},g_j\} \right)\Bigg|_{g_j=0} = 0, \qquad (1.35)$$

for $j = 1, 2, \ldots, M$, *and* $g_{M+j} = x_{j+1}$, $x_{N+1} := x_1$ *for* $j = 1, 2, \ldots, N$.

Proof of Theorem 1.6.1. First of all we determine the differential systems having the N independent partial integrals $g_j = g_j(\mathbf{x})$, $j = 1, 2, \ldots, N$. Thus we obtain the system (1.12). Clearly, this differential system admits the additional partial integrals g_j with $j = N + 1, \ldots, M$ if and only if $\mathbf{X}(g_\nu) = \Phi_\nu$, $\quad \Phi_\nu\big|_{g_\nu=0} = 0$, for $\nu = N + 1, \ldots, M$, which, using (1.11) can be written equivalently as

$$\Phi_1\{g_\nu, g_2, \ldots, g_N\} + \ldots + \Phi_N\{g_1, \ldots, g_{N-1}, g_\nu\} - \Phi_\nu\{g_1, \ldots, g_{N-1}, g_N\} = 0. \qquad (1.36)$$

Now we prove that

$$\Phi_\nu = \sum_{\alpha_1,\ldots,\alpha_{N-1}=1}^{M+N} G_{\alpha_1,\ldots,\alpha_{N-1}}\{g_{\alpha_1},\ldots,g_{\alpha_{N-1}},g_\nu\} \qquad (1.37)$$

is a solution of (1.36) for $\nu = 1, 2, \ldots, M \geq N$, where $G_{\alpha_1,\ldots,\alpha_{N-1}} = G_{\alpha_1,\ldots,\alpha_{N-1}}(\mathbf{x})$ are arbitrary functions satisfying (1.35).

Indeed, in view of (1.36) and (1.37) we have that

$$\sum_{\alpha_1,\ldots,\alpha_{N-1}=1}^{M+N} G_{\alpha_1,\ldots,\alpha_{N-1}}(\{g_{\alpha_1},\ldots,g_{\alpha_{N-1}},g_1\}\{g_\nu, g_2, \ldots, g_{N-1}, g_N\} + \cdots$$

$$+ \{g_{\alpha_1},\ldots,g_{\alpha_{N-1}},g_N\}\{g_1, g_2, \ldots, g_{N-1}, g_\nu\}$$

$$- \{g_{\alpha_1},\ldots,g_{\alpha_{N-1}},g_\nu\}\{g_1, g_2, \ldots, g_{N-1}, g_N\})$$

$$= \sum_{\alpha_1,\ldots,\alpha_{N-1}=1}^{M+N} G_{\alpha_1,\ldots,\alpha_{N-1}}\Omega\left(g_{\alpha_1},\ldots,g_{\alpha_{N-1}},g_1,\ldots,g_N,g_\nu\right),$$

which is identically zero by the identity (viii).

Inserting (1.37) into (1.12) and using the identity (v) we obtain the equation

$$
\begin{aligned}
\dot{x}_\nu &= \Phi_1 \frac{\{x_\nu, g_2, \ldots, g_N\}}{\{g_1, \ldots, g_N\}} + \cdots + \Phi_N \frac{\{g_1, \ldots, g_{N-1}, x_\nu\}}{\{g_1, \ldots, g_N\}} \\
&= \sum_{\substack{\alpha_1, \ldots, \\ \alpha_{N-1}=1}}^{M+N} \frac{G_{\alpha_1, \ldots, \alpha_{N-1}}}{\{g_1, \ldots, g_N\}} \sum_{n=1}^{N} \{g_{\alpha_1}, \ldots, g_{\alpha_{N-1}}, g_n\}\{g_1, \ldots, g_{n-1}, x_\nu, g_{n+1}, \ldots, g_N\} \\
&= \sum_{\substack{\alpha_1, \ldots, \\ \alpha_{N-1}=1}}^{M+N} G_{\alpha_1, \ldots, \alpha_{N-1}} \{g_{\alpha_1}, \ldots, g_{\alpha_{N-1}}, x_\nu\}
\end{aligned}
$$

for $j = 1, 2, \ldots, N$. Now we prove that this differential system, which coincides with (1.34), is the most general. Indeed, since $g_{M+j} = x_{j+1}$, $x_{N+1} = x_1$, for $j = 1, \ldots, N$, system (1.34) admits the representation

$$
\begin{aligned}
\dot{x}_1 &= \left(\sum_{\substack{\alpha_1, \ldots, \alpha_{N-1}=1 \\ \alpha_1, \ldots, \alpha_{N-1} \neq (M+2, \ldots, N)}}^{M+N} G_{\alpha_1, \ldots, \alpha_{N-1}} \{g_{\alpha_1}, \ldots, g_{\alpha_{N-1}}, x_1\} \right) \\
&\quad + G_{M+2, M+3, \ldots, M+N} \{x_2, \ldots, x_N, x_1\}, \\
&\vdots \qquad\qquad\qquad\qquad \vdots \\
\dot{x}_N &= \left(\sum_{\substack{\alpha_1, \ldots, \alpha_{N-1}=1 \\ \alpha_1, \ldots, \alpha_{N-1} \neq (M+1, \ldots, N-1)}}^{M+N} G_{\alpha_1, \ldots, \alpha_{N-1}} \{g_{\alpha_1}, \ldots, g_{\alpha_{N-1}}, x_N\} \right) \\
&\quad + G_{M+1, M+2, \ldots, M+N-1} \{x_1, \ldots, x_{N-1}, x_N\}.
\end{aligned}
\tag{1.38}
$$

Note that $\{x_1, \ldots, x_{j-1}, x_{j+1}, \ldots, x_N, x_j\} \in \{-1, 1\}$. Therefore, if $\dot{x}_j = \tilde{X}_j$ for $j = 1, \ldots, N$, is another differential system having the given set of partial integrals, then by choosing suitable functions $G_{M+2, M+3, \ldots, M+N}, \ldots G_{M+1, M+2, \ldots, M+N-1}$ we deduce that the constructed vector field (1.38) contain the vector field $\tilde{\mathbf{X}} = \left(\tilde{X}_1, \tilde{X}_2, \ldots, \tilde{X}_N \right)$. So the proof of Theorem 1.6.1 follows. $\qquad\square$

The following result is proved in a similar way to Theorem 1.5.1.

Theorem 1.6.2. *Under the assumptions of Theorem 1.6.1, the differential system* (1.34) *is integrable if and only if*

$$
\Phi_l = \sum_{\alpha_1, \ldots, \alpha_{N-1}=1}^{M+N} G_{\alpha_1, \ldots, \alpha_{N-1}} \{g_{\alpha_1}, \ldots, g_{\alpha_{N-1}}, g_l\} = \mu \{F_1, \ldots, F_{N-1}, g_l\},
$$

for $l = 1, \ldots, M > N$, *where* $\mu, F_1, \ldots, F_{N-1}$ *are suitable functions such that* $F_1, \ldots, F_{N-1}$ *are independent in* D *and* $\mu \{F_1, \ldots, F_{N-1}, g_l\}|_{g_l=0} = 0$.

Corollary 1.6.3. *Under the assumptions of Theorem 1.6.2, choose the functions Φ_j for $j = 1, \ldots, M$, as*

$$\Phi_j = \Upsilon_j(g_j)\Psi_j(\mathbf{x}), \quad with \quad \Upsilon_j(g_j)\big|_{g_j=0} = 0,$$

and require that for the functions $\Psi_1, \ldots, \Psi_M$ there exist constants $\sigma_j^n \in \mathbb{C}$ for $j = 1, \ldots, M$ and $n = 1, \ldots, N - 1$, not all zero, such that

$$\sum_{j=1}^{M} \sigma_j^n \Psi_j(\mathbf{x}) = 0. \tag{1.39}$$

Then the differential system (1.34) is integrable, with first integrals

$$F_n = \sum_{j=1}^{M} \sigma_j^n \int \frac{dg_j}{\Upsilon_j(g_j)}, \quad n = 1, \ldots, N - 1. \tag{1.40}$$

In particular, if $\Upsilon_j(g_j) = g_j$, then the first integrals (1.40) are (see, for instance, [43])

$$F_n = \log\left(\prod_{j=1}^{M} |g_j|^{\sigma_j^n} \right), \quad n = 1, \ldots, N - 1.$$

Proof. We have that

$$\frac{dg_j}{dt} = \Upsilon_j(g_j)\Psi_j(\mathbf{x}) \implies \frac{\dot{g}_j}{\Upsilon_j(g_j)} = \Psi_j(\mathbf{x}).$$

Consequently, if (1.39) holds, then

$$\frac{d}{dt}\left(\sum_{j=1}^{M} \sigma_j^n \int \frac{dg_j}{\Upsilon_j(g_j)} \right) = 0, \quad for \quad n = 1, \ldots, N - 1. \qquad \square$$

Corollary 1.6.4. *Under the assumptions of Theorem 1.6.1 for $N = 2$, the system (1.34) takes the form*

$$\dot{x} = \sum_{j=1}^{M} G_j\{g_j, x\} + G_{M+1}\{y, x\} = \sum_{j=1}^{M} G_j\{g_j, x\} - G_{M+1},$$

$$\dot{y} = \sum_{j=1}^{M} G_j\{g_j, y\} + G_{M+2}\{x, y\} = \sum_{j=1}^{M} G_j\{g_j, y\} + G_{M+2}, \tag{1.41}$$

where $G_j = G_j(x, y)$ for $j = 1, 2, \ldots, M + 2$ are arbitrary functions satisfying (1.35). Moreover, (1.35) becomes

$$\left(\sum_{j=1}^{M} G_j\{g_j, g_k\} + G_{M+1}\{y, g_k\} + G_{M+2}\{x, g_k\} \right)\Bigg|_{g_k=0} = 0, \tag{1.42}$$

for $k = 1, 2, \ldots, M$.

Proof. The assertions follow immediately from Theorem 1.6.1.   □

Corollary 1.6.5. *The differential system* (1.41) *and the differential system*

$$\dot{x} = \Phi_1 \frac{\{x, g_2\}}{\{g_1, g_2\}} + \Phi_2 \frac{\{g_1, x\}}{\{g_1, g_2\}},$$

$$\dot{y} = \Phi_1 \frac{\{y, g_2\}}{\{g_1, g_2\}} + \Phi_2 \frac{\{g_1, y\}}{\{g_1, g_2\}},$$

with the conditions

$$\Phi_1\{g_\nu, g_2\} + \Phi_2\{g_1, g_\nu\} - \Phi_\nu\{g_1, g_2\} = 0,$$

are equivalent.

Remark 1.6.6. We note that the conditions (1.42) hold identically if

$$G_j = \lambda_j \prod_{\substack{m=1 \\ m \neq j}}^{M} g_m, \tag{1.43}$$

where $\lambda_j = \lambda_j(x, y)$ for $j = 1, \ldots, M + 2$ are arbitrary functions.

Inserting (1.43) into (1.41) we obtain the differential system

$$\dot{x} = \sum_{j=1}^{M} \lambda_j \{g_j, x\} \prod_{\substack{m=1 \\ m \neq j}}^{M} g_m - \lambda_{M+2} \prod_{m=1}^{M} g_m,$$

$$\dot{y} = \sum_{j=1}^{M} \lambda_j \{g_j, y\} \prod_{\substack{m=1 \\ m \neq j}}^{M} g_m + \lambda_{M+1} \prod_{m=1}^{M} g_m. \tag{1.44}$$

In particular, if in (1.44) we assume that $\lambda_j = \lambda$ for $j = 1, \ldots, M$, then we obtain the differential system

$$\dot{x} = \lambda \sum_{j=1}^{M} \{g_j, x\} \prod_{\substack{m=1 \\ m \neq j}}^{M} g_m - \lambda_{M+2} \prod_{m=1}^{M} g_m,$$

$$\dot{y} = \lambda \sum_{j=1}^{M} \{g_j, y\} \prod_{\substack{m=1 \\ m \neq j}}^{M} g_m + \lambda_{M+1} \prod_{m=1}^{M} g_m.$$

Upon introducing the function $g = \prod_{m=1}^{M} g_m$, we get the differential system

$$\dot{x} = -\lambda \frac{\partial g}{\partial y} - g\lambda_{M+1}, \quad \dot{y} = \lambda \frac{\partial g}{\partial x} + g\lambda_{M+2}. \tag{1.45}$$

Indeed, since

$$\frac{\partial g}{\partial y} = \sum_{j=1}^{M} \{g_j, x\} \prod_{\substack{m=1 \\ m \neq j}}^{M} g_m, \quad \frac{\partial g}{\partial x} == \sum_{j=1}^{M} \{g_j, y\} \prod_{\substack{m=1 \\ m \neq j}}^{M} g_m,$$

(1.45) easily results from (1.44).

Remark 1.6.7. Singer in [144] showed that if a polynomial differential system has a Liouvillian first integral, i.e., roughly speaking, a first integral that can be expressed in terms of quadratures of elementary functions, then it has an integrating factor of the form

$$R(x, y) = \exp\left(\int_{(x_0, y_0)}^{(x,y)} (U(x, y)dx + V(x, y)dy)\right), \qquad (1.46)$$

where $U = U(x, y)$ and $V = V(x, y)$ are rational functions which verify $\partial U/\partial y = \partial V/\partial x$. Using this result, Christopher in [30] showed that in fact the integrating factor (1.46) can be written in the form

$$R(x, y) = \exp\left(\frac{g}{h}\right) \prod_{j=1}^{M} f_j^{\sigma_j},$$

where g and f are polynomials and $\sigma_j \in \mathbb{C}$. This type of integrability is known as *Liouvillian integrability*.

A multi-valued function of the form $\exp\left(\frac{g}{h}\right) \prod_{j=1}^{M} f_j^{\sigma_j}$ is called a *Darboux function*.

Corollary 1.6.8. *The generalized Darboux function $F = e^\tau \prod_{j=1}^{M} |g_j|^{\lambda_j}$ is a first integral of the polynomial differential system (1.44) if and only if $\lambda_{M+1} = \partial \tau/\partial y$, $\lambda_{M+2} = \partial \tau/\partial x$, where $\tau = \tau(x, y)$ is an arbitrary polynomial and $\lambda_1, \ldots, \lambda_M$ are constants.*

Proof. Assume that the λ_j's for $j = 1, \ldots, M$ of the differential system (1.44) are constant and

$$\lambda_{M+1} = \frac{\partial \tau}{\partial y} = \tau_y$$

and

$$\lambda_{M+2} = \frac{\partial \tau}{\partial x} = \tau_x,$$

where $\tau = \tau(x, y)$ is a polynomial. Then the system (1.44) becomes

$$\dot{x} = g\left(-\frac{\partial \tau}{\partial y} + \sum_{j=1}^{M} \lambda_j \{\log |g_j|, x\}\right) = \frac{g}{F}\left(-\frac{\partial \tau}{\partial y} F + F \sum_{j=1}^{M} \lambda_j \{\log |g_j|, x\}\right)$$

$$= \frac{g}{F}\left(-\frac{\partial \tau}{\partial y}F + \sum_{j=1}^{M} \lambda_j \{F, x\}\right) \quad = g\{\log F, x\},$$

$$\dot{y} = g\left(\frac{\partial \tau}{\partial x} + \sum_{j=1}^{M} \lambda_j \{\log |g_j|, y\}\right) \quad = \frac{g}{F}\left(\frac{\partial \tau}{\partial x}F + F\sum_{j=1}^{M} \lambda_j \{\log |g_j|, y\}\right)$$

$$= \frac{g}{F}\left(\frac{\partial \tau}{\partial x}F + \sum_{j=1}^{M} \lambda_j \{F, y\}\right) \quad = g\{\log F, y\},$$

where $F = e^\tau \prod_{m=1}^{M} |g_m|^{\lambda_m}$, $g = \prod_{m=1}^{M} g_m$, and we have used the formulas

$$F_x = \tau_x F + e^\tau \left(\sum_{j=1}^{M} \lambda_j g_j^{\lambda_j - 1}\{g_j, y\} \prod_{\substack{m=1 \\ m \neq j}}^{M} |g_m|^{\lambda_m}\right)$$

$$= F\left(\tau_x + \sum_{j=1}^{M} \lambda_j \{\log |g_j|, y\}\right),$$

$$F_y = \tau_y F - e^\tau \left(\sum_{j=1}^{M} \lambda_j g_j^{\lambda_j - 1}\{g_j, x\} \prod_{\substack{m=1 \\ m \neq j}}^{M} |g_m|^{\nu_m}\right)$$

$$= -F\left(-\tau_y + \sum_{j=1}^{M} \lambda_j \{\log |g_j|, y\}\right), \tag{1.47}$$

and

$$P(x, y) = -\lambda_{M+1} \prod_{m=1}^{M} g_m + \sum_{j=1}^{M} \lambda_j \left(\prod_{\substack{m=1 \\ m \neq j}}^{M} g_m\right)\{g_j, x\}$$

$$= g\left(-\lambda_{M+1} + \sum_{j=1}^{M} \lambda_j \{\log |g_j|, x\}\right),$$

$$Q(x, y) = \lambda_{M+2} \prod_{m=1}^{M} g_m + \sum_{j=1}^{M} \lambda_j \left(\prod_{\substack{m=1 \\ m \neq j}}^{M} g_m\right)\{g_j, y\}$$

$$= g\left(\lambda_{M+2} + \sum_{j=1}^{M} \lambda_j \{\log |g_j|, y\}\right) \tag{1.48}$$

Consequently, the function F is a first integral.

Let

$$F = e^\tau \prod_{m=1}^{M} |g_m|^{\nu_m},$$

where ν_m for $m = 1, \ldots, S$ are constants, be a first integral of the vector field (1.44). By applying (1.47) and (1.48) we have that

$$\dot{F} = F_x P + F_y Q$$

$$= F\left(\tau_x + \sum_{j=1}^{M} \nu_j \{\log |g|_j, y\}\right) g\left(-\lambda_{M+1} + \sum_{j=1}^{M} \lambda_j \{\log |g|_j, y\}\right)$$

$$+ F\left(\tau_y - \sum_{j=1}^{M} \nu_j \{\log |g|_j, x\}\right) + g\left(\lambda_{M+2} + \sum_{j=1}^{M} \lambda_j \{\log |g|_j, x\}\right) \equiv 0.$$

This relation holds if $\lambda_j = \nu_j$ and $\lambda_{M+1} = \tau_y$ and $\lambda_{M+2} = \tau_x$. This completes the proof of the theorem. □

Example 1.6.9. Consider the differential equation

$$\dot{z} = i\left(a_{10}z + a_{01}\bar{z} + \sum_{j+k=3} a_{jk}z^j \bar{z}^k\right),$$

where a_{jk}, $j, k = 0, 1, 2, 3$ are real constants and $z = x + iy$, $\bar{z} = x - iy$ are the complex coordinates in the plane $\mathbb{R}^2$. Denoting

$$a_{01} = \frac{1}{2}(\alpha + a), \qquad a_{10} = \frac{1}{2}(\alpha - a),$$

$$a_{03} = \frac{1}{8}(\beta + b - \gamma - c), \qquad a_{30} = \frac{1}{8}(\beta + c - \gamma - b),$$

$$a_{12} = \frac{1}{8}(\gamma + b + 3(\beta + c)), \quad a_{21} = \frac{1}{8}(\gamma - b + 3(\beta - c)),$$

we obtain the cubic planar vector field

$$\dot{x} = y(a + b\,x^2 + c\,y^2) = \tilde{X}(x), \quad \dot{y} = x(\alpha + \beta\,x^2 + \gamma\,y^2) = \tilde{X}(y). \tag{1.49}$$

We assume that

$$c(b\gamma - c\beta)((b - \gamma)^2 + 4c\beta) \neq 0. \tag{1.50}$$

Corollary 1.6.10. *System* (1.49) *has the following properties:*

(i) *It admits two invariant conics (eventually imaginary)*

$$g_j = \nu_j(x^2 - \lambda_1) - (y^2 - \lambda_2) = 0, \quad for \quad j = 1, 2,$$

where ν_1 and ν_2 are the roots of the polynomial $P(\nu) = c\nu^2 + (b - \gamma)\nu - \beta$, and

$$\lambda_1 = \frac{\gamma a - \alpha c}{b\gamma - c\beta}, \quad \lambda_2 = \frac{\alpha b - \beta a}{b\gamma - c\beta}.$$

By (1.50), $\nu_1 - \nu_2 \neq 0$,

(ii) *It admits the first integral*

$$F(x,y) = \frac{(\nu_1(x^2 - \lambda_1) - y^2 + \lambda_2)^{\gamma - \nu_2 c}}{(\nu_2(x^2 - \lambda_1) - y^2 + \lambda_2)^{\gamma - \nu_1 c}}.$$

(iii) *Its solutions can be represented as follows*

$$x^2 = \lambda_1 + X(\tau, x_0, y_0),$$
$$y^2 = \lambda_2 + Y(\tau, x_0, y_0),$$
$$t = t_0 + \int_0^\tau \frac{d\tau}{\sqrt{(\lambda_1 + X(\tau, x_0, y_0))(\lambda_2 + Y(\tau, x_0, y_0))}},$$

where X, Y are solutions of the linear differential equation of the second order with constants coefficients

$$T'' - (\gamma + b)T' + (b\gamma - c\beta)T = 0, \quad where \quad ' \equiv \frac{d}{d\tau}.$$

Proof. Since $\{g_1, g_2\} = 4xy(\nu_2 - \nu_1)$, applying Corollary 1.3.4 (see formula (1.15)), we obtain

$$\dot{g}_1 = \frac{\partial g_1}{\partial x}\dot{x} + \frac{\partial g_1}{\partial y}\dot{y} = \frac{\partial g_1}{\partial x}\left(\mu_1\{x, g_2\} + \mu_2\{g_1, x\}\right) + \frac{\partial g_1}{\partial y}\left(\mu_1\{y, g_2\} + \mu_2\{g_1, y\}\right)$$
$$= \mu_1\{g_1, g_2\} + \mu_2\{g_1, g_1\} = \{g_1, g_2\}\mu_1.$$

Hence from (1.49) we obtain

$$\{g_1, g_2\}\mu_1 = 2xy\left(\nu_1\left((a + b\,x^2 + c\,y^2)\right) - \left(\alpha + \beta\,x^2 + \gamma\,y^2\right)\right).$$

Similarly,

$$\dot{g}_2 = \{g_1, g_2\}\mu_2 = 2xy\left(\nu_2\left((a + b\,x^2 + c\,y^2)\right) - \left(\alpha + \beta\,x^2 + \gamma\,y^2\right)\right).$$

Here μ_1 and μ_2 are arbitrary functions such that $\mu_j|_{g_j=0} = 0$, for $j = 1, 2$. Now take $2\mu_j = \kappa_j\left(\nu_j(x^2 - \lambda_1) - (y^2 - \lambda_2)\right)$, where κ_1 and κ_2 are constants. Then

$$2\mu_1 = \kappa_1\left(\nu_1(x^2 - \lambda_1) - (y^2 - \lambda_2)\right) = \frac{x^2(b\nu_1 - \beta) + y^2(c\nu_1 - \gamma) + a\nu_1 - \alpha}{\nu_2 - \nu_1},$$

$$2\mu_2 = \kappa_2\left(\nu_2(x^2 - \lambda_1) - (y^2 - \lambda_2)\right) = \frac{x^2(b\nu_2 - \beta) + y^2(c\nu_2 - \gamma) + a\nu_2 - \alpha}{\nu_2 - \nu_1}.$$

Thus we obtain the relations

$$x^2\left(\kappa_1\nu_1 - \frac{b\nu_1 - \beta}{\nu_2 - \nu_1}\right) - y^2\left(\kappa_1 - \frac{c\nu_1 - \gamma}{\nu_2 - \nu_1}\right) + \left((\lambda_2 - \nu_1\lambda_1)\kappa_1 - \frac{a\nu_1 - \alpha}{\nu_2 - \nu_1}\right) = 0,$$

$$x^2\left(\kappa_1\nu_2 - \frac{b\nu_2 - \beta}{\nu_2 - \nu_1}\right) - y^2\left(\kappa_1 - \frac{c\nu_2 - \gamma}{\nu_2 - \nu_1}\right) + \left((\lambda_2 - \nu_2\lambda_1)\kappa_1 - \frac{a\nu_2 - \alpha}{\nu_2 - \nu_1}\right) = 0.$$

Hence $(\nu_2 - \nu_1)\kappa_1\nu_1 = b\nu_1 - \beta$, and $(\nu_2 - \nu_1)\kappa_1 = c\nu_1 - \gamma$, so $b\nu_1 - \beta = (c\nu_1 - \gamma)\nu_1$, therefore $c\nu_1^2 + (b - \gamma)\nu_1 - \beta = 0$.

Similarly, $(\nu_2 - \nu_1)\kappa_2\nu_2 = b\nu_2 - \beta$, and $(\nu_2 - \nu_1)\kappa_2 = -(c\nu_2 - \gamma)$, so $b\nu_2 - \beta = (c\nu_2 - \gamma)\nu_2$, therefore $c\nu_2^2 + (b - \gamma)\nu_2 - \beta = 0$.

Finally we obtain

$$(\nu_2 - \nu_1)\kappa_1(\lambda_2 - \nu_1\lambda_1) = a\nu_1 - \alpha, \qquad \nu_1\lambda_1 - \lambda_2 = \frac{a\nu_1 - \alpha}{(\nu_2 - \nu_1)\kappa_1},$$

so

$$(\nu_2 - \nu_1)\kappa_2(\lambda_2 - \nu_2\lambda_1) = a\nu_2 - \alpha \ \text{ and } \ \nu_2\lambda_1 - \lambda_2 = \frac{a\nu_2 - \alpha}{(\nu_2 - \nu_1)\kappa_2}.$$

Solving the last system with respect to λ_1 and λ_2 we obtain

$$\lambda_1(\nu_1 - \nu_2) = \frac{a\nu_1 - \alpha}{(\nu_2 - \nu_1)\kappa_1} - \frac{a\nu_2 - \alpha}{(\nu_2 - \nu_1)\kappa_2} = \frac{(a\gamma - c\alpha)(\nu_1 - \nu_2)}{c^2\nu_1\nu_2 - c\gamma(\nu_1 + \nu_2) + \gamma^2}$$
$$= \frac{(a\gamma - c\alpha)(\nu_1 - \nu_2)}{b\gamma - c\beta},$$

$$\lambda_2(\nu_2 - \nu_1) = -\nu_2\frac{a\nu_1 - \alpha}{(\nu_2 - \nu_1)\kappa_1} + \nu_1\frac{a\nu_2 - \alpha}{(\nu_2 - \nu_1)\kappa_2}$$
$$= \frac{(b\alpha - a\beta)(\nu_1 - \nu_2)}{c^2\nu_1\nu_2 - c\gamma(\nu_1 + \nu_2) + \gamma^2} = \frac{(b\alpha - a\beta)(\nu_1 - \nu_2)}{b\gamma - c\beta},$$

where we used the relations

$$\nu_j^2 + \frac{b - \gamma}{c}\nu_j - \frac{\beta}{c} = 0 \Longrightarrow \nu_1 + \nu_2 = \frac{\gamma - b}{c}, \quad \nu_1\nu_2 = -\frac{\beta}{c}.$$

Statement (i) is proved.

Now we prove statement (ii). In view of the above assumptions we obtain that

$$\mathbf{X}(g_1) = 4xy(\nu_1 - \nu_2)\kappa_1 g_1, \quad \mathbf{X}(\ln|g_2|) = 4xy(\nu_1 - \nu_2)\kappa_2 g_2.$$

Consequently, $\mathbf{X}(\ln(|g_1|^{(\nu_1 - \nu_2)\kappa_2}|g_2|^{-(\nu_1 - \nu_2)\kappa_1})) = 0.$ Thus

$$F = \frac{|g_1|^{\gamma - c\nu_2}}{|g_2|^{\gamma - c\nu_1}} = \frac{|g_1|^{b + c\nu_1}}{|g_2|^{b + c\nu_2}}$$

is a first integral of the given cubic system; here used the relation $c(\nu_1 + \nu_2) = \gamma - b$.

Finally we prove statement (iii). It is easy to check that the functions X, Y are solutions of the system

$$X' = bX + cY, \quad Y' = \beta X + \gamma Y.$$

Then by taking the second derivatives of X and Y we get that

$$X'' - (\gamma + b)X' + (b\gamma - c\beta)X = 0 \quad \text{and} \quad Y'' - (\gamma + b)Y' + (b\gamma - c\beta)Y = 0.$$

Thus (iii) is proved, which completes the proof of the corollary. $\qquad \square$

Now we shall study the particular case when (1.49) is such that

$$\dot{x} = y(\lambda b + p + x^2(\lambda + b - 2a) + y^2(\lambda - b)),$$
$$\dot{y} = x(-\lambda a - p - x^2(\lambda - a) - y^2(\lambda + a - 2b)),$$
(1.51)

where λ, b, a, p are real parameters and $b - a \neq 0$. Then it is easy to check that

$$\nu_1 = -\frac{\lambda - a}{\lambda - b}, \quad \nu_2 = -1.$$

Thus the invariant curves of the differential system are

$$g_1 = -\left(y^2 + x^2 + \lambda\right) = 0, \quad g_2 = -\left((\lambda - a)x^2 + (\lambda - b)y^2 + (1/2)(\lambda^2 + p)\right) = 0.$$

The first integral F takes the form

$$F(x,y) = \frac{(y^2 + x^2 + \lambda)^2}{(\lambda - a)x^2 + (\lambda - b)y^2 + \frac{1}{2}(\lambda^2 + p)}.$$

Consequently, all trajectories of (1.51) are algebraic curves

$$(x^2 + y^2)^2 + A(K)\,x^2 + B(K)\,y^2 + P(K) = 0,$$
(1.52)

where $F(x,y) = K$ are the level curves, and

$$A((K)) = 2\left(\lambda - \frac{K}{2}(\lambda - a)\right), \quad B(K) = 2\left(\lambda - \frac{K}{2}(\lambda - b)\right),$$

$$P(K) = \lambda^2 - \frac{1}{2}K(\lambda^2 + p).$$

It is interesting to observe that a particular case of the above family of planar curves is the locus of points (x, y) for which the product of the distance to the fixed points $(0, -c)$ and $(0, c)$ has the constant value $\kappa^2 - c^2$ (for more details see [8]). The quartic equation of this curve is

$$(x^2 + y^2)^2 + 2c^2(x^2 - y^2) = \kappa^2\left(\kappa^2 - 2c^2\right),$$
(1.53)

which is equivalent to

$$(z^2 + c^2)\overline{(z^2 + c^2)} = (\kappa^2 - c^2)^2.$$
(1.54)

Thus if $A(K) = -B(K) = 2c^2$ and $P(K) = -\kappa^2\left(\kappa^2 - 2c^2\right)$, then if $K \neq 2$, we obtain

$$K = \frac{4c^2}{a - b}, \quad p = \frac{(a - b)(\kappa^2 - c^2)^2}{2c^2} + \frac{(c^2 - 2ab)c^2}{a - b - 2c^2}, \quad \lambda = \frac{(a + b)c^2}{2c^2 - a - b},$$

while if $K = 2$

$$A(2) = 2a, \quad B(2) = 2b, \quad P(2) = -p, \quad a = -b = c^2, \quad p = \kappa^2\left(\kappa^2 - 2c^2\right),$$

for arbitrary λ.

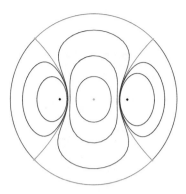

Figure 1.3: The trajectories of this phase portrait
correspond to the case when $C < 1$.

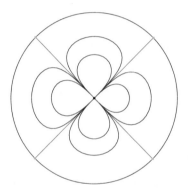

Figure 1.4: The trajectories of this phase portrait
correspond to the case when $C = 1$.

In the first case the system (1.51) takes the form

$$x' = -y[(a - b)((2a - b - 3c^2)x^2 + (c^2 + b)y^2) + p(2c^2 + b - a) + c^2 b(a + b)],$$
$$y' = x[(a - b)((a - c^2)x^2 + (3c^2 + 2b - a)y^2) + p(2c^2 + b - a) + c^2 a(a + b)],$$

where $' = d/d\tau$, with $t = (a - b - 2c^2)\tau$. This differential system admits as trajectories the family of lemniscates (1.53).

In the second case we obtain that the differential system (1.51) in complex coordinates takes the form

$$\dot{z} = i\left(\kappa^2(2c^2 - \kappa^2)z + c^2 z^3 - \lambda\bar{z}(c^2 + z^2)\right).$$

and admits as trajectories the family of lemniscates (1.54). In particular, if $c = 1$

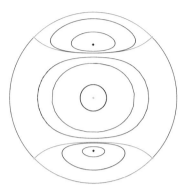

Figure 1.5: The trajectories of this phase portrait
correspond to the case when $C > 1$.

then the system
$$\dot{z} = i\left((1 - C^2)\,z + z^3 - \lambda\,\bar{z}(1 + z^2)\right).$$

The bifurcation diagrams of this differential system in the plane $(C = |\kappa^2 - 1|, \lambda)$ are given in [141]. Now consider the case when $\lambda = 0$, i.e., the differential equation
$$\dot{z} = i\left((1 - C^2)\,z + z^3\right).$$

The trajectories of this equation are given by formula (1.52) and are the lemniscates given by
$$(z^2 + 1)(\bar{z}^2 + 1) = \left(\kappa^2 - 1\right)^2 = C^2.$$

The phase portrait for the case when $C < 1$ is given in Figure 1.3, for $C = 1$ in Figure 1.4, and for $C > 1$ in Figure 1.5.

Finally, consider the following particular case of system (1.49):
$$\dot{x} = y\left(a + (r - q)x^2 + y^2\right), \quad \dot{y} = x\left(\alpha - (p^2 + q^2)x^2 + (r + q)y^2\right).$$

It is easy to show that the roots ν_1 and ν_2 are $\nu_1 = q + ip$ and $\nu_2 = q - ip$. Thus the invariant curves are complex:
$$g_1 = (q + ip)(x^2 - \lambda_1) - (y^2 - \lambda_2) = 0, \quad g_2 = (q - ip)(x^2 - \lambda_1) - (y^2 - \lambda_2) = 0.$$

Hence the first integral F is
$$F(x, y) = \left((q(x^2 - \lambda_1) - (y^2 - \lambda_2))^2 + p^2(x^2 - \lambda_1)^2\right)$$
$$\times \exp 2\arctan\left(\frac{p(x^2 - \lambda_1)}{q(x^2 - \lambda_1) - (y^2 - \lambda_2)}\right).$$

Chapter 2

Polynomial Vector Fields with Given Partial and First Integrals

2.1 Introduction

The solutions of the inverse problem in ordinary differential equations have a very high degree of arbitrariness because of the unknown functions involved. To reduce this arbitrariness we need additional conditions. In this chapter we are mainly interested in the planar polynomial differential systems which have a given set of invariant algebraic curves.

We establish the following:

(1) The normal form of planar polynomial vector fields having a given set of generic (in a suitable sense) invariant algebraic curves.

(2) The maximum degree of the non-singular invariant algebraic curves in $\mathbb{CP}^2$ of planar polynomial vector fields of a given degree.

(3) The polynomial planar differential system with non-singular and singular invariant algebraic curves.

(4) The maximum degree of an invariant algebraic curve (in function of the number of its branches) for planar quadratic vector fields. We support by several facts the conjecture that the maximum degree of this invariant algebraic curve is 12 under the condition that the vector field does not admit a Liouvillian first integral.

The invariant algebraic curves and of the integrability of planar differential systems has been also studied from other points of view. Thus, there are results related with the inverse problem using the Darboux theory of integrability that we do not consider in this book, but that can be found in [35, 36, 37, 38, 101, 102, 110, 140].

2.2   Preliminary definitions and basic results

We consider the planar polynomial differential system

$$\dot{x} = P(x,y), \qquad \dot{y} = Q(x,y), \tag{2.1}$$

of degree m, where $P, Q \in \mathbb{C}_m[x,y]$. As usual $\mathbb{C}_m[x,y]$ denotes the set of all complex polynomials of degree at most m in the variables x and y, and $m = \max\{\deg P, \deg Q\}$.

Let $g = g(x,y) \in \mathbb{C}[x,y]$, where $\mathbb{C}[x,y]$ is the ring of complex polynomials in x and y. The algebraic curve $g(x,y) = 0$ in $\mathbb{C}^2$ is called an *invariant algebraic curve* of system (2.1) if

$$Pg_x + Qg_y = Kg, \tag{2.2}$$

for some complex polynomial $K = K(x,y)$, which is called the *cofactor* of $g = 0$. Henceforth we denote by g_x and g_y the derivatives of g with respect to x and y, respectively. For simplicity, in what follows we will talk about the curve $g = 0$, only saying "the curve g."

We note that the cofactor of an invariant algebraic curve for system (2.1) has degree at most $m - 1$.

Below we shall need the following basic results.

Let $g = \prod_{i=1}^{l} g_i^{n_i}$ be the irreducible decomposition of g. Then g is an invariant algebraic curve with a cofactor K of the system (2.1) if and only if g_i is an invariant algebraic curve with a cofactor K_i of the system (2.1). Moreover, we have $K = \sum_{i=1}^{l} n_i K_i$. For a proof, see [25].

Let $F(x,y) = \exp(G(x,y)/H(x,y))$ with $G, H \in \mathbb{C}[x,y]$ coprime (or equivalently, $(G, H) = 1$). We say that F is an *exponential factor* if

$$PF_x + QF_y = LF,$$

for some polynomial $L \in \mathbb{C}_{m-1}[x,y]$, which is called the *cofactor* of F.

Let U be an open subset of $\mathbb{C}^2$. A complex function $\mathcal{H} : U \to \mathbb{C}$ is a *first integral* of the system (2.1), if it is constant on all solution curves $(x(t), y(t))$ of (2.1), i.e., $\mathcal{H}(x(t), y(t)) \equiv$ constant for all values of t for which the solution $(x(t), y(t))$ is defined on U. If the first integral $\mathcal{H}$ is differentiable, then $P\mathcal{H}_x + Q\mathcal{H}_y = 0$ in U.

If there exists a smooth function $\mathcal{R}(x,y)$ such that $(P\mathcal{R})_x + (Q\mathcal{R})_y = 0$, then $\mathcal{R}$ is called an *integrating factor* of the system (2.1).

If the system (2.1) has a first integral or an integrating factor of the form

$$g_1^{\lambda_1} \cdots g_p^{\lambda_p} F_1^{\mu_1} \cdots F_q^{\mu_q}, \tag{2.3}$$

where g_i and F_j are invariant algebraic curves and exponential factors of the system (2.1), respectively, and $\lambda_i, \mu_j \in \mathbb{C}$, then the system (2.1) is called *Darboux integrable*. This kind of function (2.3) is called a *Darboux function*.

The main result of this chapter is to construct a planar differential system with a given set of algebraic curves which can be *generic* (in the sense which we specify in the next section), *non-singular* or *singular*. In particular, we find the quadratic planar vector fields with one invariant algebraic curve. The obtained results are illustrated with concrete examples.

2.3 Planar polynomial vector fields with generic algebraic curves

In this section we are mainly interested in the polynomial differential systems which have a given set of invariant algebraic curves. Thus, first we study the normal forms of *planar polynomial vector fields* having a given set of *generic invariant algebraic curves*.

We say that a given set of invariant algebraic curves is *generic* if it satisfies the following *generic conditions*:

(i) There are no points at which C_i and its first derivatives are all zero.

(ii) The highest order terms of C_i have no repeated factors.

(iii) If two curves intersect at a point in the finite plane, they are transverse at that point.

(iv) There are at most two curves $C_i = 0$ meeting at any point in the finite plane.

(v) No two curves have a common factor in the highest order terms.

In order to distinguish a generic invariant algebraic curve $C = 0$ from a non-generic one $g = 0$, in this section we shall denote the generic curves with the capital letter C.

The first result is the following theorem (see [34]).

Theorem 2.3.1. *Let $C_i = 0$ for $i = 1, \ldots, p$, be irreducible invariant algebraic curves in $\mathbb{C}^2$, and set $r = \sum_{i=1}^{p} \deg C_i$. We assume that all the curves $C_i = 0$ are generic. Then for any polynomial vector field $\mathcal{X}$ of degree m tangent to all $C_i = 0$ one of the following statements holds true.*

(a) *If $r < m + 1$, then*

$$\mathcal{X} = \left(\prod_{i=1}^{p} C_i \right) y + \sum_{i=1}^{p} h_i \left(\prod_{\substack{j=1 \\ j \neq i}}^{p} C_j \right) \mathcal{X}_{C_i}, \qquad (2.4)$$

where $\mathcal{X}_{C_i} = (-C_{iy}, C_{ix})$ *is a Hamiltonian vector field, the* h_i *are arbitrary polynomials of degree at most* $m - r + 1$, *and* $\mathbf{Y}$ *is an arbitrary polynomial vector field of degree at most* $m - r$.

(b) *If* $r = m + 1$, *then*

$$\mathcal{X} = \sum_{i=1}^{p} \alpha_i \left(\prod_{\substack{j=1 \\ j \neq i}}^{p} C_j \right) \mathcal{X}_{C_i}, \tag{2.5}$$

with $\alpha_i \in \mathbb{C}$.

(c) *If* $r > m + 1$, *then* $\mathcal{X} = 0$.

Remark 2.3.2. This theorem, due to Christopher [29], was stated without proof in several papers, for instance, [29] and [82], and used in other papers, for instance, [17] and [103]. The proof that we present here of Theorem 2.3.1 circulated as the preprint [28], but was never published. Żołądek in [163] (see also Theorem 3 of [164]) stated a result similar to our Theorem 1, but as far as we know the paper [163] has not been published. In any case Żołądek's approach to Theorem 2.3.1 is analytic, while our approach is completely algebraic and follows the proof given in [163].

Statement (b) of this theorem has a corollary due to Christopher and Kooij [82] showing that system (2.5) has the *integrating factor* $\mathcal{R} = \left(\prod_{i=1}^{p} C_i \right)^{-1}$, and consequently the system is *Darboux integrable*.

In the proof of Theorem 2.3.1 we will make essential use of *Hilbert's Nullstellensatz* (see for instance, [58]):

Let $A, B_i \in \mathbb{C}[x, y]$ *for* $i = 1, \ldots, r$. *If* A *vanishes in* $\mathbb{C}^2$ *whenever the polynomials* B_i *vanish simultaneously, then there exist polynomials* $M_i \in \mathbb{C}[x, y]$ *and a nonnegative integer* n *such that* $A^n = \sum_{i=1}^{r} M_i B_i$. *In particular, if all* B_i *have no common zero, then there exist polynomials* M_i *such that* $\sum_{i=1}^{r} M_i B_i = 1$.

In what follows, for a polynomial A we will denote its degree by a. Also, for a polynomial C we denote by C^c the homogeneous part of degree c of C. We shall need the following result.

Lemma 2.3.3. *If* C^c *has no repeated factors, then* $(C_x, C_y) = 1$.

Proof. Suppose that $(C_x, C_y) \neq 1$. Then there exists a nonconstant polynomial A such that $A|C_x$ and $A|C_y$. ($A|C_x$ means that the polynomial A divides the polynomial C_x.) Therefore, $A^a|(C^c)_x$ and $A^a|(C^c)_y$. By the Euler theorem for homogeneous polynomials, $x(C^c)_x + y(C^c)_y = cC^c$. So $A^a|C^c$. Since A^a, $(C^c)_x$, $(C^c)_y$, and C^c are homogeneous polynomials in $\mathbb{C}[x, y]$ and A^a divides $(C^c)_x$, $(C^c)_y$, and C^c, the linear factors of A^a having multiplicity m, must be linear factors of C^c having multiplicity $m + 1$. This last statement follows easily by identifying the linear factors of the homogeneous polynomial $C^c(x, y)$ in two variables with the roots of the polynomial $C^c(1, z)$ in the variable z. Hence, A^a is a repeated factor of C^c. This contradicts with the assumption. □

We first consider the case where the system (2.1) has a given invariant algebraic curve.

Lemma 2.3.4. *Assume that the polynomial system* (2.1) *of degree m has an invariant algebraic curve $C = 0$ of degree c, and that C satisfies condition* (i) *of Theorem 2.3.1.*

(a) *If $(C_x, C_y) = 1$, then the system* (2.1) *has the normal form*

$$\dot{x} = AC - DC_y, \qquad \dot{y} = BC + DC_x, \qquad (2.6)$$

where $A, B,$ and D are suitable polynomials.

(b) *If C satisfies the genericity condition* (ii), *then the system* (2.1) *has the normal form* (2.6) *with $a, b \leq m - c$ and $d \leq m - c + 1$. Moreover, if the highest order term C^c of C does not have the factors x and y, then $a \leq p - c$, $b \leq q - c$, and $d \leq \min\{p, q\} - c + 1$.*

Proof. (a) Since there are no points at which C, C_x, and C_y vanish simultaneously, Hilbert's Nullstellensatz shows that there exist polynomials E, F, and G such that

$$EC_x + FC_y + GC = 1. \qquad (2.7)$$

From (2.2) and (2.7) it follows that

$$K = (KE + GP)C_x + (KF + GQ)C_y$$

is invariant.

Substituting K into (2.2), we get

$$[P - (KE + GP)C]C_x = -[Q - (KF + GQ)C]C_y.$$

Since $(C_x, C_y) = 1$, there exists a polynomial D such that

$$P - (KE + GP)C = -DC_y, \qquad Q - (KF + GQ)C = DC_x.$$

This proves that the system (2.1) has the form (2.6) with $A = KE + GP$ and $B = KF + GQ$. The statement (a) is proved.

(b) From (a) and Lemma 2.3.3 we get that system (2.1) has the normal form (2.6). Without loss of generality we can assume that $p \leq q$.

We first consider the case where C^c has neither as a factor x, nor y. So we have $(C^c, (C^c)_x) = 1$ and $(C^c, (C^c)_y) = 1$. In (2.6) we assume that $a > p - c$, otherwise the statement follows. Then $d = a + 1$. Moreover, examining the highest order terms of (2.6) we get

$$A^a C^c = D^{a+1} C_y^{c-1},$$

where C_y^{c-1} denotes the homogeneous part of degree $c-1$ of C_y. Since (C^c, C_y^{c-1}) $= 1$, there exists a polynomial F such that

$$A^a = FC_y^{c-1}, \qquad D^{a+1} = FC^c.$$

Now in (2.6) we replace A by $A - FC_y$ and D by $D - FC$, so the degrees of polynomials under consideration are reduced by one. We continue this process and do the same for $\dot{y}$ until we reach a system of the form

$$\dot{x} = AC - DC_y, \qquad \dot{y} = BC + EC_x, \qquad (2.8)$$

with $a \le p - c$, $d \le p - c + 1$, $b \le q - c$, and $e \le q - c + 1$. Since $C = 0$ is an invariant algebraic curve of (2.8), from (2.2) we get

$$C(AC_x + BC_y) + C_x C_y(E - D) = KC.$$

This implies that there exists a polynomial R such that $E - D = RC$, because C with C_x and C_y are coprime.

If $e \ge d$, then $r = e - c$. If we write $BC + EC_x = (B + RC_x)C + DC_x$ and denote $B + RC_x$ again by B, then the system (2.8) has the form (2.6), where A, B and D have the required degrees.

If $e < d$, then $r = d - c$. If we write $AC - DC_y = (A + RC_y)C - EC_y$ and denote $A + RC_y$ again by A, then the system (2.8) has the form (2.6), where A, B and E instead of D have the required degrees. This proves the second part of (b).

Now we prove the first part of (b). We note that even though C^c has no repeated factor, C^c and C_x^{c-1} or C_y^{c-1} may have a common factor in x or y (for example $C^3 = x(x^2 + y^2)$, $C^3 = y(x^2 + y^2)$, or $C^4 = xy(x^2 + y^2)$). In order to avoid this difficulty we rotate system (2.1) slightly such that C^c has no factors in x and y. Then, applying the above method to the new system we get that the new system has a normal form (2.6) with the degrees of A, B, and D as those of the second part of (b).

We claim that under affine changes system (2.6) preserves its form and the upper bound on degrees of the polynomials, i.e., $a, b \le m - c$ and $d \le m - c + 1$. Indeed, the *affine change of variables* $u = a_1 x + b_1 y + c_1$, $v = a_2 x + b_2 y + c_2$ with $a_1 b_2 - a_2 b_1 \ne 0$, takes the system (2.6) into

$$\dot{u} = (a_1 A + b_1 B)C - (a_1 b_2 - a_2 b_1)DC_v, \qquad \dot{v} = (a_2 A + b_2 B)C + (a_1 b_2 - a_2 b_1)DC_u.$$

Hence, the claim follows. This completes the proof of (b) and hence of the lemma. $\square$

Lemma 2.3.5. *Assume that $C = 0$ and $D = 0$ are different irreducible invariant algebraic curves of system (2.1) of degree m, and that they satisfy the genericity conditions (i) and (iii).*

(a) *If $(C_x, C_y) = 1$ and $(D_x, D_y) = 1$, then the system (2.1) has the normal form*

$$\dot{x} = ACD - EC_y D - FCD_y, \qquad \dot{y} = BCD + EC_x D + FCD_x, \qquad (2.9)$$

(b) *If C and D satisfy conditions* (ii) *and* (v), *then the system* (2.1) *has the normal form* (2.9) *with $a, b \leq m - c - d$ and $e, f \leq m - c - d + 1$.*

Proof. Since $(C, D) = 1$, the curves C and D have finitely many intersection points. By assumption (i), at each of such points there is at least one non-zero first derivative of both C and D. Similarly to the proof of the claim inside the proof of Lemma 2.3.4, we can show that under an affine change of variables, system (2.9) preserves its form and the bound for the degrees of A, B, E and F. So, we rotate system (2.1) slightly so that all first derivatives of C and D are not equal to zero at the intersection points.

By Hilbert's Nullstellensatz, there exist polynomials M_i, N_i and R_i, $i = 1, 2$ such that

$$M_1 C + N_1 D + R_1 D_y = 1, \quad M_2 C + N_2 D + R_2 C_y = 1. \tag{2.10}$$

By Lemma 2.3.4,

$$P = A_1 C - E_1 C_y = G_1 D - F_1 D_y, \tag{2.11}$$

for some polynomials A_1, E_1, G_1, and F_1. Moreover, using the first equation of (2.10) we have $F_1 = SC + TD + UC_y$ for some polynomials S, T and U. Substituting F_1 into (2.11) we obtain

$$(A_1 + SD_y) C + (-G_1 + TD_y) D + (-E_1 + UD_y) C_y = 0. \tag{2.12}$$

Using the second equation of (2.10) and (2.12) to eliminate C_y we get

$$-E_1 + UD_y = VC + WD, \tag{2.13}$$

for some polynomials V and W. Substituting (2.13) into (2.12), we have

$$(A_1 + SD_y + VC_y) C = (G_1 - TD_y - WC_y) D.$$

Since $(C, D) = 1$, there exists a polynomial K such that

$$A_1 + SD_y + VC_y = KD, \quad G_1 - TD_y - WC_y = KC. \tag{2.14}$$

Substituting E_1 from (2.13) and A_1 from (2.14) into (2.11), we get

$$P = KCD - SCD_y + WC_y D - UC_y D_y. \tag{2.15}$$

Similarly, we can prove that there exist polynomials K', S', W', and U' such that

$$Q = K'CD + S'CD_x - W'C_x D + U'C_x D_x. \tag{2.16}$$

Since C is an invariant algebraic curve of (2.1), we have that $PC_x + QC_y = K_C C$ for some polynomial K_C. Using (2.15) and (2.16) we get

$$K_C C = C \left[D \left(KC_x + K'C_y \right) - SC_x D_y + S'C_y D_x \right]$$
$$+ C_x C_y \left[D(W - W') - UD_y + U'D_x \right].$$

As C, C_x, and C_y are coprime, there exists a polynomial Z such that

$$D(W - W') - UD_y + U'D_x = ZC. \tag{2.17}$$

Substituting the expression $DW - UD_y$ into (2.15), we get

$$P = KCD - SCD_y + W'C_yD - U'C_yD_x + ZCC_y. \tag{2.18}$$

Since $D = 0$ is an invariant algebraic curve of system (2.1), we have $PD_x + QD_y = K_D D$ for some polynomial K_D. Using (2.16) and (2.18) we get

$$\begin{aligned}
K_D D = &\, D\left[C\left(KD_x + K'D_y\right) + W'\left(C_yD_x - C_xD_y\right)\right] \\
&+ D_x\left[CD_y(-S + S') + U'\left(C_xD_y - C_yD_x\right) + ZCC_y\right].
\end{aligned}$$

As D and D_x are coprime, there exists a polynomial M such that

$$CD_y(-S + S') + U'\left(C_xD_y - C_yD_x\right) + ZCC_y = MD. \tag{2.19}$$

Next, since the curves C and D are transverse, C, D and $C_xD_y - C_yD_x$ have no common zeros. From Hilbert's Nullstellensatz, there exist some polynomials M_3, N_3 and R_3 such that

$$M_3C + N_3D + R_3\left(C_xD_y - C_yD_x\right) = 1. \tag{2.20}$$

Eliminating the term $C_xD_y - C_yD_x$ from (2.19) and (2.20), we obtain that $U' = IC + JD$ for some polynomials I and J. Hence, equation (2.19) becomes

$$\begin{aligned}
C\left[I\left(C_xD_y - C_yD_x\right) + D_y(-S + S') + ZC_y\right] \\
+ D\left[J\left(C_xD_y - C_yD_x\right) - M\right] = 0.
\end{aligned}$$

Since $(C, D) = 1$, there exists a polynomial G such that

$$\begin{aligned}
M &= J\left(C_xD_y - C_yD_x\right) + GC, \\
I\left(C_xD_y - C_yD_x\right) + D_y(-S + S') + ZC_y &= GD.
\end{aligned}$$

Substituting $ZC_y - SD_y$ and U' into (2.18) we obtain that

$$P = (K + G)CD - \left(IC_x + S'\right)CD_y + \left(W' - JD_x\right)DC_y.$$

This means that P can be expressed in the form (2.15) with $U = 0$.

Working in a similar way, we can express Q in the form (2.16) with $U' = 0$. Thus, (2.17) implies that $D(W - W') = ZC$. Hence, we have $W = W' + HC$ for some polynomial H. Consequently, $Z = HD$. Therefore, from (2.19) we obtain that $CD_y(-S + S') = D(M - HCC_y)$. Since $(C, D) = 1$ and $(D, D_y) = 1$, we have $S = S' + LD$ for some polynomial L. Substituting W and S into (2.15) we obtain that P and Q have the form (2.9). This proves statement (a).

As in the proof of Lemma 2.3.4 we can show that under suitable affine changes of variables the form of system (2.9) and the bound of the degrees of the polynomials A, B, E and F are invariant. So, without loss of generality we can assume that the highest order terms of C and D are not divisible by x or y.

By the assumptions, the conditions of statement (a) hold, so we get that system (2.1) has the form (2.9). If the bounds of the degrees of A, B, E, and F are not satisfied, we have by (2.9) that

$$
\begin{aligned}
A^a C^c D^d - E^e C_y^{c-1} D^d - F^f C^c D_y^{d-1} &= 0, \\
B^b C^c D^d + E^e C_x^{c-1} D^d + F^f C^c D_x^{d-1} &= 0.
\end{aligned}
\tag{2.21}
$$

We remark that if one of the numbers $a+c+d$, $e+c-1+d$, and $f+c+d-1$ is less than the other two, then its corresponding term in the first equation of (2.21) is equal to zero. The same remark applies to the second equation of (2.21). From the hypotheses it follows that C^c and C_y^{c-1} are coprime, and so are D^d and D_y^{d-1}, and C^c and D^d, respectively. Hence, from these last two equations we obtain that there exist polynomials K and L such that $E^e = KC^c$, $F^f = LD^d$, and

$$
A^a = KC_y^{c-1} + LD_y^{d-1}, \qquad B^b = -KC_x^{c-1} - LD_x^{d-1}.
$$

We rewrite equation (2.9) as

$$
\begin{aligned}
\dot{x} &= (A - KC_y - LD_y)\,CD - (E - KC)C_yD - (F - LD)CD_y, \\
\dot{y} &= (B + KC_x + LD_x)\,CD + (E - KC)C_xD + (F - LD)CD_x.
\end{aligned}
$$

Thus, we reduced the degrees of A, B, E and F in (2.9) by one. We can continue this process until the bounds are reached. This completes the proof of statement (b). □

Lemma 2.3.6. *Let $C_i = 0$ for $i = 1, \ldots, p$ be distinct irreducible invariant algebraic curves of system (2.1) with $\deg C_i = c_i$. Assume that C_i satisfy the genericity conditions (i), (iii), and (iv). Then the following statements hold.*

(a) *If $(C_{ix}, C_{iy}) = 1$ for $i = 1, \ldots, p$, then system (2.1) has the normal form*

$$
\dot{x} = \left(B - \sum_{i=1}^{p} \frac{A_i C_{iy}}{C_i} \right) \prod_{i=1}^{p} C_i, \quad \dot{y} = \left(D + \sum_{i=1}^{p} \frac{A_i C_{ix}}{C_i} \right) \prod_{i=1}^{p} C_i, \tag{2.22}
$$

where B, D, and A_i are suitable polynomials.

(b) *If C_i satisfy conditions (ii) and (v) of Theorem 2.3.1, then system (2.1) has the normal form (2.22) with $b, d \leq m - \sum_{i=1}^{p} c_i$, and $a_i \leq m - \sum_{i=1}^{p} c_i + 1$.*

Proof. We proceed by induction. Using Lemmas 2.3.4 and 2.3.5, we assume that for any l with $2 \leq l < p$ we have

$$P = \sum_{i=1}^{l} \left(B_i - \frac{A_i C_{iy}}{C_i} \right) \prod_{i=1}^{l} C_i, \quad Q = \sum_{i=1}^{l} \left(D_i + \frac{A_i C_{ix}}{C_i} \right) \prod_{i=1}^{l} C_i,$$

where $\sum_{i=1}^{l} B_i = B$ and $\sum_{i=1}^{l} D_i = D$. Since $C_{l+1} = 0$ is an invariant algebraic curve, Lemma 2.3.4 shows that there exist polynomials E, G, and H such that

$$P = \sum_{i=1}^{l} \left(B_i - \frac{A_i C_{iy}}{C_i} \right) \prod_{i=1}^{l} C_i = E C_{l+1} - G C_{l+1,y},$$

$$Q = \sum_{i=1}^{l} \left(D_i + \frac{A_i C_{ix}}{C_i} \right) \prod_{i=1}^{l} C_i = H C_{l+1} + G C_{l+1,x}. \tag{2.23}$$

Now consider the curves

$$K_j = \prod_{\substack{i=1 \\ i \neq j}}^{l} C_i = 0, \qquad j = 1, \dots, l.$$

From the assumptions we obtain that there is no point at which all the curves $K_i = 0$ and $C_{l+1} = 0$ intersect. Moreover, at least three of the curves $C_i = 0$ for $i = 1, \dots, l+1$ intersect at some point. Hence, there exist polynomials U and V_i for $i = 1, \cdots, l$ such that

$$U C_{l+1} + \sum_{i=1}^{l} V_i K_i = 1. \tag{2.24}$$

Using this equality, we can rearrange (2.23) as

$$(E - G U C_{l+1,y}) C_{l+1} = \sum_{i=1}^{l} \left(B_i C_i - A_i C_{iy} + G V_i C_{l+1,y} \right) K_i,$$

$$(H + G U C_{l+1,x}) C_{l+1} = \sum_{i=1}^{l} \left(D_i C_i + A_i C_{ix} - G V_i C_{l+1,x} \right) K_i. \tag{2.25}$$

Using (2.24) and (2.25) to eliminate C_{l+1} we obtain that

$$E - G U C_{l+1,y} = \sum_{i=1}^{l} I_i K_i, \quad H + G U C_{l+1,x} = \sum_{i=1}^{l} J_i K_i,$$

for some polynomials I_i and J_i. Substituting these last equalities into (2.25), we have

$$\sum_{i=1}^{l} \left(B_i C_i - A_i C_{iy} + GV_i C_{l+1,y} - I_i C_{l+1} \right) K_i = 0,$$

$$\sum_{i=1}^{l} \left(D_i C_i + A_i C_{ix} - GV_i C_{l+1,x} - J_i C_{l+1} \right) K_i = 0.$$
(2.26)

It is easy to check that the expression multiplying K_i in the two sums in (2.26) is divisible by C_i. Hence, there exist polynomials L_i and F_i for $i = 1, \ldots, l$ such that

$$B_i C_i - A_i C_{iy} + GV_i C_{l+1,y} - I_i C_{l+1} = L_i C_i,$$
$$D_i C_i + A_i C_{ix} - GV_i C_{l+1,x} - J_i C_{l+1} = F_i C_i.$$
(2.27)

So, from (2.26) we get that $\sum_{i=1}^{l} L_i = 0$ and $\sum_{i=1}^{l} F_i = 0$. This implies that (2.23) can be rewritten as

$$P = \sum_{i=1}^{l} \left((B_i - L_i)C_i - A_i C_{iy} \right) K_i, \quad Q = \sum_{i=1}^{l} \left((C_i - F_i)C_i + A_i C_{ix} \right) K_i. \quad (2.28)$$

Further, we write (2.27) in the form

$$(B_i - L_i)C_i - A_i C_{iy} = I_i C_{l+1} - GV_i C_{l+1,y} = P_i,$$
$$(D_i - F_i)C_i + A_i C_{ix} = J_i C_{l+1} + GV_i C_{l+1,x} = Q_i.$$

It is easy to see that C_i and C_{l+1} are invariant algebraic curves of the system $\dot{x} = P_i, \dot{y} = Q_i$. So, from statement (a) of Lemma 2.3.5 we obtain that

$$P_i = (B_i - L_i)C_i - A_i C_{iy} = X_i C_i C_{l+1} - Y_i C_{iy} C_{l+1} - N_i C_i C_{l+1,y},$$
$$Q_i = (D_i - F_i)C_i + A_i C_{ix} = Z_i C_i C_{l+1} + Y_i C_{ix} C_{l+1} + N_i C_i C_{l+1,x}.$$

Substituting these last two equations into (2.28), we deduce that the system (2.1) has the form (2.22) with the $l + 1$ invariant algebraic curves $C_1, \ldots, C_{l+1}$. The induction is complete, and so is the proof of statement (a).

The proof of statement (b) is almost identical with that of Lemma 2.3.5(b), so we omit it here. The proof of the lemma is complete. $\qquad\square$

Proof of Theorem 2.3.1. Statement (a) of follows from Lemma 2.3.6.

By checking the degrees of polynomials A_i, B and D in statement (b) of Lemma 2.3.6 we obtain statement (b) of Theorem 2.3.1.

By statement (a) of Lemma 2.3.6, we can rearrange system (2.1) so that it has the form (2.22). But from statement (b) of Lemma 2.3.6 we must have $B = 0$, $D = 0$, and $A_i = 0$. This proves statement (c) of Theorem 2.3.1. $\qquad\square$

Remark 2.3.7. We observe that system (2.22) coincides with the system of differential equations (1.44), by introducing the corresponding notations, when the partial integrals are polynomial and generic and the arbitrary functions $\lambda_1, \ldots, \lambda_M$, λ_{M+1}, and λ_{M+2} are arbitrary rational functions such that

$$P(x,y) = \sum_{j=1}^{M} \prod_{\substack{m=1 \\ m \neq j}}^{M} g_m \lambda_j \{g_j, x\} - \prod_{m=1}^{M} g_m \lambda_{M+1},$$

$$Q(x,y) = \sum_{j=1}^{M} \prod_{\substack{m=1 \\ m \neq j}}^{M} g_m \lambda_j \{g_j, y\} + \prod_{m=1}^{M} g_m \lambda_{M+2},$$

$$m = \max\left(\deg P, \deg Q\right).$$

2.4   On the degree of invariant non-singular algebraic curves

From Joanolou's Theorem (see for instance [77, 98]) it follows that for a given polynomial differential system of degree m the maximum *degree of its irreducible invariant algebraic curves* is bounded, since either the system has a finite number $p < \frac{1}{2}m(m+1)+2$ of invariant algebraic curves, or all its trajectories are contained in invariant algebraic curves and the system admits a *rational first integral*. Thus for each polynomial system there is a natural number N which bounds the degrees of all its irreducible invariant algebraic curves. A natural problem, going back to Poincaré (for more details see [128]), is to give an effective procedure to find N. Partial answers to this question were given in [18, 20, 21]. Of course, given such a bound, it is easy to compute the algebraic curves of the system.

The purpose of this section is to give an upper bound for the degree of the invariant non-singular algebraic curves [32]. First we present the basic results which we will apply below to prove the main results.

Projectivization of polynomial vector fields in $\mathbb{C}^2$

Following Darboux, we will work in the *complex projective plane* $\mathbb{C}P^2$. Let P, Q and R be homogeneous polynomials of degree $m+1$ in the complex variables X, Y and Z. We say that the homogeneous 1-form

$$\omega = PdX + QdY + RdZ$$

of degree $m+1$ is *projective* if $XP + YQ + ZR = 0$; that is, if there exist three homogeneous polynomials L, M, and N of degree m such that

$$P = ZM - YN, \quad Q = XN - ZL, \quad R = YL - XM;$$

or, equivalently,

$$(P, Q, R) = (L, M, N) \wedge (X, Y, Z).$$

Then we can write

$$\omega = L(Y\,dZ - Z\,dY) + M(Z\,dX - X\,dZ) + N(X\,dY - Y\,dX). \qquad (2.29)$$

The triple (L, M, N) can be thought of as a *homogeneous polynomial vector field* on $\mathbb{C}P^2$ of degree m, more specifically

$$\mathcal{X} = L\frac{\partial}{\partial X} + M\frac{\partial}{\partial Y} + N\frac{\partial}{\partial Z}, \qquad (2.30)$$

where X, Y, and Z are the homogeneous coordinates on $\mathbb{C}P^2$.

If (L', M', N') also defines ω, i.e.,

$$P = ZM' - YN', \quad Q = XN' - ZL', \quad R = YL' - XM',$$

then

$$(M' - M)Z - (N' - N)Y = (N' - N)X - (L' - L)Z = (L' - L)Y - (M' - M)X = 0.$$

This implies that there exists a homogeneous polynomial H of degree $m - 1$ such that

$$L' = L + XH, \quad M' = M + YH, \quad N' = N + ZH.$$

We note that if we choose

$$H = -\frac{1}{m+2}\left(\frac{\partial L}{\partial X} + \frac{\partial M}{\partial Y} + \frac{\partial N}{\partial Z}\right),$$

then

$$\frac{\partial L'}{\partial X} + \frac{\partial M'}{\partial Y} + \frac{\partial N'}{\partial Z} = 0, \qquad (2.31)$$

where we have used that

$$X\frac{\partial H}{\partial X} + Y\frac{\partial H}{\partial Y} + Z\frac{\partial H}{\partial Z} = (m - 1)H,$$

by the Euler Theorem applied to the homogeneous polynomial H of degree $m - 1$. The vector field

$$L'\frac{\partial}{\partial X} + M'\frac{\partial}{\partial Y} + N'\frac{\partial}{\partial Z},$$

satisfying (2.31) is called the *Darboux normal form* of the projective 1-form ω. In short, the study of the *projective 1-form* $\omega = P\,dX + Q\,dY + R\,dZ$ of degree $m+1$ is equivalent to the study of the homogeneous polynomial vector fields (2.30) of degree m of $\mathbb{C}P^2$ satisfying (2.31). Here in this paper we do not need to use Darboux's normal form.

Usually in the literature a projective 1-form ω is called a *Pfaff algebraic form* of degree $m + 1$ of $\mathbb{C}P^2$; for more details Joanolou [77].

Let $F(X, Y, Z)$ be a homogeneous polynomial. Then $F = 0$ in $\mathbb{C}P^2$ is an *invariant algebraic curve* of the vector field (2.30) if $\mathcal{X}F = KF$ for some homogeneous polynomial $K(X, Y, Z)$ of degree $m - 1$, called the *cofactor* of F.

The polynomial vector field

$$\mathcal{X} = p(x, y)\frac{\partial}{\partial x} + q(x, y)\frac{\partial}{\partial y} \tag{2.32}$$

on $\mathbb{C}^2$ of degree m, or equivalently the differential system

$$\frac{dy}{dx} = \frac{q(x, y)}{p(x, y)},$$

or the 1-form

$$p(x, y)dy - q(x, y)dx = 0,$$

can be thought in $\mathbb{C}P^2$ as the projective 1-form

$$Z^{m+2}\left[p\left(\frac{X}{Z}, \frac{Y}{Z}\right)\frac{ZdY - YdZ}{Z^2} - q\left(\frac{X}{Z}, \frac{Y}{Z}\right)\frac{ZdX - XdZ}{Z^2}\right] = 0$$

of degree $m + 1$, where we have replaced x and y by X/Z and Y/Z, respectively.

Let $P(X, Y, Z) = Z^m p(X/Z, Y/Z)$ and $Q(X, Y, Z) = Z^m q(X/Z, Y/Z)$. Then the last projective 1-form becomes

$$P(X, Y, Z)(YdZ - ZdY) + Q(X, Y, Z)(ZdX - XdZ) = 0.$$

So the polynomial vector field (2.32) of $\mathbb{C}^2$ of degree m is equivalent to the following homogeneous polynomial vector field on $\mathbb{C}P^2$

$$\mathcal{X} = P\frac{\partial}{\partial X} + Q\frac{\partial}{\partial Y} \tag{2.33}$$

of degree m. This vector field is called the *projectivization* of the vector field (2.32).

We remark that the *projectivized vector field* has the third component N identically zero, and consequently the infinite straight line $Z = 0$ of the projectivization of $\mathbb{C}^2$ is a solution of the projectivized vector field.

It is easy to verify that if $f(x, y) = 0$ is an invariant algebraic curve of (2.32) of degree n with cofactor $k(x, y)$, then $F(X, Y, Z) = Z^n f(X/Z, Y/Z) = 0$ is an invariant algebraic curve of (2.33) with cofactor $K(X, Y, Z) = Z^{m-1}k(X/Z, Y/Z)$.

Singular points of homogeneous polynomial vector fields on $\mathbb{C}P^2$

In this section we present arguments and results due to Darboux [43]; they will play an essential role in proving our results. First we recall some definitions and preliminary results on planar algebraic curves in $\mathbb{C}P^2$.

Let $F(X, Y, Z)$ a homogeneous polynomial of degree n in the variables X, Y, and Z and with coefficients in $\mathbb{C}$. Then $F(X, Y, Z) = 0$ is an algebraic curve in $\mathbb{C}P^2$. Let $p = (X_0, Y_0, Z_0)$ be a point of $\mathbb{C}P^2$. Since the three coordinates of p cannot be all zero, without loss of generality we can assume that $p = (0, 0, 1)$. Then suppose that the expression of $F(X, Y, Z)$ restricted to $Z = 1$ is

$$F(X, Y, 1) = F_m(X, Y) + F_{m+1}(X, Y) + \cdots + F_n(X, Y),$$

where $0 \leq m \leq n$ and $F_j(X, Y)$ denotes a homogeneous polynomial of degree j in the variables X and Y for $j = m, \ldots, n$, with F_m different from the zero polynomial. We say that $m = m_p(F)$ is the *multiplicity of the curve* $F = 0$ at the point p. If $m = 0$, then the point p does not belong to the curve $F = 0$. If $m = 1$, we say that p is a *simple* point for the curve $F = 0$. If $m > 1$, we say that p is a *multiple* point. In particular, p is a *multiple point* of $F = 0$ if and only if

$$\frac{\partial F}{\partial X}(p) = 0, \quad \frac{\partial F}{\partial Y}(p) = 0, \quad \frac{\partial F}{\partial Z}(p) = 0.$$

Suppose that for $m > 0$ we have that $F_m = \prod_{i=1}^{r} L_i^{r_i}$, where the L_i are distinct straight lines, called *tangent lines* to $F = 0$ at the point p, and r_i is the *multiplicity* of the tangent line L_i at p. For $m > 1$ we say that p is an *ordinary multiple* point if the multiplicity of all tangent lines at p is 1, otherwise we say that p is a *non-ordinary multiple point*.

Let $F = 0$ and $G = 0$ be two algebraic curves and p a point of $\mathbb{C}P^2$. We say that $F = 0$ and $G = 0$ intersect *strictly* at p, if $F = 0$ and $G = 0$ have no common components which pass through p. We say that $F = 0$ and $G = 0$ intersect *transversally* at p if p is a simple point of $F = 0$ and of $G = 0$, and the tangents to $F = 0$ and $G = 0$ at p are distinct. The proof of the following two theorems can be found in [58].

Theorem 2.4.1 (*Intersection Number Theorem*). *There exists a unique multiplicity or intersection number, $I(p, F \cap G)$, defined for all algebraic curves $F = 0$ and $G = 0$ and for all point p of $\mathbb{C}P^2$, satisfying the following properties:*

(i) $I(p, F \cap G)$ *is a non-negative integer for all F, G and p when $F = 0$ and $G = 0$ intersect strictly at p. $I(p, F \cap G) = \infty$ if $F = 0$ and $G = 0$ do not intersect strictly at p.*

(ii) $I(p, F \cap G) = 0$ *if and only if p is not a common point to $F = 0$ and $G = 0$. $I(p, F \cap G)$ depends only on the components of $F = 0$ and $G = 0$ which pass through p.*

(iii) *If T is a change of coordinates and $T(p) = q$, then $I(q, T(F) \cap T(G)) = I(p, F \cap G)$.*

(iv) $I(p, F \cap G) = I(p, G \cap F)$.

(v) $I(p, F \cap G) \geq m_p(F) m_p(G)$, *and equality holds if and only if $F = 0$ and $G = 0$ have no common tangents at p.*

(vi) $I(p, F \cap G_1 G_2) = I(p, F \cap G_1) + I(p, F \cap G_2)$.

(vii) $I(p, F \cap G) = I(p, F \cap (G + AF))$ *for all homogeneous polynomial $A(X, Y, Z)$.*

Theorem 2.4.2 (Bézout Theorem). *Let $F = 0$ and $G = 0$ be two algebraic curves in $\mathbb{C}P^2$ of degrees r and s, respectively, and without common components. Then*

$$\sum_p I(p, F \cap G) = rs \,.$$

Let $F_i = 0$, for $i = 1, \dots, s$, be algebraic curves and p a point of $\mathbb{C}P^2$. We define the *multiplicity* or *intersection numbers* of the curves $F_1 = 0, \dots, F_s = 0$ at p as

$$I\left(p, \bigcap_{i=1}^{s} F_i\right) = \min_{i<j}\{I(p, F_i \cap F_j)\} \,.$$

The *singular* points of a projective 1-form (2.29) of degree $m + 1$, or of its associated homogeneous polynomial vector field (2.30) of degree m, are those points where the tangent is not defined; that is, the points satisfying the system of equations

$$ZM - YN = 0\,, \quad XN - ZL = 0\,, \quad YL - XM = 0\,.$$

In order to determine the number of *singular points of a homogeneous polynomial vector field* (2.30) of degree m we will use the following result, see a proof in [26].

Lemma 2.4.3 (Darboux Lemma). *Let A, A', B, B', C, and C' be homogeneous polynomials in the variables X, Y, and Z of degrees l, l', m, m', n, and n', respectively, verifying the identity $AA' + BB' + CC' \equiv 0$. Assume that the curves $A = 0$, $B = 0$, $C = 0$ and the curves $A' = 0$, $B' = 0$, $C' = 0$ have no common component, respectively. Then*

$$\sum_p I(p, A \cap B \cap C) + \sum_p I(p, A' \cap B' \cap C') = \frac{lmn + l'm'n'}{\ell} \,,$$

where $\ell = l + l' = m + m' = n + n'$.

Proposition 2.4.4 (Darboux Proposition 1). *For any homogeneous polynomial vector field (2.32) of degree m on $\mathbb{C}P^2$ having finitely many singular points we have that its number of singular points, counting multiplicities, or intersection numbers satisfies*

$$\sum_p I\left(p, (ZM - YN) \cap (XN - ZL) \cap (YL - XM)\right) = m^2 + m + 1 \,.$$

Proposition 2.4.5 (Darboux Proposition 2). *Let $f(x, y) = 0$ be an invariant algebraic curve of degree n of the polynomial vector field (2.32). Then any multiple point of $F(X, Y, Z) = Z^n f(X/Z, Y/Z) = 0$ is a singular point of its projectivization (2.33).*

Sufficient conditions for the existence of rational first integrals

The next theorem provides sufficient conditions for a polynomial vector field to have a rational first integral.

Theorem 2.4.6. *Let* $f(x, y) = 0$ *be an invariant algebraic curve of degree* $n > 1$ *with cofactor* k *for the polynomial vector field*

$$\mathcal{X} = p(x, y) \frac{\partial}{\partial x} + q(x, y) \frac{\partial}{\partial y}$$

of degree $m > 1$. *Assume that* $f(x, y) = 0$ *is not a straight line and that* k *is not identically zero. Define*

$$P(X, Y, Z) = Z^m p\left(\frac{X}{Z}, \frac{Y}{Z}\right),$$

$$Q(X, Y, Z) = Z^m q\left(\frac{X}{Z}, \frac{Y}{Z}\right),$$

$$K(X, Y, Z) = Z^{m-1} k\left(\frac{X}{Z}, \frac{Y}{Z}\right).$$

If the number of point solutions of the system

$$nP - XK = 0, \quad nQ - YK = 0, \quad ZK = 0, \tag{2.34}$$

counting multiplicities, or intersection numbers, is m^2, *then* $\mathcal{X}$ *has a rational first integral.*

Proof. Since m^2 is the number of solutions of system (2.34), then Bézout's Theorem shows that the numbers of solutions of the systems

$$nP - XK = 0, \quad nQ - YK = 0, \quad Z = 0, \tag{2.35}$$

and

$$P = 0, \quad Q = 0, \quad K = 0, \tag{2.36}$$

are m and $m^2 - m$, respectively. We note that always we take multiplicities into account.

We write

$$P(X, Y, Z) = p_0 Z^m + p_1(X, Y) Z^{m-1} + \cdots + p_m(X, Y),$$
$$Q(X, Y, Z) = q_0 Z^m + q_1(X, Y) Z^{m-1} + \cdots + q_m(X, Y),$$
$$K(X, Y, Z) = k_0 Z^{m-1} + k_1(X, Y) Z^{m-2} + \cdots + k_{m-1}(X, Y),$$

where $A_i(x, y)$ denotes a homogeneous polynomial of degree i and $A \in \{p, q, k\}$. Then, for $Z = 0$ system (2.35) becomes

$$np_m(X, Y) - X k_{m-1}(X, Y) = 0, \quad nq_m(X, Y) - Y k_{m-1}(X, Y) = 0.$$

Since this system of homogeneous polynomials of degree m in the variables X and Y has m common zeros taking into account multiplicities, there exists a homogeneous polynomial $A(X,Y)$ of degree m such that

$$p_m(X,Y) = \ell_1 A(X,Y) + \frac{1}{n} X k_{m-1}(X,Y),$$

$$q_m(X,Y) = \ell_2 A(X,Y) + \frac{1}{n} Y k_{m-1}(X,Y),$$

where $\ell_1, \ell_2 \in \mathbb{C}$ are not zero.

The polynomial system associated to the vector field $\mathcal{X}$ can be written as

$$\dot{x} = p_0 + p_1(x,y) + \cdots + p_{m-1}(x,y) + \ell_1 A(x,y) + \frac{1}{n} x\, k_{m-1}(x,y) = P(x,y,1),$$

$$\dot{y} = q_0 + q_1(x,y) + \cdots + q_{m-1}(x,y) + \ell_2 A(x,y) + \frac{1}{n} y k_{m-1}(x,y) = Q(x,y,1),$$

or equivalently

$$\dot{x} = p_0 + p_1 + \cdots + p_{m-1} - \frac{1}{n} x\, (k_0 + k_1 + \cdots + k_{m-2}) + \ell_1 A + \frac{1}{n} x\, k,$$

$$\dot{y} = q_0 + q_1 + \cdots + q_{m-1} - \frac{1}{n} y\, (k_0 + k_1 + \cdots + k_{m-2}) + \ell_2 A + \frac{1}{n} y k,$$

Since $\ell_1^2 + \ell_2^2 \neq 0$ (otherwise system (2.35) does not have m intersection points), without loss of generality we can assume that $\ell_1 \neq 0$. We change from the variables (x,y) to the variables (x,z), where $z = \ell_2 x - \ell_1 y$. In the new variables the second equation of the polynomial system reads

$$\dot{z} = \ell_2 P(x,y,1) - \ell_1 Q(x,y,1) = b(x,y) + \frac{1}{n} z k(x,y),$$

with $y = (\ell_2 x - z)/\ell_1$ and where

$$b(x,y) = \ell_2 \left(p_0 + p_1 + \cdots + p_{m-1} - \frac{1}{n} x\, (k_0 + k_1 + \cdots + k_{m-2}) \right)$$

$$- \ell_1 \left(q_0 + q_1 + \cdots + q_{m-1} - \frac{1}{n} y\, (k_0 + k_1 + \cdots + k_{m-2}) \right).$$

Since, by (2.36), the curves $\dot{x} = P(x,y,1) = 0$, $\dot{y} = Q(x,y,1) = 0$ and $K(x,y,1) = k(x,y) = 0$ have $m^2 - m$ intersection points, counting multiplicities, it follows that the curves $P(x,y,1) = 0$, $b(x,y) + k(x,y)z/n = 0$ and $k(x,y) = 0$ have the same number of intersection points, where $y = (\ell_2 x - z)/\ell_1$. Hence, the number of intersection points of the curves $b(x,y) = 0$ and $k(x,y) = 0$ is at least $m^2 - m$ points taking into account their multiplicities. But, by Bézout's Theorem, if these last two curves do not have a common component, then they have $(m-1)^2$ intersection points, counting multiplicities. Since $m^2 - m > (m-1)^2$ if $m > 1$, it follows that the curves $b = 0$ and $k = 0$ have a maximal common component $c = 0$ of degree $r \geq 1$. Therefore, $b = \bar{b}c$ and $k = \bar{k}c$ where $\bar{b}$ and $\bar{k}$ are polynomials of degree $m - r - 1$.

From (2.34) it follows that the number of intersection points of the curves $P(x, y, 1) = 0$ and $Q(x, y, 1) = 0$ with $k(x, y) = 0$ is maximal, i.e., $m^2 - m$, and then the number of intersection points of the curves $P(x, y, 1) = 0$, $Q(x, y, 1) = 0$ and $\bar{k}(x, y) = 0$ is $m(m - r - 1)$, and the number of intersection points of the curves $P(x, y, 1) = 0$, $Q(x, y, 1) = 0$ and $c(x, y) = 0$ is mr. Of course, we always compute the number of intersection points taking into account their multiplicities. Since the number of intersection points of the curves $P(x, y, 1) = 0$, $b(x, y) + k(x, y)z/n = 0$ and $\bar{k}(x, y) = 0$ is $m(m - r - 1)$, it follows that the number of intersection points of the curves $b(x, y) = 0$ and $\bar{k}(x, y) = 0$ is at least $m(m - r - 1)$. On the other hand, since the curves $b = 0$ and $\bar{k} = 0$ has no common components, by Bézout's Theorem, they intersect at $(m-1)(m-r-1)$ points, counting multiplicities. Hence, since $m > 1$ and $m(m-r-1) > (m-1)(m-r-1)$ except when $r = m-1$, we have that $r = m-1$. Therefore, $b = ak$ with $a \in \mathbb{C}$. So, $\dot{z} = k(a+z/n)$, and consequently $z + an = \ell_2 x - \ell_1 y + an = 0$ is an invariant straight line with cofactor k/n. Then, by statement (a) of Darboux' Theorem, we obtain that $H = f(x, y)(\ell_2 x - \ell_1 y + an)^{-n}$ is a rational first integral of $\mathcal{X}$. We note that since $f = 0$ is different by a straight line, we have that $f \neq (\ell_2 x - \ell_1 y + an)^n$, and consequently H is a first integral. $\square$

On the multiple points of an invariant algebraic curve

In this section we study the number of multiple points that an invariant algebraic curve of degree n of a polynomial vector field of degree m must have in function of m and n.

Theorem 2.4.7. *Let $f(x, y) = 0$ be an invariant algebraic curve of degree n of the polynomial vector field*

$$\mathcal{X} = p(x, y)\frac{\partial}{\partial x} + q(x, y)\frac{\partial}{\partial y}$$

of degree m with cofactor k. Assume that if

$$P(X, Y, Z) = Z^m p\left(\frac{X}{Z}, \frac{Y}{Z}\right),$$

$$Q(X, Y, Z) = Z^m q\left(\frac{X}{Z}, \frac{Y}{Z}\right),$$

$$K(X, Y, Z) = Z^{m-1} k\left(\frac{X}{Z}, \frac{Y}{Z}\right),$$

then

(1) *the curves $nP - XK = 0$, $nQ - YK = 0$, $ZK = 0$, do not have a common component, and*

(2) *the curve $F(X, Y, Z) = Z^n f(X/Z, Y/Z) = 0$ has finitely many multiple points in $\mathbb{C}P^2$, counting multiplicities, namely h.*

Then

$$(n-1)(n-m-1) \leq h \leq m^2 + (n-1)(n-m-1).$$

Proof. Since $F = 0$ is an invariant algebraic curve of (2.33) with cofactor K, we have that

$$\frac{\partial F}{\partial X} P + \frac{\partial F}{\partial Y} Q = KF.$$

in $\mathbb{C}P^2$. By using Euler's Theorem for the homogeneous function F of degree n, this equation is recast as

$$\frac{\partial F}{\partial X}\left(P - \frac{1}{n}XK\right) + \frac{\partial F}{\partial Y}\left(Q - \frac{1}{n}YK\right) + \frac{\partial F}{\partial Z}\left(-\frac{1}{n}ZK\right) = 0. \qquad (2.37)$$

Now we take

$$A = \frac{\partial F}{\partial X}, \quad B = \frac{\partial F}{\partial Y}, \quad C = \frac{\partial F}{\partial Z}, \quad A' = P - \frac{1}{n}XK, \quad B' = Q - \frac{1}{n}YK, \quad C' = -\frac{1}{n}ZK,$$

and

$$h = \sum_p I(p, A \cap B \cap C), \quad h' = \sum_p I(p, A' \cap B' \cap C').$$

We note that by assumption h and h' are finite.

Since A, A', B, B', C and C' satisfy equality (2.37), Darboux' Lemma yields

$$h + h' = \frac{m^3 + (n-1)^3}{m+n-1} = m^2 + (n-1)(n-m+1).$$

Therefore, the upper bound for h given in the statement of the theorem is proved. By Bézout's Theorem, the number of intersection points of the curves $A' = 0$, $B' = 0$ and $C' = 0$, counting multiplicities, is at most m^2, i.e., $h' \leq m^2$. Therefore, $h \geq (n-1)(n-m-1)$, and the theorem is proved. $\qquad \square$

The next result is immediate from Theorem 2.4.7.

Corollary 2.4.8. *Let $f(x, y) = 0$ be an irreducible invariant algebraic curve of degree n without multiple points for the polynomial vector field $\mathcal{X}$ of degree m. Then $n \leq m + 1$.*

Theorem 2.4.9. *Under the assumptions of Theorem 2.4.7, let $f(x, y) = 0$ be an invariant algebraic curve of degree $n = m + 1$ for the polynomial vector field $\mathcal{X}$ of degree $m > 1$ such that its projectivization $F(X, Y, Z) = Z^n f(X/Z, Y/Z) = 0$ has no multiple points. Then $\mathcal{X}$ has a rational first integral.*

Proof. Using the same notation that in the proof of Theorem 3 we have that $h' = m^2$, because from the assumptions we have that $n = m + 1$ and $h = 0$. Now, since we are under the hypotheses of Theorem 1 the statement of the theorem follows. $\qquad \square$

2.5 Polynomial vector fields with partial integrals

In this section the we shall apply Theorems 1.3.1, 1.6.1 and their corollaries when $N = 2$ and we will require that the constructed differential system is polynomial and the given invariant curves are irreducible algebraic curves which in general can be singular, i.e., there are points at which the curve and its first partial derivatives vanish. We shall denote $x = x_1$ and $y = x_2$.

First we prove the following corollary, which will be used in the proof of the results below.

Corollary 2.5.1. *The differential equations*

$$\dot{x} = \Psi \frac{\{x, g_2\}}{\{g_1, g_2\}} + \lambda \frac{\{g_1, x\}}{\{g_1, g_2\}} = P,$$
$$\dot{y} = \Psi \frac{\{y, g_2\}}{\{g_1, g_2\}} + \lambda \frac{\{g_1, y\}}{\{g_1, g_2\}} = Q,$$
(2.38)

where $\{g_1, g_2\} \not\equiv 0$, coincide with the differential equations

$$\dot{x} = \lambda_3 + \nu\{g_1, x\} = P,$$
$$\dot{y} = \lambda_2 + \nu\{g_1, y\} = Q,$$

if

$$\lambda = \nu\{g_1, g_2\} + \lambda_2\{x, g_2\} + \lambda_3\{g_2, y\}$$
$$\Psi = -\lambda_3\{y, g_1\} + \lambda_2\{y, g_1\}.$$

Here the functions ν, λ_2, and λ_3 are arbitrary.

Proof. We apply the Nambu bracket identity (viii) for $N = 2$, i.e.,

$$\{f_1, g_1\}\{G, g_2\} + \{f_1, g_2\}\{g_1, G\} = \{f_1, G\}\{g_1, g_2\},$$

and get

$$\{x, g_1\}\{y, g_2\} + \{x, g_2\}\{g_1, y\} = \{x, y\}\{g_1, g_2\} = \{g_1, g_2\},$$
$$\text{if} \quad f_1 = x, \quad G = y,$$
$$\{y, g_1\}\{x, g_2\} + \{y, g_2\}\{g_1, x\} = \{y, x\}\{g_1, g_2\} = -\{g_1, g_2\},$$
$$\text{if} \quad f_1 = y, \quad G = x,$$
$$\{x, g_1\}\{x, g_2\} + \{x, g_2\}\{g_1, x\} = \{x, x_1\}\{g_1, g_2\} = 0,$$
$$\text{if} \quad f_1 = x, \quad G = x,$$
$$\{y, g_1\}\{y, g_2\} + \{y, g_2\}\{g_1, y\} = \{y, y\}\{g_1, g_2\} = 0,$$
$$\text{if} \quad f_1 = y, \quad G = y.$$
(2.39)

Consequently,

$$\Psi \frac{\{x, g_2\}}{\{g_1, g_2\}} = (-\lambda_3\{y, g_1\} + \lambda_2\{x, g_1\}) \frac{\{x, g_2\}}{\{g_1, g_2\}}$$

$$= (-\lambda_3\{y, g_1\}\{x, g_2\} - \lambda_2\{y, g_1\}\{x, g_2\}) \frac{1}{\{g_1, g_2\}}.$$

In view of (2.39) we have

$$\lambda_2\{x, g_1\}\{x, g_2\} - \lambda_3\{y, g_1\}\{x, g_2\}$$
$$= \lambda_3 \left(\{g_1, g_2\} + \{y, g_2\}\{x, g_1\} + \lambda_2\{g_1, x\}\{x, g_2\}\right)$$
$$= \lambda_3\{g_1, g_2\} + (\lambda_3\{y, g_2\} - \lambda_2\{x, g_2\})\{g_1, x\}.$$

Thus we finally obtain

$$\Psi \frac{\{x, g_2\}}{\{g_1, g_2\}} = \lambda_3 + \frac{(\lambda_3\{y, g_2\} - \lambda_2\{x, g_2\})\{g_1, x\}}{\{g_1, g_2\}}.$$

Similarly,

$$\Psi \frac{\{y, g_2\}}{\{g_1, g_2\}} = \lambda_2 + \frac{(\lambda_3\{y, g_2\} - \lambda_2\{x, g_2\})\{g_1, y\}}{\{g_1, g_2\}}.$$

Inserting these relations into (2.38) we have

$$\dot{x} = \Psi \frac{\{x, g_2\}}{\{g_1, g_2\}} + \lambda \frac{\{g_1, x\}}{\{g_1, g_2\}}$$

$$= \lambda_3 + (\lambda_3\{y, g_2\} - \lambda_2\{x, g_2\}) \frac{\{g_1, x\}}{\{g_1, g_2\}} + \lambda \frac{\{g_1, x\}}{\{g_1, g_2\}}$$

$$= \lambda_3 + (\lambda_3\{y, g_2\} - \lambda_2\{x, g_2\} + \lambda) \frac{\{g_1, x\}}{\{g_1, g_2\}}$$

$$:= \lambda_3 + \nu\{g_1, x\} = P,$$

$$\dot{y} = \Psi \frac{\{y, g_2\}}{\{g_1, g_2\}} + \lambda \frac{\{g_1, y\}}{\{g_1, g_2\}}$$

$$= \lambda_2 + \frac{(\lambda_3\{y, g_2\} - \lambda_2\{x, g_2\})\{g_1, y\}}{\{g_1, g_2\}} + \lambda \frac{\{g_1, y\}}{\{g_1, g_2\}}$$

$$= \lambda_2 + \frac{(\lambda_3\{y, g_2\} - \lambda_2\{x, g_2\} + \lambda)\{g_1, y\}}{\{g_1, g_2\}}$$

$$:= \lambda_2 + \nu\{g_1, y\} = Q.$$

This completes the proof of the corollary. $\square$

2.6 Polynomial vector fields with one algebraic curve

We have the following propositions.

Proposition 2.6.1. *Assume that a polynomial differential system of degree m has an invariant irreducible algebraic curve $g_1 = 0$ of degree n, with cofactor K. Then the system is necessarily of the form*

(a)
$$\dot{x} = Kg_1 \frac{\{x, g_2\}}{\{g_1, g_2\}} + \lambda \frac{\{g_1, x\}}{\{g_1, g_2\}} = \lambda_3 g_1 + \lambda_1\{g_1, x\} = P,$$

$$\dot{y} = Kg_1 \frac{\{y, g_2\}}{\{g_1, g_2\}} + \lambda \frac{\{g_1, y\}}{\{g_1, g_2\}} = \lambda_2 g_1 + \lambda_1\{g_1, y\} = Q,$$

(2.40)

if $g_1 = 0$ is a non-singular invariant algebraic curve, where g_2 and λ are arbitrary rational functions which we choose in such a way that $\{g_1, g_2\} \neq 0$ in $\mathbb{R}^2$, λ_j, for $j = 1, 2, 3$ P and Q are polynomials and

$$\max(\deg P, \deg Q) = m, \quad \deg K \leq m - 1,$$

and

$$-\lambda_3\{y, g_1\} + \lambda_2\{x, g_1\} = K. \tag{2.41}$$

(b) *If $Kg_1 = \{g_1, g_2\}\mu$ and $\lambda = \{g_1, g_2\}\nu$, then*

$$\dot{x} = \mu\{x, g_2\} + \nu\{g_1, x\} := \mu_1 + \nu\{g_1, x\} = P(x, y),$$
$$\dot{y} = \mu\{y, g_2\} + \nu\{g_1, y\} := \mu_2 + \nu\{g_1, y\} = Q(x, y),$$

(2.42)

where g_2, μ and λ are arbitrary rational functions μ_1, μ_2, and ν are arbitrary polynomials such that P and Q are polynomials and

$$\max(\deg P, \deg Q) = m,$$
$$-\mu_1\{y, g_1\} + \mu_2\{x, g_1\} = \{g_1, g_2\}\mu = Kg_1,$$
$$\deg(\{g_1, g_2\}\mu) \leq n + m - 1.$$

Proof. By Corollary 1.3.3, the differential system

$$\dot{x} = \Phi \frac{\{x, g_2\}}{\{g_1, g_2\}} + \lambda \frac{\{g_1, x\}}{\{g_1, g_2\}} = P(x, y),$$

$$\dot{y} = \Phi \frac{\{y, g_2\}}{\{g_1, g_2\}} + \lambda \frac{\{g_1, y\}}{\{g_1, g_2\}} = Q(x, y),$$

(2.43)

with g_2 and λ arbitrary functions, $\{g_1, g_2\} \neq 0$ in $\mathbb{R}^2$ and $\Phi|_{g_1=0} = 0$, has $g_1 = 0$ as an invariant algebraic curve, we have that

$$\dot{g_1} = P\frac{\partial g_1}{\partial x} + Q\frac{\partial g_1}{\partial y} = \Phi.$$

Assuming that Φ, g_2, and λ have been chosen in such a way that P and Q in (2.43) are polynomials, we have that $\Phi = Kg_1$ if K is cofactor of the invariant curve $g_1 = 0$. Thus we obtain the system (2.40).

In view of Corollary 2.5.1 and (2.41) we deduce the equivalence of the equations in question. This completes the proof of statement (a).

By Corollary 1.3.3, the differential system

$$
\begin{aligned}
\dot{x} &= \mu\{x, g_2\} + \lambda\{g_1, x\} = P(x, y), \\
\dot{y} &= \mu\{y, g_2\} + \lambda\{g_1, y\} = Q(x, y),
\end{aligned}
\tag{2.44}
$$

with g_2, λ, and μ arbitrary functions such that $\mu|_{g_1=0} = 0$, has an invariant algebraic curve.

Now we have that

$$
\dot{g}_1 = P\frac{\partial g_1}{\partial x} + Q\frac{\partial g_1}{\partial y} = \{g_1, g_2\}\mu.
$$

Assuming that g_2, λ, and μ have been chosen in such a way that P and Q in (2.44) are polynomials, we have that $\{g_1, g_2\}\mu = Kg_1$, if K is the cofactor of $g_1 = 0$.

The equivalence of the given equations follows from Corollary 2.5.1.

We observe that if $g_1 = 0$ is a *singular invariant algebraic curve*, i.e., there are points (x_0, y_0) satisfying

$$
g(x_0, y_0) = \left.\frac{\partial g_1}{\partial x}\right|_{x=x_0, y=y_0} = \left.\frac{\partial g_1}{\partial y}\right|_{x=x_0, y=y_0} = 0,
$$

then $\{g_1, g_2\}|_{x=x_0, y=y_0} = 0$, for an arbitrary function g_2. This completes the proof of the proposition. $\square$

We illustrate these results in the following examples.

Example 2.6.2. We consider the non-singular curve $g_1 = y + x^3 + x = 0$ in $\mathbb{R}^2$. We choose in (2.40) g_2, λ_2, λ_3, and λ in such a way that

$$
\{g_1, g_2\} \neq 0 \quad \text{in} \quad \mathbb{R}^2 \Longrightarrow (3x^2 + 1)\frac{\partial g_2}{\partial y} - \frac{\partial g_2}{\partial x} \neq 0 \quad \text{in} \quad \mathbb{R}^2,
$$

$$
\lambda_3\frac{\partial g_1}{\partial x} + \lambda_2\frac{\partial g_1}{\partial y} = K \Longrightarrow \lambda_3(3x^2 + 1) + \lambda_2 = K,
$$

$$
\lambda - \lambda_3\{y, g_2\} + \lambda_2\{x, g_2\} = \lambda_1\{g_1, g_2\}.
$$

If we take $g_2 = x$, then $\{g_1, g_2\} = -1$ and $\lambda_1 = \lambda + \lambda_3$. In particular, if we choose $\lambda_3 = 0$, the differential system (2.40) takes the form

$$
\dot{x} = -\lambda, \quad \dot{y} = (1 + 3x^2)\lambda + K(y + x^3 + x),
$$

where λ and K are arbitrary functions.

Example 2.6.3. We consider the curve $g_1 = y^3 + x^3 - x^2 = 0$. Since

$$g_1(0,0) = 0, \quad \frac{\partial g_1}{\partial x}(0,0) = 0, \quad \frac{\partial g_1}{\partial y}(0,0) = 0,$$

this curve is singular.

The differential system (2.42) in this case takes the form

$$\dot{x} = -3y^2 \nu + \mu_1, \quad \dot{y} = (3y^2 - 2x) \nu + \mu_2, \tag{2.45}$$

where μ_1, μ_2, and ν are arbitrary polynomials. This is the most general differential system having $y^3 + x^3 - x^2 = 0$ as invariant curve. In particular, if

$$\mu_1 = 3y^2 \nu + 2x - 2x^3 - 3xy^2 + y^3, \quad \mu_2 = -(3y^2 - 2x) \nu + \frac{4y}{3} + x^2 - 3x^2 y - 3y^3,$$

then

$$\mu_1(3y^2 - 2x) + \mu_2 3y^2 = -(6x^2 + 2x + 9y^2 - 4)(y^3 + x^3 - x^2).$$

Therefore, (2.45) takes the form

$$\dot{x} = 2x - 2x^3 - 3xy^2 + y^3, \quad \dot{y} = \frac{4y}{3} + x^2 - 3x^2 y - 3y^3.$$

This differential system was studied in [34].

In the rest of this subsection we shall work with complex polynomial vector fields.

Now we shall study the planar polynomial vector fields with one *invariant curve with $S \geq 1$ branches* see [108] with respect to the variable y. More precisely, first we treat the planar polynomial vector field $\mathbf{X} = (P, Q)$ of degree n associated to the differential system

$$\dot{x} = P(x,y), \quad \dot{y} = Q(x,y), \tag{2.46}$$

that has the invariant algebraic curve

$$g = \sum_{j=0}^{S} a_j(x) y^{S-j} = 0, \tag{2.47}$$

where $a_j = a_j(x)$ for $j = 0, \ldots, S$ are polynomials. If $a_0(x) \neq 0$, then it is well known that

$$g = a_0(x) \prod_{j=1}^{S} (y - y_j(x)) = 0,$$

where $y_j = y_j(x)$, for $j = 1, 2, \ldots, S$ are algebraic functions. Moreover

$$a_1 = -a_0 \sum_{j=1}^{S} y_j, \quad a_2 = a_0 \sum_{1 \leq j < m \leq S} y_j y_m, \quad \ldots \quad , a_S = (-1)^S a_0 \prod_{j=1}^{S} y_j.$$

The functions $g_j = y - y_j$ for $j = 1, \ldots, S$ are called the *branches of the algebraic curve* $g = 0$ with respect to the variable y.

In what follows we shall suppose that

$$\Delta_0(x) = \begin{vmatrix} 1 & 1 & \cdots & \cdots & 1 \\ y_1 & y_2 & \cdots & \cdots & y_S \\ \vdots & \vdots & \vdots & \vdots & \vdots \\ y_1^{S-1} & y_2^{S-1} & \cdots & \cdots & y_S^{S-1} \end{vmatrix} = \prod_{1 \le m < j \le S} (y_m - y_j) \ne 0, \qquad (2.48)$$

and there are at least two functions $g_1 = y - y_1(x)$ and $g_2 = y - y_2(x)$ for which $\{g_1, g_2\} = y_2'(x) - y_1'(x) \ne 0$.

Proposition 2.6.4. *Let* (2.47) *be an invariant algebraic curve of a polynomial vector field* $\mathbf{X}$ *associated to differential system* (1.44). *Then the branches* $g_j = y - y_j(x) = 0$ *for* $j = 1, 2, \ldots, S \ge 2$ *of* $g = 0$ *are invariant curves of* $\mathbf{X}$.

Proof. Let $\mathbf{X} = (P, Q)$. We shall prove that the given orbits $g_l = y - y_l(x)$ are invariant curves of (2.46). Indeed, by Remark 1.6.6, the vector field $\mathbf{X}$ can be written as (see (1.45))

$$\mathbf{X}(*) = \sum_{j=1}^{S+2} \lambda_j \{g_j, *\} \prod_{\substack{m=1 \\ m \ne j}}^{S} g_m, \qquad (2.49)$$

$$\dot{x} = \mathbf{X}(x) = P, \quad \dot{y} = \mathbf{X}(y) = Q$$

where $\lambda_1 = \cdots = \lambda_S = \nu$, $\lambda_{S+1} = \mu_1$, $\lambda_{S+2} = -\mu_2$, are arbitrary rational functions such that P and Q are polynomials, $g_{S+1} = y$, $g_{S+2} = x$.

Hence

$$\mathbf{X}(g_l) = g_l \left(\sum_{\substack{j=1 \\ m \ne l}}^{S+2} \lambda_j \{g_j, g_l\} \prod_{\substack{m=1 \\ m \ne j, l}}^{S} g_m \right),$$

i.e., $g_l = y - y_l = 0$ is an invariant curve of $\mathbf{X}$, $l = 1, \ldots, S$.   □

Corollary 2.6.5. *The differential system associated to the vector field* (2.49) *is equivalent to the differential system*

$$\dot{x} = \begin{vmatrix} 1 & 1 & \cdots & 1 \\ y & y_1 & \cdots & y_S \\ \vdots & \vdots & \vdots & \vdots \\ y^{S-1} & y_1^{S-1} & \cdots & y_S^{S-1} \\ \tilde{\mu}_2 y^S & \tilde{\mu}_2 y_1^S + \tilde{\nu} & \cdots & \tilde{\mu}_2 y_S^S + \tilde{\nu} \end{vmatrix},$$

$$\dot{y} = \begin{vmatrix} 1 & 1 & \cdots & 1 & 1 \\ y & y_1 & \cdots & y_{S-1} & y_S \\ \vdots & \vdots & \vdots & \vdots & \vdots \\ y^{S-1} & y_1^{S-1} & \cdots & y_{S-1}^{S-1} & y_S^{S-1} \\ \tilde{\lambda}_{S+1} y^S & \tilde{\lambda}_{S+1} y_1^S + \tilde{\nu} y_1' & \cdots & \tilde{\lambda}_{S+1} y_{S-1}^S + \tilde{\nu} y_{S-1}' & \tilde{\mu}_1 y_S^S + \tilde{\nu} y_S' \end{vmatrix},$$

where

$$\tilde{\mu}_2 \prod_{1 \leq i \leq j \leq S} (y_j - y_i) = -\mu_2, \quad \tilde{\mu}_1 \prod_{1 \leq i \leq j \leq S} (y_j - y_i) = \mu_1,$$

$$\tilde{\nu}(-1)^{S-m+1} \prod_{\substack{1 \leq i \leq j \leq S \\ i, j \neq m}} (y_j - y_i) = \nu.$$

Proof. Expanding by the last row the determinants of the statement of the corollary, we get system (2.49) (for more details see [136, 139]). □

Remark 2.6.6. Let (x_0, y_0) be an intersection point of two curves $g_j = y - y_j = 0$ and $g_k = y - y_k = 0$ with $j \neq k$. Then by Proposition 2.6.4, this point is a singular point of the vector field (2.49), and on it $\Delta_0(x) = 0$. Moreover, from (2.48) it follows that these points are the unique ones where $\Delta_0(x)$ vanishes.

2.7 Quadratic vector fields with algebraic curve with one branch

Let **X** be the *quadratic vector field* associated to the quadratic system

$$\dot{x} = p_0 y^2 + p_1 y + p_2, \quad \dot{y} = q_0 y^2 + q_1 y + q_2, \tag{2.50}$$

where p_0, q_0 are constants $p_j = p_j(x) = \sum_{n=0}^{j} p_{jn} x^n$, $q_j = q_j(x) = \sum_{n=0}^{j} q_{jn} x^n$, for $j = 0, 1, 2$, and assume that for which (2.47) is its unique irreducible invariant algebraic curve.

First we shall study the case when the algebraic curve (2.47) has only one branch, i.e., it has the form

$$g = a_0(x)\, y + a_1(x) = 0, \quad \text{with} \quad a_0(x) \neq 0. \tag{2.51}$$

Then by Proposition 2.6.1 a differential system with the unique invariant algebraic curve (2.51) can be written as

$$\dot{x} = \nu\, a_0 + \mu_1 g = P, \quad \dot{y} = -\nu\, (a_0'(x)y + a_1'(x)) + \mu_2 g = Q, \tag{2.52}$$

where μ_1, μ_2 and ν are arbitrary $\mathcal{C}^1$ functions. Now we shall determine these functions in such a way that

$$\nu a_0 + \mu_1\, (a_0(x)\, y + a_1(x)) = p_2(x),$$
$$\nu\, (a_0'(x)y + a_1'(x)) + \mu_2\, (a_0(x)\, y + a_1(x)) = q_0 y^2 + q_1 y + q_2,$$

where q_j are polynomials of degree j in the variable x.

Proposition 2.7.1. *The algebraic curve $g = a_0(x)y + a_1(x) = 0$ is invariant for the quadratic system*

$$\dot{x} = p_2, \quad \dot{y} = q_0 y^2 + q_1 y + q_2, \quad with \quad q_0 \neq 0, \qquad (2.53)$$

with cofactor $K = \alpha y + \beta x + \gamma$ if and only if $\alpha a_1(x) = p_2 a_0'(x) - r a_0(x)$, where $r = (\beta - q_{11})x + \gamma - q_{10}$ and $a_0 = a_0(x)$ is a polynomial solution of the Fuchs equation

$$w'' + \varrho_1 w' + \varrho_2 w = 0, \qquad (2.54)$$

where

$$\varrho_1 = \frac{p_2' - (\beta x + \gamma) - r}{p_2}, \quad \varrho_2 = \frac{-(\beta x + \gamma)r + \alpha q_2 - r' p_2}{p_2^2}.$$

Proof. Under our assumptions the equation

$$p_2 \frac{\partial g}{\partial x} + \left(q_0 y^2 + q_1 y + q_2 \right) \frac{\partial g}{\partial y} = (\alpha y + \beta x + \gamma)g,$$

implies the relations

$$a_0(\alpha - q_0) = 0, \quad p_2 a_0' = \alpha a_1 + r a_0, \quad p_2 a_1' = (\beta x + \gamma)a_1 - q_2 a_0.$$

Upon differentiating of the second relation, and using the third we get

$$p_2^2 a_0'' + (p_2 p_2' - (\beta x + \gamma)p_2 - r p_2) a_0' - (-(\beta x + \gamma)r + \alpha q_2 - r' p_2) a_0 = 0. \quad (2.55)$$

By solving with respect to a_0'' we obtain that the function a_0 is a solution of the Fuchs equation (2.54). □

The Fuchs equation (2.55) admits a polynomial solution if

$$\alpha q_2 = (\kappa + r')p_2 + (\beta x + \gamma)r.$$

Indeed, in this case equation (2.55) takes the form

$$p_2 a_0'' + (p_2' - (\beta x + \gamma) - r)a_0' - \kappa a_0 = 0.$$

Hence a_0 can be an *orthogonal polynomial* because the degrees of $p_2 = p_2(x)$, $p_2' - (\beta x + \gamma) + r$, and κ are $2, 1, 0$ respectively. For a precise definition of a family of *orthogonal polynomials* see [1]. A very important class of orthogonal polynomials $f_0, f_1, \ldots, f_n, \ldots$ are the ones satisfying the differential equation

$$\tau_2(x)f'' + \tau_1(x)f' + \tau_0 f = 0, \qquad (2.56)$$

where $\tau_j = \tau_j(x)$ are polynomials of degree at most j, for $j = 0, 1, 2$. The solution of (2.56) is an *orthogonal polynomial* if one of the following sets of conditions holds.

(1) The polynomial $\tau_2 = \tau_2(x)$ is quadratic with two distinct real roots, the root of the polynomial $\tau_1 = \tau_1(x)$ lies strictly between the roots of τ_2, and the leading terms of τ_2 and τ_1 have the same sign. This case leads to the Jacobi-like polynomials, which are solutions of the differential equation

$$(1 - x^2)f'' + (A - B - (A + B + 2)x)f' + n(n + A + B + 1)f = 0,$$

where A, B are real constants and n is a natural number. Special cases of the Jacobi polynomials are the Gegenbauer polynomials (with parameter $\gamma = A + 1/2$), the Legendre polynomials (with $A = B = 0$), and the Chebyshev polynomials (with $A = B = \pm 1/2$).

(2) The polynomial $\tau_2(x)$ is linear. The roots of τ_2 and τ_1 are distinct, and the leading terms of τ_2 and τ_1 have the same sign if the root of τ_1 is less that the root of τ_2, or vice-versa. This case leads to the Laguerre-like polynomials, which are solutions of the differential equation

$$xf'' + (A + 1 - x)f' + nf = 0,$$

where A is a real constants and n is a natural number.

(3) τ_2 is just a nonzero constant. The leading term of τ_1 has the opposite sign of τ_2. This case leads to the Hermite-like polynomials which are solutions of the differential equation

$$f'' - xf' + nf = 0,$$

where n is a natural number.

Proposition 2.7.2. *Let f be an orthogonal polynomial satisfying equation (2.56). Then*

(a) *There exists a quadratic polynomial differential system having the invariant algebraic curve $\tilde{g} = f'(x)y + f(x) = 0$.*

(b) *There exists a polynomial differential system of degree $2, 3$, or 4 having the invariant algebraic curve $g = f(x)y + f'(x) = 0$.*

Proof. By (2.52), the vector field with the invariant algebraic curve $\tilde{g} = 0$ can be written as

$$\dot{x} = \nu(x, y)f'(x) + \mu_1(x, y)(f'(x)y + f(x)),$$
$$\dot{y} = -\nu(x, y)(f''(x)y + f'(x)) + \mu_2(x, y)(f'(x)y + f(x)), \qquad (2.57)$$

where we take

$$\nu = -y\mu_1, \quad \mu_1 f(x) = p_2(x), \quad \mu_2 = \mu_1\left(\frac{q_1 y}{p_2} - 1\right) \implies \mu_2 f = q_1 y - p_2,$$

and by considering (2.56) we have that system (2.57) takes the form

$$\dot{x} = p_2(x), \quad \dot{y} = q_0 y^2 + q_1(x)y - p_2(x).$$

The cofactor of the curve $\tilde{g} = 0$ is $q_0 y$. After the change $q_0 y \longmapsto y$ we obtain the differential system

$$\dot{x} = p_2(x), \quad \dot{y} = y^2 + q_1(x)y - q_0 p_2(x).$$

The cofactor in this case is y. This system admits three, two, or one invariant algebraic curve depending of degree of $p_2(x)$. Hence statement (a) is proved.

In particular, the quadratic differential system

$$\dot{x} = 1, \quad \dot{y} = 2n + 2xy + y^2,$$

for any $n \in \mathbb{N}$ admits a unique irreducible non-singular in the affine plane invariant algebraic curve $g = y H_n(x) + 2 H'_n(x) = 0$, where H_n are the Hermite polynomials (for more details, see [24]).

From (2.52) we obtain that the differential system

$$\begin{aligned}
\dot{x} &= \nu(x,y)f(x) + \mu_1(x,y)(f(x)y + f'(x)), \\
\dot{y} &= -\nu(x,y)(f'(x)y + f''(x)) + \mu_2(x,y)(f(x)y + f'(x)),
\end{aligned} \tag{2.58}$$

has the invariant curve $g = f(x)y + f'(x) = 0$. Upon choosing

$$\nu = -y\mu_1, \quad f'\mu_1 = p_2, \quad p_2\mu_2 = -q_0\mu_1 \implies p_2\left(f'\mu_2 + q_0\right) = 0, \tag{2.59}$$

we obtain from (2.58) the system

$$\dot{x} = p_2, \quad \dot{y} = p_2\, y^2 - \frac{y\mu_2}{q_0}p_2 f'' + \mu_2 f\, y + \mu_2 f'.$$

Moreover, in view of (2.56) and (2.59) we get that

$$\frac{\mu_2 p_2 f''}{q_0} = \frac{y\left(-q_1\mu_2 f' + \mu_2 q_0 f\right)}{q_0} = -yq_1 + \mu_2 q_0 f.$$

Hence we end up with the system

$$\dot{x} = p_2(x), \quad \dot{y} = p_2(x)y^2 - q_1(x)y - q_0.$$

This system has degree two, three, or four and admits one, two, or three invariant algebraic curves, depending on the degree of polynomial $p_2(x)$. Moreover, the cofactor of $g = 0$ is $p_2(x)y - q_1(x)$.   □

From Propositions 2.7.1 and 2.7.2 it follows that there exist polynomial differential systems with invariant algebraic curves of arbitrary high degree. The following conjecture goes back to Poincaré, see [128]: *If a polynomial vector field has an invariant algebraic curve of sufficiently large degree, then it has a rational first integral.* This conjecture turned out to be not true. Two different counterexamples were published, see the references [33] and [118]. From these counterexamples a new conjecture arose: *If a polynomial vector field has an invariant algebraic curve of sufficiently large degree, then it has a Liouvillian first integral* (see Remark 1.6.7). Until this conjecture remains open.

A particular case of the curve (2.51) is the curve

$$F(a, b, c, x)\left(y - \frac{ab}{c}x\right) + x(1 - x)F'(a, b, c, x) = 0 \tag{2.60}$$

where $F = F(a, b, c, x)$ is the hypergeometric function

$$F(a, b; c; x) = \sum_{k=0}^{\infty} \frac{(a)_k (b)_k}{(c)_k} \frac{x^k}{k!}.$$

Here we have used the notation

$$(a)_k = \begin{cases} 1, & \text{if } k = 0, \\ a(a+1)(a+2)\cdots(a+k-1), & \text{if } k > 0, \end{cases}$$

where a, b and c are real constants. The hypergeometric function $F(a, b; c; x)$ is a solution of the hypergeometric differential equation

$$x(1 - x)y'' + [c - (a + b + 1)x]y' - aby = 0,$$

and so F and F' can have a common zero only at $x = 1$.

The following theorem is proved in [33].

Theorem 2.7.3. *Consider the quadratic differential system*

$$\dot{x} = x(1 - x), \quad \dot{y} = -\lambda y + Ax^2 + Bxy + y^2, \tag{2.61}$$

where

$$\lambda = c - 1, \quad A = \frac{ab(c - a)(c - b)}{c^2}, \quad B = a + b - 1 - \frac{2ab}{c}.$$

If for any positive integer k, we choose $a = 1 - k$, $b \le a$, and c irrational, then the polynomial system (2.61) has

(a) *no rational first integral,*

(b) *the irreducible invariant curve (2.60) of degree k, and*

(c) *a Darboux integrating factor (see Remark 1.6.7).*

From Propositions 2.7.1 and 2.7.2, and also from Theorem 2.7.3, it follows that there exist polynomial differential systems with invariant algebraic curves of arbitrarily high degree.

2.8 Quadratic vector fields with algebraic curve with $S \geq 2$ branches

Theorem 2.8.1. *Let $\mathbf{X}$ be the vector field associated to the quadratic system (2.50) and assume that the algebraic curve (2.47) is its unique irreducible invariant algebraic curve. If the curve $g = 0$ has $S > 1$ branches with respect to the variable y (so $a_0 \neq 0$), then $\deg g \leq 4S$.*

Note that in view of Corollary 2.4.8, if $\deg(g) > 3$, then the curve $g = 0$ is singular.

Proof. From the relation $\mathbf{X}(g) = (\alpha y + \beta x + \gamma)g$, taking the coefficients of the powers of y we obtain the following differential system

$$p_0 \frac{da_0}{dx} = 0, \quad A\frac{d\mathbf{a}}{dx} = B\mathbf{a}, \quad p_2 \frac{da_S}{dx} = (\beta x + \gamma)a_S - q_2 a_{S-1}, \qquad (2.62)$$

where $\mathbf{a} = (a_0, a_1, \ldots, a_S)^T$, and $a_i = a_i(x)$ for $i = 0, 1, \ldots, S$, and A and B are the $(S+1) \times (S+1)$ matrices

$$A = \begin{pmatrix}
p_1 & p_0 & 0 & 0 & 0 & \cdots & 0 & 0 & 0 & 0 \\
p_2 & p_1 & p_0 & 0 & 0 & \cdots & 0 & 0 & 0 & 0 \\
0 & p_2 & p_1 & p_0 & 0 & \cdots & 0 & 0 & 0 & 0 \\
0 & 0 & p_2 & p_1 & p_0 & \cdots & 0 & 0 & 0 & 0 \\
\vdots & \vdots & \vdots & \vdots & \vdots & \vdots & \vdots & \vdots & \vdots & \vdots \\
0 & 0 & 0 & 0 & 0 & \cdots & 0 & p_2 & p_1 & p_0 \\
0 & 0 & 0 & 0 & 0 & \cdots & 0 & 0 & p_2 & p_1
\end{pmatrix},$$

$$B = \begin{pmatrix}
\tilde{a}_0 & 0 & 0 & 0 & \cdots & 0 & 0 & 0 & 0 \\
b_0 & \tilde{a}_1 & 0 & 0 & \cdots & 0 & 0 & 0 & 0 \\
c_0 & b_1 & \tilde{a}_2 & 0 & \cdots & 0 & 0 & 0 & 0 \\
0 & c_1 & b_2 & \tilde{a}_3 & \cdots & 0 & 0 & 0 & 0 \\
\vdots & \vdots & \vdots & \vdots & \vdots & \vdots & \vdots & \vdots & \vdots \\
0 & 0 & 0 & 0 & \cdots & c_{S-3} & b_{S-2} & \tilde{a}_{S-1} & 0 \\
0 & 0 & 0 & 0 & \cdots & 0 & c_{S-2} & b_{S-1} & \tilde{a}_S
\end{pmatrix},$$

where $\tilde{a}_j = \alpha + (j - S)q_0$, $b_j = \beta x + \gamma + (j - S)q_1$, $c_j = (j - S)q_2$ for $j = 0, 1, \ldots, S$. It is known (see for instance [64]) that after a linear change of variables and a rescaling of time. any quadratic system (2.50) can be written as

$$\dot{x} = P(x, y), \quad \dot{y} = q_0 y^2 + q_1 y + q_2,$$

where $P(x, y)$ is one of the following ten polynomials

$$1 + xy, \quad y + x^2, \quad y, \quad 1, \quad xy, \quad -1 + x^2, \quad 1 + x^2, \quad x^2, \quad x, \quad 0.$$

Since the last six possibilities for $P(x,y)$ force that the quadratic system has an invariant straight line (real or complex) and by assumption our quadratic system has no invariant straight lines, the polynomial $P(x,y)$ can only be $1 + xy$, $y + x^2$, y, 1.

Case 1: Assume that P is either $y + x^2$, or y. We consider the quadratic system

$$\dot{x} = y + p_2(x), \quad \dot{y} = q_0 y^2 + q_1 y + q_2,$$

with $p_2(x) = x^2$ or $p_2(x) = 0$. After the recursive integration of system (2.62), since the a_j's are polynomials, we deduce that

$$\alpha = S q_0,$$
$$a_0 = a_{00},$$
$$a_1 = a_{12} x^2 + a_{11} x + a_{10},$$
$$a_2 = a_{24} x^4 + a_{23} x^3 + a_{22} x^2 + a_{21} x + a_{20},$$
$$\vdots$$
$$a_{S-1} = a_{S-1,2(S-1)} x^{2(S-1)} + \cdots,$$
$$a_S = a_{S,2S} x^{2S} + \cdots,$$

where all the a_{ij} are constants. Therefore $\deg g \leq 2S$.

Case 2: Assume that $P = 1$, so we are dealing with the quadratic system

$$\dot{x} = 1, \quad \dot{y} = q_0 y^2 + q_1 y + q_2. \tag{2.63}$$

We note that the differential system (2.63) can be written as a Ricatti differential equation. Since this system has no singular points, the algebraic invariant curve $g = 0$ must be *non-singular in the affine plane*. If the curve is *non-singular in* $\mathbb{C}P^2$, then the degree of g is at most two (see Corollary 2.4.8). So if the algebraic curve $g = 0$ of (2.63) has degree larger than two, it is non-singular in the affine plane and singular at infinity, i.e., in $\mathbb{C}P^2$. We shall determine the curve $g = 0$ solution of (2.62) with degree > 2.

Assume that $q_{11} \neq 0$. After the change of variables $(q_{11}x, y) \to (y, x)$ and introducing the notations

$$\frac{q_{22}}{q_{11}^2} = p_0, \quad \frac{q_{21}}{q_{11}} = p_{11}, \quad p_2(x) = q_0 x^2 + q_{10} x + q_{20},$$

we obtain the system

$$\dot{x} = p_0 y^2 + xy + p_{11} y + p_2(x), \quad \dot{y} = q_{11}. \tag{2.64}$$

We consider the differential system (2.62) associated to system (2.64). If $p_0 \neq 0$, then without loss of generality we can take $p_0 = 1$, and (2.62) takes the form

$$
\begin{aligned}
a'_0 &= 0, \\
a'_1 &= ma_0, \\
a'_2 &= (m^2 + m(q_0 - q_{11}) + \beta - sq_{11})x \\
&\quad + (m(C_1 - p_{10}C_0) + q_0C_1 + (\gamma - Sq_{11})C_0, \\
&\vdots
\end{aligned}
$$

where $m = \alpha - Sq_0$. Hence we obtain that $\deg a_j \leq j$, for $j = 0, 1, \ldots, S$, and consequently $\deg g \leq S$.

We study the case $p_0 = 0$. After the change $(x + p_{11}, y) \to (x, y)$ and

$$
\left(q_0, q_{10} - 2q_0p_{11}, q_{20} - p_{11}q_{10} + p_{11}^2 q_0 \right) \to (p_{22}, p_{21}, p_{20}),
$$

the differential system (2.62) becomes

$$
\begin{aligned}
xa'_0 &= \alpha a_0, \\
xa'_{j+1} &= \alpha a_{j+1} + (\beta x + \gamma)a_j - (p_{22}x^2 + p_{21}x + p_{20})a'_j - q_{11}(S + 1 - j)a_{j-1},
\end{aligned}
$$

for $j = 0, \ldots, S$, where $a_{-1} = 0$. Solving the first differential equation we get $a_0 = C_0 x^\alpha$, hence α must be a non-negative integer, and without loss of generality we can take $C_0 = 1$. Now substituting this into the differential equation of a'_1 we obtain

$$
\begin{aligned}
xa'_1 &= \alpha a_1 + (\beta x + \gamma)x^\alpha - \alpha(p_{22}x^2 + p_{21}x + p_{20})x^{\alpha-1} \\
&= \alpha a_1 + (\beta - \alpha p_{22})x^{\alpha+1} + (\gamma - \alpha p_{21})x^\alpha - \alpha p_{20}x^{\alpha-1}.
\end{aligned}
$$

Solving this linear differential equation we have

$$
a_1 = (\beta - \alpha p_{22})x^{\alpha+1} + C_1 x^\alpha + \alpha p_{20}x^{\alpha-1} + (\gamma - \alpha p_{21})x^\alpha \log x.
$$

Since a_1 must be a polynomial, we get that

$$
\gamma = \alpha p_{21}.
$$

Solving the differential equation of a'_2 we obtain

$$
\begin{aligned}
a_2(x) &= \alpha p_{20}^2(\alpha - 1)x^{\alpha-2} + \alpha p_{20}(p_{21} - C_1)x^{\alpha-1} + C_2 x^\alpha \\
&\quad - (C_1 - p_{21})(\alpha p_{22} - \beta)x^{\alpha+1} + \frac{1}{2}(\alpha p_{22} - \beta)((\alpha + 1)p_{22} - \beta)x^{\alpha+2} \\
&\quad - (Sq_{11} - (2\alpha p_{22} - \beta)p_{20}x^\alpha \log x.
\end{aligned}
$$

Again, since a_2 must be a polynomial, we get that

$$S = \frac{p_{20}(2\alpha p_{22} - \beta)}{q_{11}}.$$

Arguing similarly and considering that we can write

$$(\beta x + \gamma)a_j - (p_{22}x^2 + p_{21}x + p_{20})a'_j - q_{11}(S + 1 - j)a_{j-1}$$
$$= (-q_{11}(S - j)C_{j-1} + \cdots)x^\alpha + \cdots ,$$

for $j \geq 3$, we can obtain solving the linear differential equation for a'_j that all a_j for $j \geq 3$ are polynomials, choosing the arbitrary constant C_{j-1} conveniently.

After the recursive integrations we finally deduce that

$$a_j = \prod_{m=1}^{j} (\beta - (-1 + m + \alpha)p_{22}) \frac{x^{\alpha+j}}{j!} + x^{\alpha-j}P_{2j-1}(x)$$
$$= x^{\alpha-j}\left(\prod_{m=1}^{j} (\beta - (-1 + m + \alpha)p_{22}) \frac{x^{2j}}{j!} + P_{2j-1}(x) \right),$$

where $P_m(x)$ is a polynomial of degree m in x, and by definition $P_{-1}(x) = 0$. The invariant algebraic curve in this case admits the representation

$$g = x^{\alpha-S} \sum_{j=0}^{S} (xy)^{S-j} \left(\prod_{m=1}^{j} (\beta - (-1 + m + \alpha)p_{22}) \frac{x^{2j}}{j!} + P_{2j-1}(x) \right).$$

Hence $\deg g \leq \alpha + S$.

If $\alpha - S \geq 0$, then by considering that the curve is irreducible, we have $\alpha = S$, and as a consequence $\deg g \leq 2S$. If $\alpha - S < 0$, then $\deg g \leq \alpha + S < 2S$. In short, in Case 2 and when $q_{11} \neq 0$ we have that $\deg g \leq 2S$. Substituting a_S and a_{S-1} in the last equation of (2.62) and taking the leading coefficient in x (i.e., the coefficient of $x^{\alpha+S+1}$), we deduce that

$$\prod_{m=1}^{S+1} (\beta - (1 - m + \alpha)p_{22}) = 0.$$

The particular Case 2 with $q_{11} \neq 0$ when $S > \alpha = 1$ and $\beta = p_{22} \neq 0$, $p_{20} = \gamma = 0$ is interesting. The solutions of (2.62) are polynomial of degree one of the form $a_j = c_j x + r_j$, for $j = 0, 1, \ldots, S$, where c_j and r_j are suitable constants, satisfying the equations

$$r_{j+1} = p_{20}c_j + (S + 1 - j)q_{11}r_{j-1}, \quad p_{22}r_j = q_{11}(S + 1 - j)c_{j-1}, \quad r_{-1} = c_{-1} = 0,$$

for $j = 0, 1, \ldots, S$. Hence, we obtain that

$$p_{22}p_{20} = Sq_{11},$$

$$r_{2j} = 0, \quad r_1 = p_{20}, \quad r_{2j+1} = \frac{p_{20}(S - 2j)}{S}c_{2j},$$

$$c_{2j+1} = 0, \quad c_0 = 1, \quad c_{2j} = \frac{(-q_{11})^j S!}{2^j j!(S - 2j)!}.$$

Consequently, the curve $g = 0$ takes the form

$$g = \sum_{k=0}^{S} a_k(x)y^{S-k} = x\sum_{k=0}^{S} c_k y^{S-k} + \sum_{k=0}^{S} r_k y^{S-k},$$

or equivalently

$$g = x\sum_{k=0}^{[S/2]} c_{2k}y^{S-2k} + \sum_{k=0}^{[(S-1)/2]} r_{2k+1}y^{S-2k-1},$$

where $[x]$ denotes the integer part of the real number x.

If we denote

$$H_S(y) = \sum_{k=1}^{[S/2]} c_{2k}(q_{11}y)^{S-2k} = \sum_{k=0}^{[S/2]} \frac{(-q_{11})^k S!}{2^k k!(S - 2k)!}(q_{11}y)^{S-2k},$$

then we obtain for g the representation

$$g(x, y) = xH_S(y) + \frac{p_{20}}{S}H'_S(y).$$

It is easy to check that if if $q_{11} = 2$, then H_S coincide with the Hermite polynomial. Clearly, $\deg g = S + 1$.

Now we assume that in (2.63) $q_{11} = 0$ and $q_{21} \neq 0$. Then performing the change of variables $(q_{21}x, y) \mapsto (y, x)$ we obtain

$$\dot{x} = p_0 y^2 + y + p_2(x), \quad \dot{y} = q_{21},$$

where $p_0 = q_{22}/q_{21}^2$, $p_2(x) = q_0 x^2 + q_{10}x + q_{20}$. If $p_0 = q_{22} \neq 0$, then system (2.62) admits the polynomial solutions

$$a_0 = a_{00}, \quad a_1 = a_{11}x + a_{10}, \ldots, a_S = a_{SS}x^S + \cdots + a_{S0},$$

so $\deg g \leq S$. If $p_0 = q_{22} = 0$, then the integration of equation (2.62) is analogous to Case 1. Hence $\deg g \leq 2S$.

Case 3: Assume that $P = xy + 1$. Hence, we must study the quadratic systems

$$\dot{x} = xy + 1, \quad \dot{y} = q_0 y^2 + (q_{11}x + q_{10})y + q_{22}x^2 + q_{21}x + q_{20}. \tag{2.65}$$

If $\alpha - Sq_0 = m$ we shall show that the functions $a_j(x)$ in (2.62) are polynomials of degree $\deg a_j \leq q_0 j + m$ if $q_0 \neq 0$, and of $\deg a_j \leq j + m$ if $q_0 = 0$, for $j = 0, 1, \ldots, S$.

We denote $q_0 = k$. Since $p_0 = 0$, $p_1 = x$, $p_2 = 1$, from the first nonzero differential equation of system (2.62) we obtain that $a_0 = x^m C_0$ with $C_0 \in \mathbb{C} \setminus \{0\}$. Since $a_0(x)$ must be a polynomial, m must be a non-negative integer, i.e., $\alpha - Sq_0 = m \geq 0$ as a consequence $Sq_0 \leq \alpha$.

Solving the second nonzero differential equation of system (2.62) we obtain

$$a_1(x) = C_0 \left(\frac{Sq_{11} - \beta}{k-1} x^{m+1} + \frac{Sq_{10} - \gamma}{k} x^m + \frac{m}{1+k} x^{m-1} \right) + C_1 x^{m+k}$$
$$= x^{m-1}(P_2 + C_1 x^{k+1}),$$

if $k(k^2 - 1) \neq 0$.

From the computation of $a_1(x)$ we obtain that $k \in \mathbb{N} \setminus \{1\}$, because we are looking for the algebraic curve $g = 0$ given in (2.47) with the maximum degree, and consequently $a_1(x)$ must be a polynomial with $m + k$ a positive integer larger than or equal to $m + 1$, but since some denominator contains $k - 1$ it follows that $k > 1$. (Recall that $\mathbb{N}$ denotes the set of positive integers.)

By the recursive integration of system (2.62) we compute $a_2(x), \ldots, a_S(x)$, and using induction after a tedious but elementary computation we deduce that the components $a_j = a_j(x)$ of the vector $\mathbf{a}$ are of the form

$$a_0 = x^m C_0,$$
$$a_1 = x^{m-1}(P_2 + C_1 x^{k+1}) = x^{m-1} P_{k+1},$$
$$a_2 = x^{m-2}(P_{k+3} + C_2 x^{2k+2}) = x^{m-2} P_{2k+2},$$
$$\vdots$$
$$a_{S-1} = x^{m-S+1}(P_{(S-2)k+S} + C_{S-1} x^{(S-1)k+S-1})$$
$$= x^{m-S+1} P_{(S-1)k+S-1},$$
$$a_S = x^{m-S}(P_{(S-1)k+S+1} + C_S x^{Sk+S}) = x^{m-S} P_{Sk+S}.$$

Here $P_j = P_j(x)$ denotes a polynomial of degree j, and C_j are arbitrary constants coming from the integration, that we must assume different from zero in order to obtain the maximum possible degree for the polynomial g defining the algebraic curve (2.47).

Since P_j are polynomials of degree j in x, we have that the $\deg a_j \leq kj + m$, for $j = 0, 1, 2, \ldots, S$. Therefore, $\deg g \leq kS + m = q_0 S + \alpha - q_0 S = \alpha$. On the other hand, since

$$g = \sum x^{m-j} y^{S-j} P_{j(k+1)} = x^{m-S} \sum (xy)^{S-j} P_{j(k+1)} = 0,$$

and since the curve must be irreducible, we obtain that $m = \alpha - kS = S$, therefore $\alpha = S(k+1)$. Clearly, if $m - S < 0$, then $kS + m < S(k+1) = \alpha$. So $\deg g \le (k+1)S$. We are interesting in determining the biggest finite upper bound of the degree of the polynomial g.

We shall study the last equation of system (2.62). We prove that if $C_S \ne 0$, then the curve $g = 0$ has the cofactor $K = \alpha y$. Indeed, inserting a_S and a_{S-1} in the last equation of system (2.62), we obtain that

$$\beta C_S x^{Sk+m+1} + \gamma C_S x^{Sk+m} + P_{Sk+m-1}(x) = 0, \quad \text{if} \quad k \ge 3.$$

Hence $\beta = \gamma = 0$ and the cofactor is αy.

We claim that if $C_S C_{S-1} \ne 0$, then $k = 3$. Indeed, from the last equation of system (2.62) we obtain that the polynomials a_S and a_{S-1} are such that

$$\frac{da_S}{dx} + q_2 a_{S-1} = 0. \tag{2.66}$$

Integration yields

$$a_S = q_{22} \left(\frac{C_{S-1}}{k(S-1)+m} x^{(S-1)k+3+m} + \cdots \right).$$

On the other hand, the polynomial a_S has degree $kS + m$, therefore

$$C_S x^{Sk+m} + \cdots = q_{22} \left(\frac{C_{S-1}}{k(S-1)+3+m} x^{(S-1)k+3+m} + r_0 x^{k(S-2)+m+3} \cdots \right),$$

where r_0 is a real constant. Hence if $C_S C_{S-1} \ne 0$, then $k = q_0 = 3$. Consequently, $\deg g \le (k+1)S = 4S$.

If $C_{S-1} = 0$ and $r_0 \ne 0$, then $k = 3/2$, and consequently $\deg g \le (3/2+1)S \le 3S$. Clearly, if $C_S = 0$ then from (2.66) it follows that $C_{S-1} = 0$, thus $\deg g < 4S$. In this case arguing as in the case where the constants C_j are different from zero with $m = 0$ we get for the curve

$$g = y^S + a_1 y^{S-1} + a_2 y^{S-2} + \cdots + a_s = 0, \tag{2.67}$$

that $\deg g \le 3S$.

Now we assume that $k = q_0 = 0$. The recursive integration of system (2.62) produces the following polynomial solutions

$$a_0 = x^\alpha, \quad a_j = r_j x^{\alpha+j} + x^{\alpha-j} P_{2j-1} = x^{\alpha-S} \left(r_j x^{S+j} + x^{S-j} P_{2j-1} \right),$$

for $j = 1, 2, \ldots, S$, where r_j are rational functions in the variables $q_{11}, q_{12}, q_{21}, q_{22}, q_{20}, q_{10}, \alpha, \beta$, and $P_m(x)$ is a polynomial of degree m in the variable x. Note that $\deg a_j \le \alpha + j$. The polynomial g becomes

$$g = x^{\alpha-S} \sum_{j=0}^{S} (r_j x^{S+j} + x^{S-j} P_{2j-1}) y^{S-j} = x^{\alpha-S} \sum_{j=0}^{S} (xy)^{S-j} (r_j x^{2j} + P_{2j-1}),$$

where $r_0 = 1$ and $P_{-1}(x) = 0$. Since the curve $g = 0$ is irreducible, we have $\alpha = S$. If $\alpha - S < 0$, then $\deg g \leq \alpha + S < 2S$.

If $k = 1$, then system (2.62) becomes

$$xa_0' = ma_0,$$
$$xa_{j+1}' = (m + j + 1)a_{j+1} + ((\beta - (S - j)q_{11})x + \gamma - (S - j)q_{10})a_j$$
$$- a_j' - (S - j)q_2 a_{j-1},$$

for $j = 1, 2, \ldots, S$, where $a_{-1} = 0$. Hence after integration it is easy to see that

$$a_0 = C_0 x^m = x^{m-1}P_1,$$
$$a_1 = C_1 x^{m+1} + (Sq_{10} - \gamma)x^m + \frac{m}{2}x^{m-1} = x^{m-1}P_2,$$
$$a_2 = x^{m-1}P_3,$$
$$\vdots$$
$$a_S = x^{m-1}P_{S+1},$$

where $P_j = P_j(x)$ is a polynomial of degree j in the variable x. Hence

$$g = x^{m-1}\sum_{j=0}^{S} P_{j+1}(x)y^{S-j} = 0.$$

Since we are requiring that this curve be irreducible, we have that $m = \alpha - S = 1$. As a consequence $S = \alpha - 1 < \alpha$ and $\deg g \leq S + m = S + 1$.

For the case when $k = -1$ system (2.62) takes the form

$$xa_0' = (\alpha + S)a_0,$$
$$xa_1' = (\alpha + S - 1)a_1 + ((\beta - Sq_{11})x + \gamma - Sq_{10})\,a_0 - a_0',$$
$$xa_2' = (\alpha + S - 2)a_2 + ((\beta - (S-1)q_{11})x + \gamma - (S-1)q_{10}))\,a_1 - a_1',$$
$$\vdots$$

After the recursive integrations we obtain that the polynomial solutions exist in particular if $\alpha + S = 0$. In this case we obtain that $\deg a_j \leq j$, and so $\deg g \leq S$.

Theorem 2.8.1 is proved. □

Remark 2.8.2. Let

$$R(x) = p_2 \frac{da_S}{dx} - ((\beta x + \gamma)a_S - q_2 a_{S-1}) = \sum_{j=0}^{4S+1} A_j x^j,$$

be a polynomial of degree at most $4S + 1$ in the variable x, where

$$A_j = A_j\,(q_0, q_{11}, q_{10}, q_{22}, q_{21}, q_{20}, \alpha, \beta, \gamma, C_0, C_1, \ldots, C_S),$$

for $j = 0, 1, \ldots, 4S + 1$. To determine the exact degree of the invariant curve $g = 0$ in all the cases studied in the proof of Theorem 2.8.1, it is necessary that the polynomial $R(x)$ be zero. This holds if and only if all the coefficients are zero, i.e., $A_j = 0$ for $j = 0, 1, \ldots, 4S+1$. The compatibility of all these equations is required. Working a little it is possible to reduce the system $A_j = 0$ to a polynomial system in the variables $q_0, q_{11}, q_{10}, q_{22}, q_{21}, q_{20}, \alpha, \beta, \gamma, C_0, C_1, \ldots, C_S$. The polynomials thus obtained in general have high degree and it is not easy to work with them for proving that they have no common solutions. One would guess that $\deg g$ (which from the proof of Theorem 2.8.1 must be a multiple of S smaller than or equal to $4S$) must be $\leq 3S$.

In view of this remark and comments later on we introduce the following two conjectures

Conjecture 2.8.3. *If a quadratic polynomial differential system* (2.50) *admits a unique invariant irreducible algebraic curve $g = 0$ given in* (2.47), *then* $\deg g \leq 3S$.

Conjecture 2.8.4. *If a quadratic polynomial differential system* (2.50) *with an invariant irreducible algebraic curve $g = 0$ given by* (2.47) *does not admit a Liouvillian first integral, then* $\deg g \leq 12$.

These conjectures are supported mainly by the following facts. First, we are able to show that for $S = 1, 2, \ldots, 5$ there are irreducible invariant algebraic curves $g = 0$ of degree $3S$ for suitable quadratic system (2.65). This curve has a cofactor $K = 3Sy$. On the other hand, without loss of generality we can suppose that the given invariant curve has the form (2.67) for which the $\deg g \leq 3S$.

We illustrate this in the following example.

Example 2.8.5. The quadratic polynomial differential system

$$\dot{x} = 1 + xy, \quad \dot{y} = 3y^2 - \frac{8q}{13}yx - \frac{24q^2}{169}x^2 + q, \tag{2.68}$$

with $q \in \mathbb{R}$ admits the following family of invariant algebraic curves $g = 0$ of degree 18 with cofactor $18y$:

$$\left(y^6 + x^3 y^5 + \left(-\frac{20q}{13}x^4 + \frac{3}{4}x^2 + \frac{4q^2}{169}x^6 \right) y^4 \right) C$$

$$+ \left(\frac{136q^2}{169}x^5 + \frac{3}{16}x - \frac{64q^3}{2197}x^7 - \frac{1}{2}x^3 q \right) y^3 C$$

$$+ \left(\frac{241}{4394}x^4 q^3 + \frac{1}{208}q - \frac{1728}{4826809}x^{10}q^6 - \frac{23}{832}x^2 q^2 + \frac{2592}{371293}x^8 q^5 - \frac{90}{2197}x^6 q^4 \right) C$$

$$+ \left(-\frac{69}{338}q^2 x^4 + \frac{3}{26}qx^2 - \frac{264}{2197}q^3 x^6 \frac{1}{64} + \frac{288}{28561}q^4 x^8 \right) y^2 C$$

$$+ \left(\frac{474}{2197}q^3 x^5 + \frac{23}{416}qx - \frac{720}{28561}q^4 x^7 - \frac{11}{52}q^2 x^3 \right) yC - \frac{24}{13}y^5 qx$$

$$+ \left(\frac{9q}{13} + \frac{12288}{28561} x^6 q^4 + \frac{204}{169} q^2 x^2 \right) y^4$$

$$+ \left(\frac{20840448}{815730721} x^{11} q^8 + \frac{12582912}{137858491849} x^{15} q^{10} \right) y$$

$$+ \left(-\frac{196608}{371293} x^7 q^5 + \frac{360448}{4826809} x^9 q^6 - \frac{608}{2197} x^3 q^3 + \frac{18432}{28561} x^5 q^4 - \frac{180}{169} x q^2 \right) y^3$$

$$+ \left(\frac{3342336 q^8}{815730721} x^{12} - \frac{196608 q^5}{371293} x^6 + \frac{1695744 q^6}{4826809} x^8 + \frac{24 q^2}{169} - \frac{4325376 q^7}{62748517} x^{10} \right.$$

$$+ \frac{1224 q^3}{2197} x^2 + \frac{10896 q^4}{28561} x^4 \left. \right) y^2 + \left(-\frac{540672}{4826809} x^9 q^7 + \frac{1104}{28561} x^3 q^4 - \frac{59136}{371293} x^5 q^5 \right.$$

$$-\frac{354}{2197} x q^3 - \frac{26738688}{10604499373} q^9 x^{13} + \frac{964608}{4826809} x^7 q^6 \left. \right) y - \frac{5712}{371293} x^4 q^5$$

$$+ \frac{60555264}{137858491849} x^{14} q^{10} + \frac{15}{2197} q^3 - \frac{2818048}{815730721} q^9 x^{12} - \frac{1744896}{62748517} x^8 q^7$$

$$+ \frac{16777216}{23298085122481} x^{18} q^{12} - \frac{50331648}{1792160394037} q^{11} x^{16} + \frac{11452416}{815730721} x^{10} q^8$$

$$+ \frac{132544}{4826809} x^6 q^6 + \frac{177}{2197} x^2 q^4 = 0,$$

where C is an arbitrary constant. Consequently, the given quadratic system admits a rational first integral.

We observe that the quadratic vector fields (2.68) have two critical points. For $q = 13$ the coordinates of these points are $M_0 = (-1, 1)$, $M_1 = (1, -1)$ which are node.

Example 2.8.6. The quadratic polynomial differential system

$$\dot{x} = 1 + xy, \quad \dot{y} = 3y^2 - \frac{322}{179} axy - \frac{17052}{32041} a^2 x^2 + a,$$

where a is an arbitrary nonzero constant, admits the invariant algebraic curve of degree 15

$$y^5 + \left(\frac{34300 a^2}{32041} x^3 - \frac{805 a}{179} x \right) y^4 + \left(\frac{263620 a^2}{32041} x^2 - \frac{22089200 a^3}{5735339} x^4 + \frac{15 a}{179} \right) y^3$$

$$+ \frac{414124480 a^4}{1026625681} x^6 y^3 + \left(-\frac{200022123840 a^5}{183765996899} x^7 + \frac{5370844920 a^4}{1026625681} x^5 \right) y^2$$

$$+ \left(-\frac{31815 a^2}{128164} x - \frac{43924580 a^3}{5735339} x^3 + \frac{2414759842880 a^6}{32894113444921} x^9 \right) y^2$$

$$+ \left(\frac{32340533610000 a^6}{32894113444921} x^8 - \frac{584433172400 a^5}{183765996899} x^6 - \frac{777552669407360 a^7}{5888046306640859} x^{10} \right) y$$

$$+ \left(\frac{135 a^2}{512656} + \frac{2829015 a^3}{11470678} x^2 + \frac{3724251125 a^4}{1026625681} x^4 \right) y$$

$$+ \frac{6927287418356480a^8}{1053960288888713761}x^{12}y + \frac{7874940337187646464a^{10}}{3376994161628327761620 1}x^{15} - \frac{135a^3}{512656}x$$

$$+ \frac{24017066863300a^6}{32894113444921}x^7 - \frac{1115293274355393280a^9}{18865889171107976321 9}x^{13} - \frac{128504153897a^5}{183765996899}x^5$$

$$- \frac{337403955a^4}{4106502724}x^3 - \frac{1750523330217200a^7}{5888046306640859}x^9 + \frac{62786609721967040a^8}{1053960288888713761}x^{11} = 0$$

with cofactor $15y$.

To study the problem of the existence of rational first integral we shall use the following result due to Poincaré: *If a planar complex polynomial system has a rational first integral, then the eigenvalues λ and μ associated to any critical point of the system must be resonant, in the sense that there exist integers m and n with $|m| + |n| > 1$ such that $m\lambda + n\mu = 0$.* By applying this result to the above quadratic system we obtain that the critical points are

$$M_1 = \left(\sqrt{\frac{179}{29|a|}} - \sqrt{\frac{29|a|}{179}} \right), \quad M_2 = \left(-\sqrt{\frac{179}{29|a|}}, \sqrt{\frac{29|a|}{179}} \right)$$

The eigenvalues are

$$\lambda = 30\, i \sqrt{\frac{|a|}{179}}, \quad \mu = 45i \sqrt{\frac{|a|}{179}},$$

so in this case if we take $m = 45$, $n = -30$ we have that $|m| + |n| = 15 > 0$ and $m\lambda + n\mu = 0$. So by Poincaré's result the system probably has a rational first integral for all $a \neq 0$. If $a = 0$, then the system admits the rational first integral

$$F = \frac{(1 + 4xy)^3}{y^4}.$$

Second, we consider the more general quadratic systems (2.65) having some foci. Thus we get

$$q_0 = 3, \quad q_{22} = \frac{84ae^2 - 36e^2 - 25e^4 - h^2}{288}, \quad q_{10} = 0,$$

$$q_{11} = a, \quad q_{20} = \frac{36a^2 - 36ae^2 + e^4 + h^2}{48e^2}, \tag{2.69}$$

where $eh \neq 0$. The points $(\sqrt{6}/e, -e/\sqrt{6})$ and $(-\sqrt{6}/e, e/\sqrt{6})$ are critical points of the corresponding quadratic system, with eigenvalues $(6a - 7e^2 \pm ih)/(2\sqrt{6}\,e)$ and $(-6a + 7e^2 \pm ih)/(2\sqrt{6}\,e)$ respectively, so they are strong *foci* if $6a - 7e^2 \neq 0$, and consequently these quadratic system do not admit a rational first integral (see for instance [97]).

We study the particular systems of (2.62) satisfying (2.69) with $S = 4$, that is the family of quadratic systems

$$\dot{x} = xy + 1, \quad \dot{y} = 3y^2 - 10axy - 150a^2x^2 + 59a, \tag{2.70}$$

where a is a nonzero parameter. This system admits the following family of invariant algebraic curves of degree 12:

$$- 781250000a^8x^{12} + 312500000a^7x^{10} - 62500000a^6yx^9 - 159375000a^6x^8$$
$$- 3750000a^5yx^7 + 230375000a^5x^6 + 375000a^4y^2x^6 - 9975000a^4yx^5$$
$$- 82923125a^4x^4 - 4215000a^3y^2x^4 + 281000a^2y^3x^3 + 5291500a^3yx^3$$
$$+ 3833820a^3x^2 + 210750a^2y^2x^2 - 6860ay^3x - 129960a^2yx + 343y^4$$
$$+ 110592a^2 + 12348ay^2 = 0.$$

The critical points of the system (2.70) are foci, hence it has no rational first integrals. This example shows that the degree of the invariant algebraic curve of the studied quadratic systems without rational first integral is greater than or equal to 12.

2.9 Polynomial vector fields with $M \geq 2$ algebraic curves

First we study the case when $M = 2$.

Proposition 2.9.1. *Assume that a polynomial system of degree m has two invariant irreducible invariant algebraic curves $g_1 = 0$ and $g_2 = 0$ of degree n_1 and n_2, respectively, and such that $\{g_1, g_2\} \not\equiv 0$. Then the system is of the form*

$$\dot{x} = K_1 g_1 \frac{\{x, g_2\}}{\{g_1, g_2\}} + K_2 g_2 \frac{\{g_1, x\}}{\{g_1, g_2\}} = P(x, y),$$
$$\dot{y} = K_1 g_1 \frac{\{y, g_2\}}{\{g_1, g_2\}} + K_2 g_2 \frac{\{g_1, y\}}{\{g_1, g_2\}} = Q(x, y),$$
$$(2.71)$$

where K_1 and K_2 are the cofactors of the curve $g_1 = 0$ and $g_2 = 0$, respectively, and such that $\max(\deg P, \deg Q) = m$. Moreover, upon setting $K_j = \{g_1, g_2\}\mu_j$ for $j = 1, 2$, we have

$$\dot{x} = g_1\mu_1\{x, g_2\} + g_2\mu_2\{g_1, x\} = P(x, y),$$
$$\dot{y} = g_1\mu_1\{y, g_2\} + g_2\mu_2\{g_1, y\} = Q(x, y),$$
$$(2.72)$$

where μ_1 and μ_2 are arbitrary rational functions such that

$$\max(\deg P, \deg Q) = m, \quad \deg(\{g_1, g_2\}\mu_j) \leq n_j + m - 1, \quad j = 1, 2.$$

Systems (2.71) and (2.72) are the most general planar systems that have the invariant curve $g_j = 0$, for $j = 1, 2$.

Proof. The proof follows from Corollary 1.3.4 for $N = M = 2$. Indeed, from the system (1.14), (1.15) we obtain the following equations and in view of the relations

$$\dot{g}_j = \Phi_j = K_j g_j, \quad \dot{g}_j = \{g_1, g_2\}\mu_j g_j = K_j g_j, \quad \text{for} \quad j = 1, 2.$$

we obtain the proof of the proposition.

We observe that from the condition $\{g_1, g_2\} \neq 0$ it follows that the algebraic curves $g_j = 0$ for $j = 1, 2$ are non-singular and not transverse. $\square$

Now we study the case when $M > 2$.

Proposition 2.9.2. *Assume that a polynomial system of degree m has $M > 2$ irreducible invariant algebraic curves $g_j = 0$ of respective degrees n_j and such that $\{g_1, g_2\} \not\equiv 0$. Then the system is of the form*

$$\dot{x} = \sum_{j=1}^{M} G_j \{g_j, x\} - G_{M+1} = P(x, y),$$

$$\dot{y} = \sum_{j=1}^{M} G_j \{g_j, y\} + G_{M+2} = Q(x, y),$$

(2.73)

where $G_j = G_j(x, y)$ for $j = 1, 2, \ldots, M + 2$ are arbitrary functions satisfying

$$\max(\deg P,\ \deg Q) = m,$$

$$\sum_{k=1}^{M+2} G_k \{g_k, g_j\} = K_j\, g_j,$$

(2.74)

for $j = 1, 2, \ldots, M$, where $g_{M+1} = x$, $g_{M+2} = y$, and K_j are the cofactors of the curve $g_j = 0$ for $j = 1, \ldots, M$. A particular case of (2.73) is the system

$$\dot{x} = \sum_{j=1}^{M} \prod_{\substack{m=1 \\ m \neq j}}^{M} g_m \lambda_j \{g_j, x\} - \prod_{m=1}^{M} g_m \lambda_{M+1} = P(x, y),$$

$$\dot{y} = \sum_{j=1}^{M} \prod_{\substack{m=1 \\ m \neq j}}^{M} g_m \lambda_j \{g_j, y\} + \prod_{m=1}^{M} g_m \lambda_{M+2} = Q(x, y),$$

(2.75)

where $\lambda_1, \ldots, \lambda_{M+2}$ are arbitrary rational functions such that $\max(\deg P,\ \deg Q) = m$.

System (2.73) under the conditions (2.74) is the most general planar system that has the invariant curves $g_j = 0$, $j = 1, \ldots, M$.

Proof. This proposition is a simple consequence of Corollary 1.6.4. $\square$

Example 2.9.3. We determine the differential equations having the following invariant straight lines:

$$g_1 = x - 1, \quad g_2 = x + 1, \quad g_3 = x - (\sqrt{5} - 2), \quad g_4 = x + (\sqrt{5} - 2)$$
$$g_5 = y - 1, \quad g_6 = y + 1, \quad g_7 = y - (\sqrt{5} - 2), \quad g_8 = y + (\sqrt{5} - 2).$$

Since

$$\{g_1, g_5\} = 1, \quad \{g_j, x\} = 0 \quad \text{for} \quad j = 1, 2, 3, 4, \quad \{g_j, y\} = 0 \quad \text{for} \quad j = 5, 6, 7, 8,$$

differential system (2.75) takes the form

$$\dot{x} = -\sum_{j=5}^{8} \lambda_j \prod_{\substack{m=j \\ m \neq j}}^{8} g_m - \lambda_{10} \prod_{j=1}^{8} g_m = \left(-\prod_{j=1}^{4} g_m\right)\left(\sum_{j=5}^{8} \lambda_j \prod_{\substack{m=5 \\ m \neq j}}^{8} g_m + \lambda_{10} \prod_{j=5}^{8} g_m\right)$$
$$:= (x^2 - 1)(x^2 - (\sqrt{5} - 2)^2)\mu_1$$

$$\dot{y} = \sum_{j=1}^{4} \lambda_j \prod_{\substack{m=j \\ m \neq j}}^{8} g_m - \lambda_9 \prod_{j=1}^{8} g_m = \left(\prod_{j=5}^{8} g_m\right)\left(\sum_{j=1}^{4} \lambda_j \prod_{\substack{m=1 \\ m \neq j}}^{4} g_m - \lambda_9 \prod_{j=1}^{4} g_m\right)$$
$$:= (y^2 - 1)(y^2 - (\sqrt{5} - 2)^2)\mu_2,$$

where μ_1 and μ_2 are arbitrary functions. In particular if

$$\mu_1 = x + \sqrt{5}y, \quad \mu_2 = y + \sqrt{5}x,$$

(see page 225 of [6]) then the previous system has 14 invariant straight lines (for more details see [6]).

Example 2.9.4. We determine the most general differential planar system having the invariant curves

$$g_1 = x^2 + y^2 - 1 = 0, \quad g_2 = y - 1.$$

In this case we have that $\{g_1, g_2\} = 2x$. The curves are transverse at the point $(0, 1)$. The differential system (2.72) becomes

$$\dot{x} = \mu_1 - 2y\mu_2, \quad \dot{y} = 2x\mu_2,$$

where μ_1 and μ_2 are arbitrary rational functions such that

$$\dot{g}_1 = 2x\mu_1, \quad \mu_1|_{g_1=0} = 0 \quad \dot{g}_2 = 2x\mu_2, \quad \mu_2|_{g_2=0} = 0.$$

In particular, if

$$\mu_1 = \frac{(x + y^2 + xy)(x^2 + y^2 - 1)}{x}, \quad \mu_2 = \frac{(y + x^2 + y^2)(y - 1)}{2x},$$

then we obtain the differential system

$$\dot{x} = -1 - x + x^2 + xy + y^2 + yx^2 + y^3, \quad \dot{y} = (y + x^2 + y^2)(y - 1),$$

which was introduced in [34].

Chapter 3

Hilbert's 16th Problem for Algebraic Limit Cycles

3.1 Introduction

In this chapter we state Hilbert's 16th problem restricted to algebraic limit cycles. Namely, consider the set Σ'_n of all real polynomial vector fields $\mathcal{X} = (P, Q)$ of degree n having real irreducible (on $\mathbb{R}[x, y]$) invariant algebraic curves. A simpler version of the second part of Hilbert's 16th problem restricted to algebraic limit cycles can be stated as: *Is there an upper bound on the number of algebraic limit cycles of polynomial vector fields in Σ'_n?* We solve this simpler version of Hilbert's 16th problem in two cases. Specifically, when the invariant algebraic curves are generic in the sense given in the previous chapter, or when they are non-singular in $\mathbb{CP}^2$. We state the following conjecture: *The maximum number of algebraic limit cycles for polynomial planar vector fields of degree n is $H(n) := 1 + \dfrac{(n-1)(n-2)}{2}$.* We prove this conjecture for the case when n is even and the algebraic curves are generic M-curves, and then for the case when all the curves are non-singular in $\mathbb{R}^2$ and the sum of their degrees are less than $n + 1$.

3.2 Preliminary results

We recall that a *limit cycle* of a polynomial vector field $\mathbf{X}$ is an isolated periodic orbit in the set of all periodic orbits of $\mathbf{X}$. An *algebraic limit cycle* of degree m of $\mathbf{X}$ is an oval of a real irreducible (on $\mathbb{R}[x, y]$) invariant algebraic curve $f = 0$ of degree m which is a limit cycle of $\mathbf{X}$.

Consider the set Σ of all real polynomial vector fields of degree n associated to the differential system

$$\dot{x} = P(x, y), \quad \dot{y} = Q(x; y). \tag{3.1}$$

Hilbert in [73] asked: *Is there an upper bound on the number of limit cycles of every real polynomial vector field of* Σ? This is a version of the *second half of Hilbert's* 16*th problem*. As Stephen Smale stated in [145], except for the Riemann hypothesis, it seems to be the most elusive of Hilbert's problems.

Consider the set Σ_n' of all real polynomial vector fields (3.1) of degree n having real irreducible (on $\mathbb{R}[x,y]$) invariant algebraic curves. A simpler version of the second part of Hilbert's 16th problem is: *Is there an upper bound on the number of algebraic limit cycles of any polynomial vector field of* Σ_n'? At this time we cannot provide an answer to this question for general real algebraic curves. In this chapter we give the answer for the generic and non-singular algebraic curves and for the case when all the curves are non-singular in $\mathbb{R}^2$ and the sum of their degrees are less than $n + 1$.

We shall need the following basic results.

The next well-known result states that we can restrict our attention to the irreducible invariant algebraic curves; for a proof see, for instance, [98].

Proposition 3.2.1. *We suppose that* $f \in \mathbb{C}[x,y]$ *and let* $f = f_1^{n_1} \cdots f_r^{n_r}$ *be its factorization into irreducible factors over* $\mathbb{C}[x,y]$*. Then for a polynomial vector field* $\mathcal{X}$*,* $f = 0$ *is an invariant algebraic curve with cofactor* K_f *if and only if* $f_i = 0$ *is an invariant algebraic curve for each* $i = 1, \ldots, r$ *with cofactor* K_{f_i}*. Moreover* $K_f = n_1 K_{f_1} + \cdots + n_r K_{f_r}$*.*

Theorem 2.3.1 will be play a fundamental role in the proof of the main result of this section.

Theorem 3.2.2. *If one of the genericity conditions* (i)–(v) *stated just before Theorem 2.3.1, is not satisfied, then the statements of Theorem 2.3.1 do not hold.*

The next result is due to Christopher and Kooij [82]. For more information on Darboux integrability see, e.g., [98].

Proposition 3.2.3. *The polynomial vector field* (2.5) *has the integrating factor* $\mathcal{R} = \left(\prod_{i=1}^{k} f_i \right)^{-1}$*, and consequently the system is Darboux integrable.*

Let U be an open subset of $\mathbb{R}^2$. A function $V : U \to \mathbb{R}$ is an *inverse integrating factor* of a vector field (P, Q) defined on U if V verifies the linear partial differential equation

$$P\frac{\partial V}{\partial x} + Q\frac{\partial V}{\partial y} = \left(\frac{\partial P}{\partial x} + \frac{\partial Q}{\partial y} \right) V \tag{3.2}$$

in U. We note that V satisfies (3.2) in U if and only if $R = 1/V$ is an integrating factor in $U \setminus \{(x, y) \in U : V(x, y) = 0\}$.

The following result can be found in [66]; for another proof, see [109].

Theorem 3.2.4. *Let* **X** *be a* C^1 *vector field defined in the open subset* U *of* $\mathbb{R}^2$. *Let* $V : U \to \mathbb{R}$ *be an inverse integrating factor of* **X**. *If* γ *is a limit cycle of* **X**, *then* γ *is contained in* $\{(x, y) \in U : V(x, y) = 0\}$.

3.3 Hilbert's 16th problem for generic algebraic limit cycles

Our main result is the following.

Theorem 3.3.1. *For a polynomial vector field* **X** *of degree* n *all of whose irreducible invariant algebraic curves are generic (i.e., satisfying the genericity properties* (i)–(v) *stated just before Theorem 2.3.1), the maximum number of algebraic limit cycles is at most* $1 + (n-1)(n-2)/2$ *if* n *is even, and* $(n-1)(n-2)/2$ *if* n *is odd. Moreover, this upper bound is attained.*

The authors who have worked on algebraic limit cycles for polynomial vector fields have formulated the following two more or less well-known conjectures (see for instance [99, 111].

Conjecture 3.3.2. *Is* 1 *the maximum number of algebraic limit cycles that a quadratic polynomial vector field can have?*

Conjecture 3.3.3. *Is* 2 *the maximum number of algebraic limit cycles that a cubic polynomial vector field can have?*

Here we add a next new conjecture, which contains the previous two conjectures as particular cases.

Conjecture 3.3.4. *Is* $1 + (n - 1)(n - 2)/2$ *the maximum number of algebraic limit cycles that a polynomial vector field of degree* n *can have?*

Note that from Theorem 3.3.1 Conjecture 1 follows for the class of quadratic polynomial vector fields for which all irreducible invariant algebraic curves are generic. But it remains open for quadratic systems having non-generic invariant algebraic curves.

On the other hand, for cubic polynomial vector fields for which all irreducible invariant algebraic curves are generic, Theorem 3.3.1 says that the maximum number of algebraic limit cycles is 1. However, there are examples of cubic polynomial vector fields having two algebraic limit cycles, of course when non-generic invariant algebraic curves are present. Thus, the system

$$\dot{x} = 2y(10 + xy), \quad \dot{y} = 20x + y - 20x^3 - 2x^2y + 4y^3,$$

has two algebraic limit cycles contained in the invariant algebraic curve $2x^4 - 4x^2 + 4y^2 + 1 = 0$, see Proposition 19 of [111].

Another example is

$$\dot{x} = y(a^2 - r^2 - 3ax + 3x^2 - ay + 2xy + y^2),$$
$$\dot{y} = -a^2x + 3ax^2 - 2x^3 - r^2y + axy - x^2y + y^3,$$

with $r < a/2$. This system has the algebraic limit cycles $x^2 + y^2 = r^2$ and $(x - a)^2 + y^2 = r^2$, see [139].

We shall need the following technical result.

Lemma 3.3.5. *Let* $D \subset \mathbb{R}^s$ *be the compact set*

$$D = \{(x_1, x_2, \ldots, x_s) \in \mathbb{R}^s : x_j \geq 1, \quad j = 1, 2, \ldots, s, \quad \sum_{j=1}^{s} x_j \leq l\}$$

with $l \geq s$ *a positive integer, and let* $k : D \to \mathbb{R}$ *be the function defined by*

$$k(x_1, x_2, \ldots, x_s) = \frac{1}{2} \sum_{j=1}^{s} (x_j - 1)(x_j - 2) = \frac{1}{2} \sum_{j=1}^{s} \left(x_j - \frac{3}{2}\right)^2 - \frac{s}{8}.$$

Then the maximum value of k *is* $(l - s)(l - s - 1)/2 \geq 0$, *which is attained at the vertex* $(l + 1 - s, 1, \ldots, 1)$ *of the simplex* D.

Proof. Note that the function $2k(x_1, x_2, \ldots, x_s)$ is the square of the distance between the point $(x_1, x_2, \ldots, x_s)$ of the simplex D and the point $(3/2, 3/2, \ldots, 3/2)$ plus the constant $s/4$. Therefore, it is clear that the maximum of the function $k(x_1, x_2, \ldots, x_s)$ is attained at some of the vertices of the simplex D.

If $s = 1$, then $D = [1, l]$. It is clear that the maximum value of the function $(x_1 - 1)(x_1 - 2)/2$ on the interval $[1, l]$ is $(l - 1)(l - 2)/2$ and is attained at the endpoint $x_1 = l$. Therefore, the lemma is proved for $s = 1$.

If $s = 2$, then D is the triangle $T \subset \mathbb{R}^2$ of vertices $(1, 1)$, $(l - 1, 1)$, and $(1, l-1)$. It is easy to check that the maximum value $(l-2)(l-3)/2$ of the function $\sum_{j=1}^{2} (x_j - 1)(x_j - 2)/2$ on T is attained at the vertices $(l - 1, 1)$ and $(1, l - 1)$. So the lemma is proved for $s = 2$.

For $s > 2$, the set D is a simplex $S \subset \mathbb{R}^s$ with vertices $(1, \ldots, 1)$, $(l - s + 1, 1, \ldots, 1)$, $(1, l - s + 1, 1, \ldots, 1)$, $\ldots$, $(1, \ldots, 1, l - s + 1)$. It is not difficult to verify that the maximum value $(l - s)(l - s - 1)/2 \geq 0$ of the function k on S is attained at the vertices of the simplex S different from $(1, \ldots, 1)$. Hence the lemma follows. $\square$

Several proofs of the next results are known, see for instance [158, 165, 41].

Theorem 3.3.6 (*Harnack's Theorem 1*). *A real irreducible curve of degree m cannot have more than*

$$1 + \frac{1}{2}(n-1)(n-2) - \sum_p s_p(s_p - 1),$$

where s_p is the order of the singular point of the curve.

Theorem 3.3.7 (*Harnack's Theorem 2*). *The number of ovals of a real algebraic curve of degree m is at most $1 + (m-1)(m-2)/2$ when m is even, and $(m-1)(m-2)/2$ when m is odd. Moreover, these upper bounds are attained for certain non-singular algebraic curves of degree m called M-curve .*

Let $\mathbf{X}$ be a real polynomial vector field and let its complex irreducible invariant algebraic curves of $\mathbf{X}$ be $f_j = 0$ for $j = 1, \ldots, k$. Then we consider the set of all real irreducible invariant algebraic curves $g_i = 0$ of $\mathbf{X}$ which are formed by the curves $f_j = 0$ if f_j is a real polynomial, or $f_j \overline{f}_j = 0$ if f_j is not a real polynomial. In what follows we shall call the set of curves $g_i = 0$ for $i = 1, \ldots, s$ the set of *real invariant algebraic curves* associated to $f_j = 0$ for $j = 1, \ldots, k$.

We denote by $A(l, s)$ the maximum number of algebraic limit cycles contained in the irreducible invariant algebraic curves $g_i = 0$ for $i = 1, \ldots, s$ of a real polynomial vector field having exactly s invariant algebraic curves with $l = \sum_{i=1}^{s} \deg(g_i)$. Here the irreducibility of a polynomial is considered in $\mathbb{R}[x, y]$. In the next proposition we provide an upper bound for $A(l, s)$. This result plays an important role in the proof of the main theorem of this section.

Proposition 3.3.8. *For $j = 1, \ldots, s$ let $g_j = 0$ be irreducible (in $\mathbb{R}[x, y]$) real algebraic curves such that*

$$\deg(g_j) = m_j \geq 1, \quad l \geq \sum_{j=1}^{s} m_j \geq s,$$

and let k_j be the maximum possible number of ovals of the curve $g_j = 0$. For a polynomial vector field $\mathbf{X}$ of degree n for which the curves $g_j = 0$, for $j = 1, \ldots, s$, are the only invariant algebraic curves we have

$$A(l, s) \leq \frac{1}{2}(l-1)(l-2) + \sum_{j=1}^{s} a_j,$$

where $a_1 = 1$ if m_1 is even and $a_1 = 0$ if m_1 is odd.

Proof. From Harnack's Theorem 2 we know that

$$k_j' = \begin{cases} 1 + \frac{1}{2}(m_j - 1)(m_j - 2), & \text{if } m_j \text{ is even}, \\ \frac{1}{2}(m_j - 1)(m_j - 2), & \text{if } m_j \text{ is odd}. \end{cases}$$

The maximum number of ovals of the invariant algebraic curves of the polynomial vector field $\mathbf{X}$ is

$$A(l, s) \leq \sum_{j=1}^{s} k_j = k(m_1, \ldots, m_s) + \sum_{j=1}^{s} a_j, \tag{3.3}$$

where $a_j = 1$ if m_j is even, and $a_j = 0$ if m_j is odd. By Lemma 3.3.5,

$$k(m_1, \ldots, m_s) + \sum_{j=1}^{s} a_j \tag{3.4}$$

Relations (3.3) and (3.4) complete the proof.   □

The next result is due to Christopher [31].

Theorem 3.3.9. *Let* $g = 0$ *be a real non-singular algebraic curve of degree* n, *and* h *a first-degree polynomial, chosen so that the real straight line* $h = 0$ *lies outside all ovals of* $g = 0$. *Choose the real numbers* a *and* b *so that* $ah_x + bh_y \neq 0$. *Then the polynomial vector field of degree* n,

$$\dot{x} = ag - hg_y, \qquad \dot{y} = bg + hg_x,$$

has all the ovals of $g = 0$ *as hyperbolic limit cycles. Furthermore, this vector field has no other limit cycles.*

Proposition 3.3.10. *For a real polynomial vector field* $\mathbf{X}$ *of degree* n *with only one real non-singular irreducible invariant algebraic curve of degree* $n \geq 2$, *we have*

$$A(n, 1) = \begin{cases} 1 + \frac{1}{2}(n-1)(n-2), & \text{if } n \text{ is even,} \\ \frac{1}{2}(n-1)(n-2), & \text{if } n \text{ is odd.} \end{cases}$$

We note that the upper bound $A(n, 1)$ for the maximum number of algebraic limit cycles of $\mathbf{X}$ is reached.

Proof. The statement follows immediately from Theorem 3.3.9 and Harnack's Theorem 2 stated.   □

Example 3.3.11. Let

$$g = \left(\frac{x^2}{4} + y^2 - 1 \right) \left(\frac{y^2}{4} + x^2 - 1 \right) + \frac{1}{10} = 0,$$

be the *M-curve* of degree four. The planar polynomial differential system of degree four

$$\dot{x} = a \left(\left(\frac{x^2}{4} + y^2 - 1 \right) \left(\frac{y^2}{4} + x^2 - 1 \right) + \frac{1}{10} \right)$$
$$- \frac{y}{8}(Ax + By + C)(17x^2 + 8y^2 - 20),$$

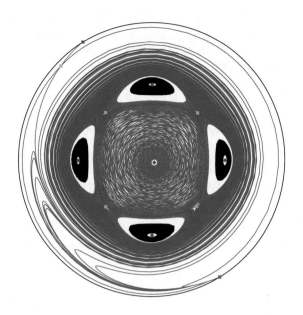

Figure 3.6: The phase portrait of Example 3.3.11
with its four limit cycles.

$$\dot{y} = b\left(\left(\frac{x^2}{4} + y^2 - 1\right)\left(\frac{y^2}{4} + x^2 - 1\right) + \frac{1}{10}\right)$$
$$+ \frac{x}{8}(Ax + By + C)(17y^2 + 8x^2 - 20),$$

where a, b, A, B, and C are arbitrary constants, admits four algebraic limit cycles
if the straight line $Ax + By + C = 0$ does not intersect the curve $g = 0$, see
Theorem 3.3.9. The phase portrait is the following one given in Figure 3.6.

Proposition 3.3.12. *Assume that $f_j = 0$ for $j = 1, \ldots, k$ are real algebraic curves
satisfying the assumptions of Theorem 2.3.1, and that*

$$l = \sum_{j=1}^{k} \deg f_j = n + 1.$$

*Let $g_i = 0$ for $i = 1, \ldots, s$ be the set of real invariant algebraic curves associated
to the $f_j = 0$ for $j = 1, \ldots, k$. Then, for $s = 2, \ldots, n$ the polynomial vector field
(2.5) of degree n satisfies that*

$$A(n+1, s) \leq \begin{cases} 1 + \frac{1}{2}(n-1)(n-2), & \text{if } n \text{ is even,} \\ \frac{1}{2}(n-1)(n-2), & \text{if } n \text{ is odd.} \end{cases}$$

If $s = 1$, then $A(n+1, 1) = 0$.

Proof. By Proposition 3.2.3 we know that $\prod_{i=1}^{s} g_i$ is an inverse integrating factor of the polynomial vector field (2.5). Therefore, by Theorem 3.2.4 all the limit cycles of (2.5) must be contained in the algebraic curves $g_i = 0$ for some $i = 1, \ldots, s$. In particular, all the limit cycles of (2.5) are algebraic.

Using Proposition 3.3.8,

$$A(n+1, s) \leq \frac{1}{2}(n+1-s)(n-s) + \sum_{i=1}^{s} a_j$$

for $s = 2, \ldots, n$. Since the maximum of the right-hand side expression is attained for $s = 2$ and the degrees of the two invariant algebraic curves are n and 1, we conclude that

$$A(n+1, s) \leq \begin{cases} 1 + \frac{1}{2}(n-1)(n-2), & \text{if } n \text{ is even,} \\ \frac{1}{2}(n-1)(n-2), & \text{if } n \text{ is odd.} \end{cases}$$

So the proposition is proved for $s = 2, \ldots, n$.

If $s = 1$ the polynomial vector field (2.5) is Hamiltonian, and consequently it has no limit cycles.   □

The solution of Hilbert's 16th problem for algebraic limit cycles which are contained in generic invariant algebraic curves is given by Theorem 3.3.1.

Proof of Theorem 3.3.1. Since all the irreducible invariant algebraic curves $f_j = 0$ for $j = 1, \ldots, k$ of **X** are generic, the assumptions of Theorem 2.3.1 hold. So **X** is the polynomial vector field given by either (2.4), or (2.5), or 0. Clearly this last case has no limit cycles.

For the polynomial vector field (2.5) the proof follows from Proposition 3.3.12.

Let $g_i = 0$ for $i = 1, \ldots, s$ be the set of all irreducible (on $\mathbb{R}[x, y]$) invariant algebraic curves associated to $f_j = 0$ for $j = 1, \ldots, k$.

Now we assume that **X** is the polynomial vector field given by (2.4) (so $l < n+1$), and we apply Proposition 3.3.8 to it (hence $l \geq s$). The l defined in the statement of that proposition takes the maximum value when $l = n$. Therefore we have that $l = n$ and $l > s$, because if $l = s$, then all the n invariant algebraic curves are straight lines, and consequently **X** has no algebraic limit cycles.

Applying Proposition 3.3.8 to the vector field **X** with $l = n > s$, and using that **X** has exactly s irreducible (on $\mathbb{R}[x, y]$) invariant algebraic curves, we get that

$$A(n, s) \leq \frac{1}{2}(n-s)(n-s-1) + \sum_{j=1}^{s} a_j.$$

The rest of the proof is divided into two cases.

Case 1: Assume that n is even. Then we have that

$$A(n, s) \leq \frac{1}{2}(n - s)(n - s - 1) + \sum_{j=1}^{s} a_j \leq \frac{1}{2}(n - s)(n - s - 1) + s. \qquad (3.5)$$

We claim that

$$\frac{1}{2}(n - s)(n - s - 1) + s \leq \frac{1}{2}(n - 1)(n - 2) + 1 \quad \text{for } n \geq 3. \qquad (3.6)$$

Now we shall prove the claim. We define the function

$$h(s) = \frac{1}{2}(n - s)(n - s - 1) + s - \frac{1}{2}(n - 1)(n - 2) - 1.$$

This function is a parabola in s satisfying that

$$h(s) \leq 0 \quad \text{if and only if } s \in [1, 2(n - 2)].$$

Therefore, if $n \geq 3$ then $1 \leq s \leq n - 1 \leq 2(n - 2)$. So $h(s) \leq 0$ if $1 \leq s \leq n - 1$, and consequently the claim is proved.

From (3.5) and (3.6) we have for $n \geq 4$ even that

$$A(n, s) \leq \frac{1}{2}(n - 1)(n - 2) + 1.$$

Hence the theorem is proved in this case. For n even it remains to prove the theorem when $n = 2 > s = 1$. This case is trivial because from Proposition 3.3.8 we have that $A(2, 1) \leq 1$. In short the theorem is proved for $n \geq 2$ even.

Case 2: Assume that n is odd. Then, since $l = n > s$, there must be at least one invariant algebraic curve g_i of odd degree, so

$$A(n, s) \leq \frac{1}{2}(n - s)(n - s - 1) + \sum_{j=1}^{s} a_j \leq \frac{1}{2}(n - s)(n - s - 1) + s - 1.$$

From the inequality (3.6) we have that

$$A(n, s) \leq \frac{1}{2}(n - s)(n - s - 1) + s - 1 \leq \frac{1}{2}(n - 1)(n - 2).$$

Hence the theorem is proved for $n \geq 3$ odd. So the first part of the theorem is proved.

By Proposition 3.3.10, the bound for $A(n, s)$ is attained. This completes the proof of the theorem. $\qquad\square$

3.4   Hilbert's 16th problem for non-singular algebraic limit cycles

In this section we give an upper bound for the maximum number N of algebraic limit cycles that a planar polynomial vector field of degree n can exhibit if the vector field has exactly k non-singular in $\mathbb{CP}^2$ and only irreducible invariant algebraic curves. Additionally, we provide sufficient conditions in order for all the algebraic limit cycles to be hyperbolic.

First we introduce the following concept: Let $g = g(x, y) = 0$ be an irreducible invariant algebraic curve of degree m of a polynomial vector field. We say that $g = 0$ is *non-singular* in $\mathbb{CP}^2$ if its projection

$$G = G_\nu(X, Y, Z) := Z^{m_\nu} g_\nu\left(X/Z, Y/Z\right) = 0,$$

is nonsigular in $\mathbb{CP}^2$, i.e., there are no points at which G and its first derivatives are all zero.

We say that $g = g(x, y) = 0$ is non-singular in $\mathbb{R}^2$ if there are no points in $\mathbb{R}^2$ at which g and its first derivatives are all zero.

We establish the following results.

Theorem 3.4.1. *Let $g_\nu = g_\nu(x, y) = 0$ for $\nu = 1, \ldots, s$ be all the real algebraic curves of degree m_ν with k_ν ovals and invariant under the polynomial vector field* $\mathbf{X}$ *of degree n. Let $\displaystyle\sum_{\nu=1}^{s} m_\nu \le l$.*

If all the curves $g_\nu = 0$ are non-singular in $\mathbb{CP}^2$ and irreducible, then

$$A(l, s) \le \sum_{\nu=1}^{s} k_\nu \le s\left(\frac{(n-1)(n-2)}{2} + 1\right) \le \frac{n^4 - 2n^3 + 3n^2 - 2n + 8}{4}.$$

Note that this theorem provides an upper bound for the maximum number of algebraic limit cycles that a polynomial vector field of degree n can have if all its invariant algebraic curves are non-singular in $\mathbb{CP}^2$ and irreducible. Hence it provides an answer to Hilbert's 16th problem restricted to the algebraic limit cycles and restricted to all polynomial vector fields having only invariant algebraic curves that are non-singular in $\mathbb{CP}^2$ and irreducible.

Proof of Theorem 3.4.1. We shall need the following result.

Theorem 3.4.2 (Joanolou's Theorem). *Let $\mathbf{X}$ be a polynomial vector field in $\mathbb{C}^2$ of degree n having irreducible invariant algebraic curves $g_\nu = g_\nu(x, y) = 0$, of multiplicity q_ν, for $\nu = 1, \ldots, s$. Then*

$$\sum_{\nu=1}^{s} q_\nu \ge \frac{n(n+1)}{2} + 2,$$

if and only if $\mathbf{X}$ has a rational first integral.

In fact Joanolou in [77] proved this theorem without taking into account the multiplicity of the curves $g_\nu = 0$, i.e., he considered $q_\nu = 1$. The improvement that accounts for the multiplicity is done in [39]. For a definition of the multiplicity of an invariant algebraic curve see [112].

The proof continues as follows. We denote by m_ν and K_ν the degree and the maximum number of ovals of the curve $g_\nu = 0$. Then by Harnack's Theorem 2 (see Theorem 3.3.7)) $k_\nu \leq \dfrac{(m_\nu - 1)(m_\nu - 2)}{2} + 1$.

If $\mathbf{X}$ has a rational first integral, then clearly $\mathbf{X}$ has no limit cycles. Therefore we can assume that $\mathbf{X}$ has no rational first integrals. Then if $g_\nu = 0$ is non-singular in $\mathbb{CP}^2$ and irreducible, from Corollary 2.4.8 and Theorems 2.4.7 and 2.4.9 we get that $m_\nu < n + 1$, hence

$$k_\nu \leq \frac{(m_\nu - 1)(m_\nu - 2)}{2} + 1 \leq \frac{(n-1)(n-2)}{2} + 1.$$

Therefore,

$$A(l, s) \leq \sum_{\nu=1}^{s} k_\nu \leq s\left(\frac{(n-1)(n-2)}{2} + 1\right).$$

So the first inequality of Theorem 3.4.1 is proved.

Next, Joanolous's Theorem implies that if the number k of irreducible invariant algebraic curves is large than $\dfrac{(n+1)n}{2} + 2$, then $\mathbf{X}$ has rational first integrals. So $s \leq \dfrac{(n+1)n}{2} + 1$. Consequently,

$$A(l, s) \leq s\left(\frac{(n-2)(n-1)}{2} + 1\right)$$
$$\leq \frac{(n^2 + n + 2)(n^2 - 3n + 4)}{4} = \frac{n^4 - 2n^3 + 3n^2 - 2n + 8}{4}.$$

Hence the theorem is proved. $\qquad\square$

Theorem 3.4.3. *Let $g_\nu = g_\nu(x, y) = 0$ for $\nu = 1, \ldots, s$ be all the real irreducible invariant algebraic curves of degree m_ν with k_ν ovals for the polynomial vector field $\mathbf{X}$ of degree n. Assume that $\displaystyle\sum_{\nu=1}^{s} m_\nu \leq n + 1$. If all the curves $g_\nu = 0$ are non-singular in $\mathbb{R}^2$, then*

(a) *the maximum number of algebraic limit cycles of $\mathbf{X}$ is at most $1 + \frac{1}{2}(n - 1)(n - 2)$, and*

(b) *this bound is attained if n is even.*

Proof. This proof is analogous to the last part of the proof of Theorem 3.3.1.

We assume that $\mathbf{X}$ is a polynomial vector field of degree n given by Propositions 2.6.1, 2.9.1, or 2.9.2, depending on s. We shall apply Proposition 3.3.8 to it (hence $l \geq s$). The l defined in the statement of that proposition takes the maximum value when $l = n$. Therefore we have that $l = n$ and $l > s$, because if $l = s$, then all the n invariant algebraic curves are straight lines, and consequently $\mathbf{X}$ has no algebraic limit cycles.

Applying Proposition 3.3.8 to the vector field $\mathbf{X}$ with $l = n > s$, and using the fact that $\mathbf{X}$ has exactly s irreducible (on $\mathbb{R}[x, y]$) invariant algebraic curves, we get

$$A(n, s) \leq \frac{1}{2}(n - s)(n - s - 1) + \sum_{j=1}^{s} a_j.$$

The rest of the proof is divided into two cases.

Case 1: Assume that n is even and $n \geq 3$. Then

$$A(n, s) \leq \frac{1}{2}(n - s)(n - s - 1) + \sum_{j=1}^{s} a_j \leq \frac{1}{2}(n - 1)(n - 21) + 1. \qquad (3.7)$$

For $n = 2$, and $s = 1$, Proposition 3.3.8 yields

$$A(2, 1) \leq 1.$$

In short the theorem is proved for $n \geq 2$ even.

Case 2: Assume that n is odd. Then, since $l = n > s$ it must be at least one invariant algebraic curve g_i of odd degree, so

$$A(n, s) \leq \frac{1}{2}(n-s)(n-s-1) + \sum_{j=1}^{s} a_j \leq \frac{1}{2}(n-s)(n-s-1)+s-1 \leq \frac{1}{2}(n-1)(n-2).$$

Hence the assertion is established for $n \geq 3$ odd, and the first part of the theorem is proved.

By Proposition 3.3.10, the bound for $A(n, s)$ is attained. This completes the proof of the theorem. □

From Theorem 3.4.3 it follows that the Conjecture 3.3.4 is valid for non-singular invariant curves under the assumptions of Theorem 3.4.3; see the next result, obtained by Dolov [46].

Corollary 3.4.4. *Under the assumptions of Theorem 3.4.3 for $s = 2$, and if $g_1 = x - a = 0$ and $g = g_2 = 0$ is an M-curve of degree $n - 1$, it holds that*

(a) *the maximum number of algebraic limit cycles of $\mathbf{X}$ is at most $1 + \frac{1}{2}(n - 2)(n - 3)$, and*

(b) *this bound is attained if n is odd.*

Proof. The proof is a simple consequence of Theorem 3.4.3 and of the fact that all the M-curves are non-singular (see Theorem 3.3.7). The upper bound is attained for the following polynomial vector field of degree n:

$$\dot{x} = (x - a)\frac{\partial g}{\partial y}, \quad \dot{y} = -(x - a)\frac{\partial g}{\partial x} + (Ax + By + C)g, \tag{3.8}$$

where a, A, B and C are constants, and the straight line $x - a = 0$ lies outside of the ovals of the M-curve $g = 0$ of degree $n - 1$. From the Harnack Theorem 2 we know that the number of ovals of the algebraic curve $g = 0$ of degree $m = n - 1$ is at most $1 + (n - 2)(n - 3)/2$ when m is even, and $(n - 2)(n - 3)/2$ when m is odd. In particular, for $n = 4$ we have that the maximum number of algebraic limit cycles is 1. This bound is attained in particular for the curve of degree 3

$$g = y^2 - (x - a_1)(x - a_2)(x - a_3) = 0, \quad a_1 < a_2 < a_3. \qquad \square$$

Theorem 3.4.5. *Let* $g(x, y) = 0$ *be the unique real algebraic invariant curve of degree* $m = \deg g \leq n + 1$ *of the polynomial vector field* $\mathbf{X}$ *of degree* n. *Assume that* g *is non-singular in* $\mathbb{R}^2$ *and irreducible. Then* $\mathbf{X}$ *can be written as*

$$\mathbf{X} = (\lambda_3 g - \lambda_1 g_y, \lambda_2 g + \lambda_1 g_x), \tag{3.9}$$

where $\lambda_\nu = \lambda_\nu(x, y)$ *for* $\nu = 1, 2, 3$ *are arbitrary polynomials. Assume that the following conditions are satisfied.*

(i) *The ovals of* $g = 0$ *do not intersect the algebraic curve* $\lambda_1 = 0$.
(ii) *If* $\gamma \subset \{g = 0\}$ *is an isolated periodic solution of* $\mathbf{X}$ *which does not intersect the curve* $\lambda_1 = 0$, *then*

$$\oint_\gamma \frac{\lambda_3 dy - \lambda_2 dx}{\lambda_1} = \iint_\Gamma \left(\left(\frac{\lambda_3}{\lambda_1}\right)_x + \left(\frac{\lambda_2}{\lambda_1}\right)_y \right) dxdy \neq 0,$$

where Γ *is the bounded region limited by* γ.
(iii) *The polynomial* $(\lambda_3\lambda_{1x} + \lambda_2\lambda_{1y})|_{\lambda_1=0}$ *does not vanish in* $\mathbb{R}^2 \setminus \{g = 0\}$.

Then the following statements hold.

(a) *All ovals of* $g = 0$ *are hyperbolic limit cycles of* $\mathbf{X}$. *Furthermore,* $\mathbf{X}$ *has no other limit cycles.*
(b) *Assume that* $a \in \mathbb{R}\setminus\{0\}$ *and* $G(x, y)$ *is an arbitrary polynomial of degree* $n - 1$ *such that the algebraic curve*

$$g = ax^{n+1} + G(x, y) = 0 \tag{3.10}$$

is non-singular and irreducible. We denote by $B_1(n)$ *(respectively* $B_2(n))$ *the maximum number of ovals of the curve (3.10) when* n *is odd (respectively* *even). If $\bar{A}(n+1,1)$ denotes the maximum number of algebraic limit cycles of the vector fields $\mathbf{X}$ given by (3.9), then*

$$\max\left(\frac{(n-1)(n-2)}{2},\, B_1(n)\right) \leq \bar{A}(n+1,1),$$

when n is odd, and

$$\max\left(\frac{(n-1)(n-2)}{2}+1,\, B_2(n)\right) \leq \bar{A}(1,n),$$

when n is even.

Proof. Let $g(x,y)=0$ be the unique real algebraic curve invariant of degree m of the polynomial vector field $\mathbf{X}$ of degree n. If $g=0$ is non-singular and irreducible, then $\mathbf{X}$ can be written as in (3.9) using (2.40).

One can check that the curve $g=0$ is invariant under $\mathbf{X}$. From (3.9) we have $\mathbf{X}g = (\lambda_3 g_x + \lambda_2 g_y)g$, consequently $g=0$ is an invariant algebraic curve with cofactor $K = \lambda_3 g_x + \lambda_2 g_y$.

Clearly a singular point of $\mathbf{X}$ on $g=0$ satisfies $-\lambda_1 g_x = \lambda_1 g_y = 0$, thus, since the given curve is non-singular, we have $\lambda_1 = 0$. Due to assumption (i) this can not occur. Thus each oval of $g=0$ is a periodic orbit of $\mathbf{X}$. Now we shall show that these periodic orbits are hyperbolic limit cycles.

Consider an oval γ of $g=0$. In order to verify that γ is a hyperbolic algebraic limit cycle we must show that

$$I = \oint_\gamma u(t)dt = \oint_\gamma \left(\frac{\partial P}{\partial x} + \frac{\partial Q}{\partial y}\right)(x(t),y(t))dt \neq 0, \tag{3.11}$$

where $u(t) := \operatorname{div}\mathbf{X}(x(t),y(t))$, $\mathcal{X} = (P,Q)$ and $(x(t),y(t))$ is the periodic solution corresponding to the periodic orbit γ; for more details see Theorem 1.2.3 of [49].

In (i) we assumed that γ does not intersect the curve $\lambda_1 = 0$. After straightforward calculations and using the fact that

$$g_y = -\frac{\dot{x} - \lambda_3 g}{\lambda_1}, \quad g_x = \frac{\dot{y} - \lambda_2 g}{\lambda_1},$$

on the points of γ we have that

$$u(t) = \lambda_{3x}g + \lambda_3 g_x - \lambda_{1x}g_y - \lambda_1 g_{xy} + \lambda_{2y}g + \lambda_2 g_y + \lambda_{1y}g_x + \lambda_1 g_{xy}$$
$$= (\lambda_{3x} + \lambda_{2y})g + (\lambda_2 + \lambda_{1x})g_y + (\lambda_3 + \lambda_{1y})g_x$$
$$= (\lambda_{3x} + \lambda_{2y})g - (\lambda_2 + \lambda_{1x})\frac{\dot{x} - \lambda_3 g}{\lambda_1} + (\lambda_3 + \lambda_{1y})\frac{\dot{y} - \lambda_2 g}{\lambda_1}$$
$$= (\lambda_{3x} + \lambda_{2y})g + \frac{\lambda_{1x}\dot{x} + \lambda_{1y}\dot{y}}{\lambda_1} + \frac{\lambda_3\dot{y} - \lambda_2\dot{x}}{\lambda_1} + \frac{g}{\lambda_1}(\lambda_3\lambda_{1x} - \lambda_2\lambda_{1y}).$$

Therefore

$$u(t)dt = d(\log|\lambda_1|) + \frac{\lambda_3 dy - \lambda_2 dx}{\lambda_1} + \frac{g}{\lambda_1}(\lambda_3\lambda_{1x} - \lambda_2\lambda_{1y} + \lambda_1\lambda_{3x} + \lambda_1\lambda_{2y})dt.$$

Hence, by assumptions (i) and (ii) and since $\gamma \subset \{g = 0\}$ we have that

$$\oint_\gamma u(t)dt = \oint_\gamma \frac{\lambda_3 dy - \lambda_2 dx}{\lambda_1} \neq 0.$$

Thus every oval of the algebraic curve $g = 0$ is a hyperbolic algebraic limit cycle.

Now suppose that there is a limit cycle $\tilde{\gamma}$ which is not contained in $g = 0$. Then this limit cycle is not algebraic. On the algebraic curve $\lambda_1 = 0$ we have

$$\dot{\lambda}_1 = \lambda_{1x}\dot{x} + \lambda_{1y}\dot{y}$$
$$= \lambda_{1x}(\lambda_3 g - \lambda_1 g_y) + \lambda_{1y}(\lambda_2 g + \lambda_1 g_x)$$
$$= \lambda_1(\lambda_{1y}g_x - \lambda_{1x}g_y) + g(\lambda_3\lambda_{1x} + \lambda_2\lambda_{1y}).$$

Hence, by assumption (iii) we have that

$$\dot{\lambda}_1|_{\lambda_1=0} = g(\lambda_3\lambda_{1x} + \lambda_2\lambda_{1y})|_{\lambda_1=0} \neq 0, \qquad (3.12)$$

in $\mathbb{R}^2 \setminus \{g = 0\}$. Thus the curve $\lambda_1 = 0$ is transverse to the flow associated to the vector field $\mathbf{X}$. Thus $\tilde{\gamma}$ cannot intersect the curve $\lambda_1 = 0$. Therefore, $\tilde{\gamma}$ lies in a connected component U of $\mathbb{R}^2 \setminus \{\lambda_1 g = 0\}$, so that g and λ_1 have constant sign on U. Since

$$\dot{g} = g_x\dot{x} + g_y\dot{y} = \frac{\dot{y} - \lambda_2 g}{\lambda_1}\dot{x} - \frac{\dot{x} - \lambda_3 g}{\lambda_1}\dot{y} = \frac{g(\lambda_3\dot{y} - \lambda_2\dot{x})}{\lambda_1},$$

we have $d(\log|g|) = (\lambda_3 dy - \lambda_2 dx)/\lambda_1$, and by (ii) we get the contradiction

$$0 = \oint_{\tilde{\gamma}} d(\log|g|) = \oint_{\tilde{\gamma}} \frac{\lambda_3 dy - \lambda_2 dx}{\lambda_1} \neq 0.$$

Statement (a) is proved.

The lower bounds for $\bar{A}(n+1, 1)$ are deduced as follows. We consider the polynomial differential system

$$\dot{x} = \alpha g - (Ax + By + C)g_y, \quad \dot{y} = \beta g + (Ax + By + C)g_x, \qquad (3.13)$$

where $A, B, C, \alpha, \beta \in \mathbb{R}$, $\alpha A + \beta B \neq 0$, and g is an arbitrary polynomial of degree n such that the ovals of $g = 0$ do not intersect $Ax + By + C = 0$. This system was already studied in [29]. Of course, (3.13) is a particular case of (3.9) taking

$$\lambda_1 = Ax + By + C, \quad \lambda_3 = \alpha, \quad \lambda_2 = \beta.$$

Then from (3.12) we get that

$$\dot{\lambda}_1|_{\lambda_1=0} = g(\alpha A + \beta B)|_{\lambda_1=0} \neq 0 \quad \text{in } \mathbb{R}^2 \setminus \{g = 0\}.$$

By direct computation we get

$$\oint_\gamma \frac{\lambda_3 dy - \lambda_2 dx}{\lambda_1} = -(\alpha A + \beta B) \iint_\Gamma \frac{dx dy}{(Ax + By + C)^2} \neq 0.$$

Hence conditions (ii) and (iii) of the statement of Theorem 1.3.1 hold for the system (3.13).

Hence, applying Proposition 3.3.9, we obtain for that system (3.13) that its number of algebraic limit cycles is $1 + (n - 1)(n - 2)/2$ when n is even, and $(n - 1)(n - 2)/2$ when n is odd. Consequently, $\bar{A}(n + 1, 1) \geq 1 + (n - 1)(n - 2)/2$ when n is even, and $\bar{A}(n + 1, 1) \geq (n - 1)(n - 2)/2$ when n is odd.

Now we consider the polynomial differential system of degree n (also studied in [134])

$$\dot{x} = (a + bxy)g_y, \quad \dot{y} = (n + 1)byg - (a + bxy)g_x, \tag{3.14}$$

where $a, b \in \mathbb{R} \setminus \{0\}$ and g is given in (3.10). The coefficient a which appears in the expression of g is the same a which appears in system (3.14). It easy to check that this system has the invariant algebraic curve $g = 0$. This system is a particular case of (3.9) taking

$$\lambda_1 = -(a + bxy), \quad \lambda_2 = (n + 1)by, \quad \lambda_3 = 0.$$

For system (3.14) we have from (3.12) that

$$\dot{\lambda}_1|_{\lambda_1=0} = -(n + 1)b^2 xyg|_{\lambda_1=0} = (n + 1)b(\lambda_1 + a)g|_{\lambda_1=0} = (n + 1)abg|_{\lambda_1=0}$$

is not zero in $\mathbb{R}^2 \setminus \{g = 0\}$. Let γ be an oval of the curve (3.10). We compute

$$\oint_\gamma \frac{\lambda_3 dy - \lambda_2 dx}{\lambda_1} = -(n + 1)ab \iint_\Omega \frac{dx dy}{(a + bxy)^2} \neq 0.$$

Moreover

$$(\lambda_3 \lambda_{1x} + \lambda_2 \lambda_{1y})|_{\lambda_1=0} = -(n + 1)b^2 xy|_{\lambda_1=0} = (n + 1)ab \neq 0.$$

Hence system (3.14) satisfies conditions (i), (ii) and (iii). So we obtain that $B_1(n) \leq \bar{A}(n+1, 1)$ if n is odd, and $B_2(n) \leq \bar{A}(n+1, 1)$ if n is even. Thus, (b) of the theorem is proved.   □

Corollary 3.4.6. *For the number $B_1(n)$ the following estimates hold:*

$$\frac{(n - 1)(n - 2)}{2} + 1 \leq B_1(n) \leq \left(\frac{(n - 1)(n - 2)}{2} + 1 \right) + \frac{n - 3}{2},$$

in $\mathbb{RP}^2$.

Proof. From Theorem 3.4.1 and for n odd we know that

$$\bar{A}(n+1,1) \geq \frac{(n-1)(n-2)}{2}.$$

We conjecture that the following result holds, but we do not have a proof for it. *Under the assumptions of Theorem 3.4.1, for all positive integer n we have that*

$$\bar{A}(n+1,1) \geq \frac{(n-1)(n-2)}{2} + 1. \tag{3.15}$$

If n is even, then of course (3.15) follows directly from statement (b) of Theorem 3.4.1. So assume that n is odd.

We study the curve

$$f = ax^{2m} + G(x,y) = 0, \quad \deg G(x,y) = 2m - 2.$$

After some computations one can check that the maximum *genus* G of this curve is $\mathrm{G} = 2(m-1)^2 - 1$ see for more details [58]. Hence the maximum number of ovals of the given curve (which we denote by $B_1(n)$) is not greater than $\mathrm{G}+1 = 2(m-1)^2$ in $\mathbb{RP}^2$ [165]. Oleg Viro (personal communication) proved that the number of ovals of this curve is at least $1 + (2m-2)(2m-3)/2 = 2(m-1)^2 - (m-2)$ in $\mathbb{RP}^2$.

To prove this lower bound Viro considers an algebraic curve $G(x,y) = 0$ of degree $2m-2$ having $\frac{1}{2}(2m-2)(2m-4)$ ovals in the affine plane and an unbounded component having at infinity $2m-2$ different points distinct from the infinity point of the y-axis. Then, with a a suitable choice of the sign of the polynomial $G(x,y)$, the algebraic curve

$$\varepsilon\, x^{2m} + G(x,y) = 0, \tag{3.16}$$

with $\varepsilon > 0$ sufficiently small has all the perturbed ovals of $G(x,y) = 0$, plus $2m-2$ ovals coming from the infinity and the unbounded component of $G(x,y) = 0$. So, the algebraic curve (3.16) has

$$\frac{1}{2}(2m-2)(2m-4) + 2m - 2 \tag{3.17}$$

ovals in the *projective plane*. If we change $n = 2m - 1$ in (3.17) we obtain $\frac{1}{2}(n-1)(n-2)+1$. Note that if one adds to (3.17) the number $m - 2 = \frac{n-3}{2}$, then we obtain Harnack's upper bound

$$\left(\frac{1}{2}(2m-2)(2m-4) + 2m - 2\right) + m - 2 = 2(m-1)^2 = \mathrm{G} + 1.$$

As a consequence, we have the following estimates for $B_1(n)$ in the projective plane:

$$2(m-1)^2 - (m-2) \leq B_1(n) \leq 2(m-1)^2,$$

or, equivalently,

$$\frac{(n-1)(n-2)}{2} + 1 \leq B_1(n) \leq \left(\frac{(n-1)(n-2)}{2} + 1\right) + \frac{n-3}{2}, \qquad (3.18)$$

because $n = 2m - 1$. We are interested in realizing the lower bound in (3.18) in the affine plane, because then the inequality (3.15) would hold, and in particular we should obtain the number of algebraic limit cycles stated in the Conjecture 3.3.4 for n odd. But for the moment we are not able to realize (and we do not know if it is possible thus realization in general) it in the affine plane for $n > 3$.

For the particular case $n = 3$, i.e., $m = 2$, we know that $B_1(3) = 2$. This number of ovals is realized in the affine plane for the algebraic curve which was already studied in [111],

$$y^2 + (x^2 - 1/2)^2 - 1/5 = 0.$$

We observe that the curve

$$\varepsilon\, x^4 + y^2 - x^2 = 0, \qquad (3.19)$$

has the same number of ovals; here ε is a parameter. □

Corollary 3.4.7. *Under the assumptions of Theorems 3.4.1 and 3.4.5,*

(a) $\bar{A}(n+1, 1) = \dfrac{(n-1(n-2)}{2} + 1$, *for n even. In particular, for $n = 2$ we have* $\bar{A}(3, 1) = 1$,

(c) $\bar{A}(4, 1) = 2$, *and*

(d) $6 \leq \bar{A}(6, 1) \leq 7$.

Proof. (a) The proof of the equalities for $\bar{A}(n+1, 1)$ trivially follows from Theorem 3.4.1 with $s = 1$ and from statement (b) of Theorem 3.4.5.

(c) The proof that $\bar{A}(4, 1) = 2$ is easily obtained from Theorems 3.4.1 and 3.4.5 and the fact that $B_1(3) = 2$.

(d) follows from Theorems 3.4.1 and 3.4.5, and from the fact that $B_1(5) \geq 6$ because the algebraic curve

$$\varepsilon\, x^6 + \left(y^2 - \frac{\sqrt{19}}{3}x^2 - \frac{77 - 11\sqrt{19}}{144}\right)\left(y^2 + \frac{\sqrt{19}}{3}x^2 - \frac{77 + 11\sqrt{19}}{144}\right) + \frac{367}{380}$$

$$:= \varepsilon x^6 + G(x, y) = 0, \qquad (3.20)$$

where $G(x, y)$ is a polynomial of degree four, has genus seven and, in particular, if $\epsilon = 1$, then this curve can be rewritten as

$$\left(x^2 - \frac{1}{9}\right)(x^2 - 1)^2 + \left(y^2 - \frac{1}{2}\right)^2 - \frac{1}{10} = 0, \qquad (3.21)$$

and has six ovals in $\mathbb{R}^2$. □

Figure 3.7: The phase portrait of system (3.14) for $a = b = 1$ and $n = 5$, and with the curve $g = 0$ given by the equation (3.21) with its 6 limit cycles. In the picture we only have drawn the separatrices of the system.

The polynomial differential system (3.14) for $a = b = 1$ and $n = 5$, and with the curve $g = 0$ given by the equation (3.21) has the six ovals as limit cycles, see the phase portrait of Figure 3.7.

Remark 3.4.8. We observe that the curves (3.19) and (3.20) are particular case of the curve

$$f = a_0 \prod_{j=1}^{S_1} (x - a_j)^{\alpha_j} + b_0 \prod_{j=1}^{S_2} (y - b_j)^{\beta_j}, \tag{3.22}$$

where $a_0 \neq 0$, $b_0 \neq 0$, $a_1 < \cdots < a_{S_1}$, $b_1 < \cdots < b_{S_2}$, and $\alpha_1, \ldots, \alpha_{S_1}$, $\beta_1, \ldots, \beta_{S_2}$ are positive integers. (For more details see [134].)

We study the following subcases of (3.22):

(a) $n = 2m - 1$, $S_1 = m + 1$, $S_2 = m$, $\alpha_j = 1$, $\beta_j = 1$, and $a_1, \ldots, a_{m+1}, b_1, \ldots, b_n$ are such that

$$f := f_{(m)} = a_0 \prod_{j=1}^{m+1} \left(x^2 - a_j^2\right) + b_0 \prod_{j=1}^{m} \left(y^2 - b_j^2\right) := x^{2m} + G(x, y) = 0,$$

where $a_1 < \cdots < a_m$ and $b_1 < \cdots < b_m$, a_0 and b_0 are nonzero constants, and $G(x, y)$ is a polynomial of degree $2m - 2$, has the genus $\mathrm{G} = 2(m - 1)^2 - 1$. Consequently, the maximum number $B_1(n)$ of ovals of this curve in $\mathbb{RP}^2$ obeys the estimate (3.18).

A particular case of this curve is the curve

$$f := f_{(2m+2)} = \prod_{n=1}^{m+1} \left(\frac{4x^2}{((2n-1)\pi)^2} - 1 \right) + \prod_{n=1}^{m} \left(\frac{4y^2}{((2n-1)\pi)^2} - 1 \right) - a_m$$
$$= x^{2m+2} + G(x, y) = 0,$$

where a_m is a suitable constant. The analytic curve

$$f_\infty = \prod_{j=1}^{\infty} \left(\frac{4x^2}{((2n-1)\pi)^2} - 1 \right) + \prod_{j=1}^{\infty} \left(\frac{4y^2}{((2n-1)\pi)^2} - 1 \right) - 1$$
$$= \cos x + \cos y - 1 = 0,$$

has infinitely many ovals. This result follows from the fact that the function $f_\infty = \cos x + \cos y$ has infinitely many maxima which is reached at the points $(\pi/2 + 2\pi k, \pi/2 + 2\pi j)|_{(j,k)\in\mathbb{Z}^2}$. The value of the function at these points is $f_\infty(\pi/2 + 2\pi k, \pi/2 + 2\pi j) = 2$.

(b) $n = 2m$, $S_1 = m + 1$, $S_2 = m$, $\alpha_j = 1$, $\beta_j = 1$, and $a_1 = 0$, $a_2, \ldots, a_{m+1}$, $b_1 = 0$, $b_2, \ldots, b_n$ are such that

$$f := f_{(2m+1)} = x \prod_{j=1}^{m} \left(\left(\frac{x}{j\pi} \right)^2 - 1 \right) + y \prod_{j=1}^{m-1} \left(\left(\frac{y}{j\pi} \right)^2 - 1 \right) - 1$$
$$= x^{2m+1} + G(x, y),$$

where $G(x, y)$ is a polynomial of degree $2m-1$, has the genus $\mathrm{G} = 2m(m-1)$, and consequently the maximum number of ovals of this curve in $\mathbb{RP}^2$ is $\mathrm{G}+1 = 2m(m-1)+1$. The analytic curve

$$f_\infty = x \prod_{j=1}^{\infty} \left(\left(\frac{x}{j\pi} \right)^2 - 1 \right) + y \prod_{j=1}^{\infty} \left(\left(\frac{y}{j\pi} \right)^2 - 1 \right) - 1$$
$$= \sin x + \sin y - 1,$$

has infinitely many ovals.

The following question arise: how many ovals in the affine plane does the curve (3.22) have?

Consider the polynomial of degree $n + 1$

$$H := \int_{x_0}^{x} \prod_{j=1}^{n} (x - a_j) dx + \int_{y_0}^{y} \prod_{j=1}^{n} (y - b_j) dy, \qquad (3.23)$$

where

$$a_1 < \ldots < a_n, \quad b_1 < \ldots < b_n. \tag{3.24}$$

This curve is non-singular in $\mathbb{RP}^2$ for it, consequently its genus is $\mathrm{G} = n(n-1)/2$. Thus the maximum number of ovals in $\mathbb{RP}^2$ is not greater than $\mathrm{G}+1 = \dfrac{n(n-1)}{2}+1$ (for more details see [59, 130]).

Figure 3.8: The phase portrait of a Hamiltonian system of degree three having five centers.

Corollary 3.4.9. *The Hamiltonian system*

$$\dot{x} = -\frac{\partial H}{\partial y}, \quad \dot{y} = -\frac{\partial H}{\partial x},$$

with H given by (3.23) has $\dfrac{n^2}{2}$ *centers if n is even and* $\dfrac{n^2+1}{2}$ *centers if n is odd.*

Proof. The Hamiltonian system has the *analytic first integral* $H = h$. Expanding H at the point (a_j, b_k) we obtain

$$H = \frac{y^2}{2} \prod_{\substack{m=1 \\ m \neq k}}^{n} (b_k - b_m) + \frac{x^2}{2} \prod_{\substack{m=1 \\ m \neq j}}^{n} (a_j - a_m) + H^*(x, y),$$

where H^* is a polynomial of degree greater than 2 and less than or equal to $n+1$. Consequently, this first integral is positive definite if and only if

$$\left(\prod_{\substack{m=1 \\ m \neq k}}^{n} (b_k - b_m) \right) \left(\prod_{\substack{m=1 \\ m \neq j}}^{n} (a_j - a_m) \right) > 0. \tag{3.25}$$

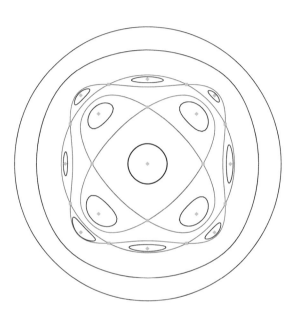

Figure 3.9: The phase portrait of a Hamiltonian system
of degree five having 13 centers.

Theorem 3.4.10 (Lyapunov's Theorem). *A necessary and sufficient condition for a singular point of the planar differential system*

$$\dot{x} = P(x,y), \quad \dot{y} = Q(x,y),$$

to be a center is the existence of a positive definite analytic first integral in a neighborhood of this point.

From (3.24), (3.25) and Lyapunov's Theorem we obtain after some computations the claimed numbers of centers.   □

Theorem 3.4.11. *Let $g_\nu(x,y) = 0$ for $\nu = 1,2$ be the unique algebraic curves of degree m_ν such that $l = m_1 + m_2 \leq n+1$ invariant for the polynomial vector field*

$$\mathbf{X} = \left(\lambda_4 g_1 g_2 - r_1 g_{1y} - r_2 g_{2y}, \lambda_3 g_1 g_2 + r_1 g_{1x} + r_2 g_{2x} \right), \tag{3.26}$$

of degree n, *where* $r_1 = \lambda_1 g_2$, $r_2 = \lambda_2 g_1$, *and* $\lambda_j = \lambda_j(x, y)$ *for* $j = 1, 2, 3, 4$ *are polynomials. Assume that the two curves* $g_\nu = 0$ *are non-singular and irreducible and that the following conditions are satisfied.*

(i) *For* $\nu = 1, 2$ *the ovals of* $g_\nu = 0$ *do not intersect the algebraic curve* $r_\nu = 0$.

(ii) *If* γ *is an isolated periodic orbit of* $\mathbf{X}$ *which does not intersect the curve* $r_\nu = 0$ *for* $\nu = 1, 2$, *then*

$$I_1 = \oint_\gamma \left(\frac{\lambda_4 dy - \lambda_3 dx}{\lambda_1} - \frac{\lambda_2}{\lambda_1} d(\log |g_2|) \right) \neq 0,$$

$$I_2 = \oint_\gamma \left(\frac{\lambda_4 dy - \lambda_3 dx}{\lambda_2} - \frac{\lambda_1}{\lambda_2} d(\log |g_1|) \right) \neq 0.$$

(iii) *The polynomials*

$$(\lambda_4 r_{1x} + \lambda_3 r_{1y} + \lambda_2 \{g_2, \lambda_1\}) |_{r_1=0},$$
$$(\lambda_4 r_{2x} + \lambda_3 r_{2y} + \lambda_1 \{g_1, \lambda_2\}) |_{r_2=0},$$

do not vanish in $\mathbb{R}^2 \backslash \{g_1 g_2 = 0\}$.

Then the following statements hold.

(a) *All ovals of* $g_\nu = 0$ *for* $\nu = 1, 2$ *are hyperbolic limit cycles of* $\mathbf{X}$. *Furthermore* $\mathbf{X}$ *has no other limit cycles.*

(b) *Assume that* $a \in \mathbb{R} \backslash \{0\}$ *and* $G(x, y)$ *is an arbitrary polynomial of degree* $n - 1$ *such that the algebraic curve*

$$f = ax^n + G(x, y) = 0 \tag{3.27}$$

is non-singular, irreducible, and $\{Ax + By + C, f\}$ *does not change sign in the region* $\in \mathbb{R}^2 \backslash \{0 < Ax + By + C < -K/n\}$ *where* A, B, C, K *are real parameters such that* $aBK \neq 0$. *We denote by* $b_1(n)$ *(respectively* $b_2(n)$*) the maximum number of ovals of all curves* $f = 0$ *when* n *is odd (respectively even). If* $\hat{A}(n + 1, 2)$ *denotes the maximum number of algebraic limit cycles of all the vector fields* $\mathbf{X}$ *given by* (3.26), *then*

$$b_1(n) \leq \hat{A}(2, n),$$

when n *is odd, and*

$$b_2(n) \leq \hat{A}(2, n), \quad \text{for} \quad n \geq 4, \quad b_2(2) = 0,$$

when n *is even. Here*

$$1 + \frac{(n-2)(n-3)}{2} \leq b_1(n) \leq \frac{(n-1)(n-2)}{2},$$
$$\frac{(n-2)(n-3)}{2} \leq b_2(n) \leq \frac{(n-1)(n-2)}{2} + 1. \tag{3.28}$$

Proof. The proof is obtained from the one of Theorem 3.4.5 by observing that the vector field $\mathbf{X}$ given by (3.26) can be written in the two forms

$$\dot{x} = l_\nu g_\nu - r_\nu g_{\nu y}, \quad \dot{y} = s_\nu g_\nu + r_\nu g_{\nu x}, \quad \nu = 1, 2,$$

where

$$
\begin{aligned}
r_1 &= \lambda_1 g_2, & r_2 &= \lambda_2 g_1, \\
l_1 &= g_2 \lambda_4 - \lambda_2 g_{2y}, & s_1 &= g_2 \lambda_3 + \lambda_2 g_{2x}, \\
l_2 &= g_1 \lambda_4 - \lambda_1 g_{1y}, & s_2 &= g_1 \lambda_3 + \lambda_1 g_{1x}.
\end{aligned}
$$

It is easy to check that $g_\nu = 0$ for $\nu = 1, 2$ are invariant algebraic curves of $\mathbf{X}$ because we have

$$\mathbf{X}(g_1) = (g_2(\lambda_4 g_{1x} + \lambda_3 g_{1y}) + \lambda_2 \{g_2, g_1\}) g_1,$$
$$\mathbf{X}(g_2) = ((g_1(\lambda_4 g_{2x} + \lambda_3 g_{2y}) + \lambda_2 \{g_1, g_2\}) g_2.$$

Clearly a singular point on $g_1 = 0$ satisfies either $r_1 = 0$, or $g_{1x} = g_{1y} = 0$. Due to assumption (i) and the fact that $g_1 = 0$ is non-singular, any of these two cases can occur. Thus each oval of $g_1 = 0$ must be a periodic orbit of $\mathbf{X}$. Analogously one proves that each oval of $g_2 = 0$ must be a *periodic solution* of $\mathbf{X}$. Now we shall show that these periodic orbits are in fact hyperbolic limit cycles.

Consider an oval γ_ν of $g_\nu = 0$ for $\nu = 1, 2$. From assumption (i) we know that γ_ν does not intersect the curve $r_\nu = 0$. In order to see that γ_ν is a hyperbolic algebraic limit cycle we must show that condition (3.11) holds for γ_ν and our vector field $\mathbf{X}$ given by (3.26).

Since

$$g_{\nu y} = -\frac{\dot{x} - l_\nu g_\nu}{r_\nu}, \qquad g_{\nu x} = \frac{\dot{y} - s_\nu g_\nu}{r_\nu}, \qquad \nu = 1, 2,$$

we obtain as in the proof of Theorem 3.4.5 that

$$u(t)dt = d(\log |r_\nu|) + \frac{l_\nu dy - s_\nu dx}{r_\nu} + \frac{g_\nu}{r_\nu}(l_\nu r_{\nu x} - s_\nu r_{\nu y} + r_\nu l_{\nu x} + r_\nu s_{\nu y})dt,$$

for $\nu = 1, 2$. Hence we have that

$$\oint_{\gamma_\nu} u(t)dt = \oint_{\gamma_\nu} \frac{l_\nu dy - s_\nu dx}{r_\nu},$$

for $\nu = 1, 2$. Since

$$
\begin{aligned}
\frac{l_1 dy - s_1 dx}{r_1} &= \frac{\lambda_4 dy - \lambda_3 dx}{\lambda_1} - \frac{\lambda_2}{\lambda_1} d(\log |g_2|), \\
\frac{l_2 dy - s_2 dx}{r_2} &= \frac{\lambda_4 dy - \lambda_3 dx}{\lambda_2} - \frac{\lambda_1}{\lambda_2} d(\log |g_1|),
\end{aligned}
\tag{3.29}
$$

we obtain that

$$\oint_{\gamma_1} u(t)dt = \oint_{\gamma_1} \left(\frac{\lambda_4 dy - \lambda_3 dx}{\lambda_1} - \frac{\lambda_2}{\lambda_1} d(\log|g_2|) \right),$$

$$\oint_{\gamma_2} u(t)dt = \oint_{\gamma_2} \left(\frac{\lambda_4 dy - \lambda_3 dx}{\lambda_2} - \frac{\lambda_1}{\lambda_2} d(\log|g_1|) \right),$$

and both integrals are not zero by assumption (ii). Thus every oval of the algebraic curve $g_\nu = 0$ is a hyperbolic algebraic limit cycle.

Now let us assume that there is a limit cycle $\tilde{\gamma}$ which is not contained in $g_1 g_2 = 0$. Then this limit cycle cannot be algebraic. On the algebraic curve $r_\nu = 0$ for $\nu = 1, 2$ we have

$$\dot{r}_\nu = r_{\nu x}\dot{x} + r_{\nu y}\dot{y}$$
$$= r_{\nu x}(l_\nu g_\nu - r_\nu g_{\nu y}) + r_{\nu y}(s_\nu g_\nu + r_\nu g_{\nu x})$$
$$= r_\nu(r_{\nu y}g_{\nu x} - r_{\nu x}g_{\nu y}) + g_\nu(l_\nu r_{\nu x} + s_\nu r_{\nu y}).$$

Therefore

$$\dot{r}_1 = r_1\{g_1, r_1\} + g_1 g_2(\lambda_4 r_{1x} + \lambda_3 r_{1y} + \lambda_2\{g_2, \lambda_1\}),$$
$$\dot{r}_2 = r_2\{g_2, r_2\} + g_1 g_2(\lambda_4 r_{2x} + \lambda_3 r_{2y} + \lambda_1\{g_1, \lambda_2\}).$$

Therefore by assumption (iii) we have that

$$\dot{r}_1|_{r_1=0} = g_1 g_2(\lambda_4 r_{1x} + \lambda_3 r_{1y} + \lambda_2\{g_2, \lambda_1\})|_{r_1=0} \neq 0,$$
$$\dot{r}_2|_{r_2=0} = g_1 g_2(\lambda_4 r_{2x} + \lambda_3 r_{2y} + \lambda_1\{g_1, \lambda_2\})|_{r_2=0} \neq 0,$$

in $\mathbb{R}^2 \setminus \{g_1 g_2 = 0\}$. Thus $\tilde{\gamma}$ cannot intersect the curve $r_\nu = 0$ for $\nu = 1, 2$. So $\tilde{\gamma}$ lies in a connected component U_ν of $\mathbb{R}^2 \setminus \{g_1 g_2 r_\nu = 0\}$, consequently g_1, g_2 and r_ν have constant sign on U_ν. Therefore, from

$$\dot{g}_\nu = g_{\nu x}\dot{x} + g_{\nu y}\dot{y} = \frac{\dot{y} - s_\nu g_\nu}{r_\nu}\dot{x} - \frac{\dot{x} - l_\nu g_\nu}{r_\nu}\dot{y} = \frac{g_\nu}{r_\nu}(l_\nu \dot{y} - s_\nu \dot{x}),$$

and (3.29) we have

$$d(\log|g_1|) = \frac{l_1 dy - s_1 dx}{r_1} = \frac{\lambda_4 dy - \lambda_3 dx}{\lambda_1} - \frac{\lambda_2}{\lambda_1} d(\log|g_2|),$$

$$d(\log|g_2|) = \frac{l_2 dy - s_2 dx}{r_2} = \frac{\lambda_4 dy - \lambda_3 dx}{\lambda_2} - \frac{\lambda_1}{\lambda_2} d(\log|g_1|).$$

By assumption (ii) we get the contradiction

$$0 = \oint_{\tilde{\gamma}} d(\log|g_\nu|) = I_\nu \neq 0, \quad \nu = 1, 2.$$

So statement (a) is proved.

The lower bounds are deduced as follows. In [139] the differential system

$$\dot{x} = \lambda(x - b)g_{2y}, \quad \dot{y} = (n\lambda + K)g_2 - \lambda(x - b)g_{2x}, \tag{3.30}$$

was studied, where $\lambda = Ax + By + C$ and

$$g_1(x, y) = x - b, \quad g_2(x, y) = ax^n + G(x, y),$$

with G an arbitrary polynomial of degree $n - 1$ such that the curve $g_2 = 0$ is irreducible, non-singular and $\{Ax + By + C, g_2\}$ does not change sign in the region $\mathbb{R}^2 \setminus D$, where

$$D = \left\{ 0 < Ax + By + C < -\frac{K}{n} \right\}.$$

Note that system (3.30) has the form as system (3.26) with

$$\lambda_1 = n\lambda + K, \quad \lambda_2 = \lambda = Ax + By + C, \quad \lambda_3 = 0, \quad \lambda_4 = 0,$$

where A, B, C, K, a, b are real parameters with $aBK \neq 0$. We remark that if $B = 0$, then system (3.30) has three invariant algebraic curves $x - b = 0$, $Ax + C = 0$ and $g_2 = 0$. Therefore, using the Darboux theory of integrability (see for instance Chapter 8 of [49]) it is easy to show that system (3.30) has a first integral.

The parameter b is chosen in such a way that the straight line $(x - b) = 0$ does not intersect any oval of $g_2 = 0$ and any periodic orbit of the system (3.30). Then the system (3.30) satisfies condition (i) of Theorem 3.4.11.

Now we prove that (3.30) satisfies condition (ii) of Theorem 3.4.11.

Let γ be an *isolated periodic solution* of (3.30) which does not intersect the algebraic curves $r_1 = \lambda_1 g_2 = 0$, $r_2 = \lambda g_1 = 0$. First we show that $I_j \neq 0$ for $j = 1, 2$ in D. In this region D we have $0 < \lambda < -K/n$, hence it is easy to deduce that

$$n + \frac{K}{\lambda} < 0, \quad 1 - \frac{K}{n\lambda + K} < 0.$$

In view of the equalities

$$\frac{\lambda_1}{\lambda_2} = \frac{n\lambda + K}{\lambda} = n + \frac{K}{\lambda} < 0, \quad \frac{\lambda_2}{\lambda_1} = \frac{\lambda}{n\lambda + K} = \frac{1}{n}\left(1 - \frac{K}{n\lambda + K}\right) < 0$$

we obtain that

$$I_1 = -\oint_\gamma \frac{\lambda_2}{\lambda_1} d(\log|g_2|) = -\oint_\gamma \frac{K}{n\lambda + K} d(\log|g_2|) < -\oint_\gamma d(\log|g_2|) = 0,$$

$$I_2 = -\oint_\gamma \frac{\lambda_1}{\lambda_2} d(\log|g_1|) = -\oint_\gamma \frac{K}{\lambda} d(\log|g_1|) \quad > n\oint_\gamma d(\log|g_1|) = 0.$$

Hence we obtain that $I_j \neq 0$ for $j = 1, 2$ in the region D. Now we show that these two integrals are non-zero in $\mathbb{R}^2 \setminus D$. If we denote by Γ the bounded region limited by γ, then

$$
\begin{aligned}
I_1 &= -\oint_\gamma \frac{\lambda_2}{\lambda_1} d(\log |g_2|) = -\oint_\gamma \left(\frac{\lambda_2 g_{2x}}{\lambda_1 g_2} \, dx + \frac{\lambda_2 g_{2y}}{\lambda_1 g_2} \, dy \right) \\
&= \iint_\Gamma \frac{1}{g_2} \left(g_{2x} \frac{\lambda_{2y}\lambda_1 - \lambda_2 \lambda_{1y}}{\lambda_1^2} - g_{2y} \frac{\lambda_{2x}\lambda_1 - \lambda_2 \lambda_{1x}}{\lambda_1^2} \right) dxdy \\
&= \iint_\Gamma \frac{1}{g_2} \left\{ g_2, \frac{\lambda_2}{\lambda_1} \right\} dxdy = -\frac{K}{n} \iint_\Gamma \frac{1}{g_2} \left\{ g_2, \frac{1}{\lambda_1} \right\} dxdy \\
&= \frac{K}{n} \iint_\Gamma \frac{1}{\lambda_1^2 g_2} (\lambda_{1x} g_{2y} - \lambda_{1y} g_{2x}) dxdy \\
&= \frac{K}{n} \iint_\Gamma \frac{\{\lambda_1, g_2\}}{\lambda_1^2 g_2} dxdy = K \iint_\Gamma \frac{\{\lambda, f\}}{\lambda_1^2 f} dxdy.
\end{aligned}
$$

By assumption, $g_2 = f$ and $\lambda_1^2 g_2$ has constant sign on Γ because γ does not intersect the curve $\lambda_1 g_2 = 0$, and $\{\lambda, g_2\}$ does not change sign in $\mathbb{R}^2 \setminus D$. Therefore we get that $I_1 \neq 0$.

Working in a similar way we have that

$$
\begin{aligned}
I_2 &= \iint_\Gamma \frac{1}{g_1} \left\{ g_1, \frac{\lambda_1}{\lambda_2} \right\} dxdy = K \iint_\Gamma \frac{1}{g_1} \left\{ g_1, \frac{1}{\lambda} \right\} dxdy \\
&= -K \iint_\Gamma \frac{\{\lambda, g_1\}}{\lambda^2 g_1} dxdy = \iint_\Gamma \frac{KB}{(Ax + By + C)^2 (x - b)} dxdy \neq 0,
\end{aligned}
$$

because $x - b = 0$ and $\lambda_2^2 g_1 = 0$ do not intersect any periodic orbit of system (3.30). Thus, condition (ii) of Theorem 1.6.1 is satisfied by system (3.30).

In $\mathbb{R}^2 \setminus \{g_1 g_2 = 0\}$ we have that

$$
\alpha = (\lambda_4 r_{1x} + \lambda_3 r_{1y} + \lambda_2 \{g_2, \lambda_1\})|_{r_1=0} = (\lambda_2 \{g_2, \lambda_1\})|_{\lambda_1=0}.
$$

Since $\{g_2, \lambda\}$ does not change sign in $\mathbb{R}^2 \setminus D$ and consequently $\{g_2, \lambda_1\}|_{\lambda_1=0} \neq 0$ in $\mathbb{R}^2 \setminus \{g_1 g_2 = 0\}$, because $\lambda > 0$ in D. In a similar way,

$$
\beta = (\lambda_4 r_{2x} + \lambda_3 r_{2y} + \lambda_1 \{g_1, \lambda_2\})|_{r_2=0} = (\lambda_1 \{g_1, \lambda_2\})|_{\lambda_2=0} = B\lambda_1|_{\lambda_2=0}.
$$

Since $\lambda_1 = K \neq 0$ if $\lambda_2 = 0$ because $K \neq 0$, we get that $\beta \neq 0$ in $\mathbb{R}^2 \setminus \{g_1 g_2 = 0\}$. Hence system (3.30) also satisfies condition (iii) of Theorem 3.4.11.

Since system (3.30) satisfies conditions (i), (ii), and (iii) of Theorem 3.4.11, the ovals of the curve $g_2 = 0$ are hyperbolic limit cycles of system (3.30), and this system has no other limit cycles. Then $\bar{A}(2, n) \geq b_1(n)$ when n is odd, and $\bar{A}(2, n) \geq b_2(n)$ when n is even.

The upper and lower bounds for $b_\nu(n)$ of (3.28) follow from Harnack's Theorem, taking $\alpha \neq 0$ and $\alpha = 0$, respectively. This completes the proof of Theorem 3.4.11. $\qquad\square$

Remark 3.4.12. Statement (b) of Theorem 3.4.11 improves Dolov's result given in Corollary 3.4.4. In particular, for $n = 4$ the differential system (3.8) admits 1 limit cycle, while the differential system (3.30) for $n = 4$ admits 2 limit cycles. This bound is reached in particular when the curve (3.27) is

$$g = (x - a_1)(x - a_2)(x - a_3)(x - a_4) - y^2 := x^4 + G(x, y) = 0,$$

where $a_1 < a_2 < a_3 < a_4$ and $G(x, y)$ is a polynomial of degree 3.

Corollary 3.4.13. *We have that*

$$\hat{A}(4, 2) \geq 2.$$

Proof. This follows from Example 1.3.7. The system is a particular case of (3.26) with $\lambda_1 = x + y - a$, $\lambda_2 = -(x + y)$, $\lambda_3 = 0$, $\lambda_4 = 0$.

The curves $g_1 = 0$ and $g_2 = 0$ are invariant circles with cofactor $K_1 = 4ay(x + y)$ and $K_2 = 4ay(x + y - a)$, respectively. The system has in $\mathbb{R}^2$ three singular points, $(0, 0)$, $(a, 0)$, and $(a/2, 0)$. The first two are strong foci and the third is a *saddle*. Since the invariant circles contains singular points, they are periodic orbits, and due to the fact that they surround a focus, they are algebraic limit cycles. Moreover, since the unique algebraic curves of system, are $g_1 = 0$ and $g_2 = 0$, the system has exactly two algebraic limit cycles. Hence $\hat{A}(4, 2) \geq 2$ and the corollary follows (for more details see [139]). □

Theorem 3.4.14. *Let $g_\nu(x, y) = 0$ for $\nu = 1, 2, \ldots, s$ be the unique invariant algebraic curves of the polynomial vector field*

$$\mathbf{X} = \left(\lambda_{k+2} g - \sum_{\nu=1}^{s} r_\nu g_{\nu y}, \; \lambda_{s+1} g + \sum_{\nu=1}^{k} r_\nu g_{\nu x} \right), \tag{3.31}$$

of degree n, where $g = \prod_{\nu=1}^{s} g_\nu$, $\lambda_j = \lambda_j(x, y)$ for $j = 1, \ldots, s+2$ are polynomials, and $r_\nu = \lambda_\nu \prod_{j \neq \nu} g_j$ for $\nu = 1, \ldots, s$. Assume that the curves $g_\nu = 0$ are non-singular and irreducible and that the following conditions are satisfied.

(i) *For $\nu = 1, \ldots, s$ the ovals of $g_\nu = 0$ and $r_\nu = 0$ do not intersect.*

(ii) *If γ is an isolated periodic orbit of $\mathbf{X}$ which does not intersect the curve $r_\nu = 0$ for $\nu = 1, \ldots, s$, then*

$$\oint_\gamma \frac{\lambda_{k+2} dy - \lambda_{s+1} dx}{\lambda_\nu} - \sum_{\substack{j=1 \\ j \neq \nu}}^{s} \oint_\gamma \frac{\lambda_j}{\lambda_\nu} d \log |g_j| \neq 0.$$

(iii) *For $\nu = 1, \ldots, s$ the polynomials*

$$\left. \left(\lambda_{s+2} r_{\nu x} + \lambda_{s+1} r_{\nu y} + \sum_{\substack{j=1 \\ j \neq \nu}}^{s} \lambda_j \{\lambda_\nu, g_j\} \prod_{\substack{m=1 \\ m \neq j, \nu}}^{k} g_m \right) \right|_{r_\nu = 0}$$

do not vanish in $\mathbb{R}^2 \setminus \{\prod_{j=1}^{s} g_j = 0\}$.

Then all ovals of $g_\nu = 0$ are hyperbolic limit cycles of **X**. *Furthermore* **X** *has no other limit cycles.*

We omit the proof of Theorem 3.4.14 because it is analogous to that of Theorem 3.4.11.

Corollary 3.4.15. *If $\widehat{A}(l, s)$ denotes the maximum number of algebraic limit cycles of all the vector fields* **X** *given by* (3.31), *then we have that*

$$\widehat{A}(2(n-1), n-1) \geq n-1.$$

Proof. To obtain this lower bound we introduce the following polynomial differential system of degree n already considered in [139] :

$$\dot{x} = 2y\left((x+y)\prod_{j=1}^{l} g_j - (x+y-a)\prod_{j=1}^{l} g_{j+l}\right),$$

$$\dot{y} = -2x\left((x+y)\prod_{j=1}^{l} g_j - (x+y-a)\prod_{j=1}^{l} g_{j+l}\right) + 2a(x+y)\prod_{j=1}^{l} g_j,$$

where

$$g_j = x^2 + y^2 - R_j^2, \quad g_{j+l} = (x-a)^2 + y^2 - R_j^2,$$

for $j = 1, \ldots, l$, and $0 < R_1 < R_2 < \cdots < R_l < a/2$, which is a particular case of system (3.31). One can easily verify that the circles $g_j = 0$ and $g_{j+l} = 0$ for $j = 1, 2, \ldots, l$ are invariant of this polynomial differential system of degree $n = 2l + 1$, with respective cofactors

$$K_j = 4ay(x+y)\prod_{m\neq j}^{l} g_m, \quad K_{j+l} = 4ay(x+y-a)\prod_{m\neq j}^{l} g_{m+l},$$

for $j = 1, 2, \ldots, l$.

It is easy to show that the singular points of the given polynomial system in $\mathbb{R}^2$ are the strong foci $(0,0)$ and $(a,0)$ and the saddle $(a/2, 0)$.

In this case we have that

$$\lambda_\nu = -g_\nu(x+y)\left(\prod_{j=l}^{l} g_j\right)^{-1}, \quad \lambda_{\nu+l} = g_{\nu+l}(x+y-a)\left(\prod_{j=l+1}^{2l} g_j\right)^{-1}, \quad \nu = 1, 2, \ldots, l,$$

and as a consequence

$$r_\nu = -(x+y)\prod_{j=l+1}^{2l} g_j, \quad r_{\nu+l} = (x+y-a)\prod_{j=l}^{l} g_j, \quad \nu = 1, 2, \ldots, l.$$

As in the proof of Corollary 3.4.13, the $2l = n-1$ circles are algebraic limit cycles. This yields the claimed lower bound $\widehat{A}(2(n-1), n-1) \geq n-1$. The proof is complete. $\qquad\square$

Remark 3.4.16. As it follows from the results established in [104] and [105], in order to solve Hilbert's problem for algebraic limit cycles it is necessary to determine the maximum degree of the invariant algebraic curves (Poincaré's problem). It is well known that if the invariant algebraic curve $g = 0$ of a polynomial vector field of degree n is non-singular in $\mathbb{C}P^2$, then $\deg g < n + 1$ (see for instance [32]). Moreover, in [17] the authors gave the following result: if all the singularities on the invariant algebraic curve $g = 0$ are double and ordinary, then $\deg g \leq 2n$. If the algebraic curve is of nodal type, i.e., it is singular and all its singularities are of normal crossing type (that is, at any singularity of the curve there are exactly two branches of $g = 0$ which intersect transversally), then $\deg g \leq n + 2$. There exist polynomial planar systems with an invariant singular *algebraic curve of arbitrarily high degree* (see for instance Proposition 2.7.2).

In general, what is an upper bound for the degree of a singular invariant algebraic curve is an open problem.

The solution of Hilbert's 16th problem for limit cycles in the case of generic non-singular, and *nodal algebraic curves* are given in [104], [105] and [162], respectively. For limit cycles on singular curves the problem remains open.

Chapter 4

Inverse Problem for Constrained Lagrangian Systems

4.1 Introduction

The aim of this chapter is to provide a solution of the *inverse problem of the constrained Lagrangian mechanics* which can be stated as follows: Determine for a given natural mechanical system with N degrees of freedom the most general field of forces depending only on the positions and satisfying a given set of constraints with are linear in the velocities. This statement of the inverse problem for constrained Lagrangian systems is new.

As we can see from the first chapter (see for instance Theorem 1.3.1) the solutions of the inverse problem for ordinary differential equations have a very high degree of arbitrariness due to the undetermined functions which appear in them. To obtain specific solutions we need additional conditions to reduce this arbitrariness. In this chapter we will obtain these additional conditions in order to derive the equations of motion for constrained Lagrangian mechanics.

One of the main objectives in this inverse problem is to study the behavior of the nonholonomic systems with constraints linear in the velocity in a way different from the classical approach deduced from the d'Alembert–Lagrange principle. We explain this in more detail in Remark 4.2.5.

In particular, solving the inverse problem for the constrained Lagrangian systems we obtain a solution for the *inverse problem of dynamics* (for more details [60]). The first inverse problem in dynamics appeared in Celestial Mechanics; it was stated and solved by Newton (1687) [122] and concerns the determination of the potential force field that ensures the planetary motion in accordance to the observed properties, namely the Kepler laws.

Bertrand (1877) in [10] proved that the expression for Newton's force of attraction can be obtained directly from the Kepler first law. He stated also a more general problem of determining a positional force, under which a particle describes a conic section for any initial conditions. Bertrand's ideas were developed further in the works [42, 149, 78, 51, 60, 132, 139], among others.

In the modern scientific literature the importance of the inverse problem in Celestial Mechanics was already recognized by Szebehely (see [152]).

Clearly, in view of the second Newton law, that acceleration is equal to force, we obtain that the above inverse problems are equivalent to determining the second-order differential equations from given properties on the right-hand side.

4.2   Solution of the inverse problem for constrained Lagrangian systems

Let us introduce the notations and definitions that we need for presenting our applications of Theorem 1.3.1.

We denote by Q an N-dimensional *smooth manifold* and by TQ the tangent bundle of Q with local coordinates $\mathbf{x} = (x_1, \ldots, x_N)$, and $(\mathbf{x}, \dot{\mathbf{x}}) = (x_1, \ldots, x_N, \dot{x}_1, \ldots, \dot{x}_N)$, respectively (see for instance [67]).

The following definitions can be found in [4].

A *Lagrangian system* is a pair $(Q, \tilde{L})$ consisting of a smooth manifold Q, a function $\tilde{L} : TQ \to \mathbb{R}$. The points $\mathbf{x} \in Q$ specify the *position* of the system and the tangent vectors $\dot{\mathbf{x}} \in T_{\mathbf{x}}Q$ specify the *velocity* of the system at the point $\mathbf{x}$. A pair $(\mathbf{x}, \dot{\mathbf{x}})$ is called a *state* of the system. In Lagrangian mechanics it is usual to call Q the *configuration space*, and the tangent bundle TQ the *phase space*, $\tilde{L}$ is the *Lagrange function* or *Lagrangian*, and the dimension N of Q is the number of *degrees of freedom*.

Consider M equations

$$h_j = h_j(\mathbf{x}, \dot{\mathbf{x}}) = 0, \quad \text{for} \quad j = 1, \ldots, M \leq N, \tag{4.1}$$

with rank $\left(\dfrac{\partial(h_1, \ldots, h_M)}{\partial(\dot{x}_1, \ldots, \dot{x}_N)} \right) = M$ in all the points of Q, except in a set of Lebesgue measure zero. We say that equations (4.1) define M independent *constraints* for the Lagrangian systems $(Q, \tilde{L})$ if the orbits $(\mathbf{x}(t), \dot{\mathbf{x}}(t))$ of the mechanical system are required to satisfy (4.1).

Let $\mathcal{M}^*$ be the submanifold of TQ defined by the equations (4.1), i.e.,

$$\mathcal{M}^* = \{(\mathbf{x}, \dot{\mathbf{x}}) \in TQ : h_j(\mathbf{x}, \dot{\mathbf{x}}) = 0, \quad \text{for} \quad j = 1, \ldots, M \leq N\}$$

The triplet $(Q, \tilde{L}, \mathcal{M}^*)$ is called a *constrained Lagrangian system*.

We call the *inverse problem for a constrained Lagrangian system* the problem of determining for a given constrained Lagrangian system $(Q, \tilde{L}, \mathcal{M}^*)$, the force field $\mathbf{F} = \mathbf{F}(\mathbf{x}) = (F_1(\mathbf{x}), \ldots, F_N(\mathbf{x}))$ in such a way that the given submanifold $\mathcal{M}^*$ is invariant under the flow of the second-order differential equations

$$\frac{d}{dt}\left(\frac{\partial \tilde{L}}{\partial \dot{x}_j}\right) - \frac{\partial \tilde{L}}{\partial x_j} = F_j(\mathbf{x}) \quad \text{for} \quad j = 1, \ldots, N.$$

We shall study the case when the constraints are linear in the velocities in $\mathcal{M}^*$, i.e.,

$$h_j(\mathbf{x}, \dot{\mathbf{x}}) = \sum_{k=1}^{N} a_{jk}(\mathbf{x})\dot{x}_k + \alpha_j(\mathbf{x}) = 0, \quad \text{for} \quad j = 1, \ldots, M.$$

Our first main result provides the equations of motion of a constrained mechanical system with Lagrangian function

$$\tilde{L} = T = \frac{1}{2}\sum_{n,j=1}^{N} G_{jn}(\mathbf{x})\dot{x}_j\dot{x}_n := \frac{1}{2}\langle \dot{\mathbf{x}}, \dot{\mathbf{x}}\rangle = \frac{1}{2}||\dot{\mathbf{x}}||^2, \tag{4.2}$$

where T is a *Riemannian metric* on Q (representing the *kinetic energy* of the system), and $M = N$ *linear constraints* given by

$$g_j = \sum_{n=1}^{N} G_{jn}(\mathbf{x})\left(\dot{x}_n - v_n(\mathbf{x})\right) = 0 \quad \text{for} \quad j = 1, \ldots, N, \tag{4.3}$$

where $\mathbf{v}(\mathbf{x}) = (v_1(\mathbf{x}), \ldots, v_N(\mathbf{x}))$ is a given vector field.

Theorem 4.2.1. *Let Σ be a constrained Lagrangian mechanical system with configuration space Q, kinetic energy T given by (4.2), and constraints given by (4.3). The equations of motion of Σ are the Lagrange differential equations*

$$\frac{d}{dt}\left(\frac{\partial L}{\partial \dot{x}_j}\right) - \frac{\partial L}{\partial x_j} = 0 \quad \text{for} \quad j = 1, \ldots, N, \tag{4.4}$$

with $L = \frac{1}{2}||\dot{\mathbf{x}} - \mathbf{v}||^2 = T - \langle \dot{\mathbf{x}}, \mathbf{v}\rangle + \frac{1}{2}||\mathbf{v}||^2$, which are equivalent to

$$\frac{d}{dt}\left(\frac{\partial T}{\partial \dot{x}_j}\right) - \frac{\partial T}{\partial x_j} = \frac{\partial}{\partial x_j}\left(\frac{1}{2}||\mathbf{v}||^2\right) + \sum_{n=1}^{N} \dot{x}_n\left(\frac{\partial p_j}{\partial x_n} - \frac{\partial p_n}{\partial x_j}\right)$$

$$= \frac{\partial}{\partial x_j}\left(\frac{1}{2}||\mathbf{v}||^2\right) + \sum_{n=1}^{N} v_n\left(\frac{\partial p_j}{\partial x_n} - \frac{\partial p_n}{\partial x_j}\right), \tag{4.5}$$

where

$$p_j = \sum_{n=1}^{N} G_{jn}v_n, \quad \text{for} \quad j = 1, 2, \ldots, N. \tag{4.6}$$

Proof. We consider the differential system (1.4) with N replaced by $2N$ and with invariant hypersurfaces $g_j(x_1, \ldots, x_{2N}) = 0$ for $j = 1, \ldots, N_1 \leq N$. Take the functions g_m for $m = N_1, \ldots, 2N$ as follows: $g_\alpha = g_\alpha(x_1, \ldots, x_{2N})$, $g_{N+j} = x_j$, for $\alpha = N_1 + 1, \ldots, N$ if $N_1 < N$ and $j = 1, \ldots, N$, . We assume that $\{g_1, g_2, \ldots, g_N, x_1, \ldots, x_N\} \neq 0$. Hence the system (1.4) takes the form

$$\dot{x}_j = \lambda_{N+j},$$

$$\dot{x}_{j+N} = \sum_{k=1}^{N_1} \Phi_k \frac{\{g_1, \ldots, g_{k_1}, x_{j+N}, g_{k+1}, \ldots, g_N, x_1, \ldots, x_N\}}{\{g_1, g_2, \ldots, g_N, x_1, \ldots, x_N\}} + \cdots \qquad (4.7)$$

$$+ \sum_{k=N_1+1}^{2N} \lambda_k \frac{\{g_1, \ldots, g_{N_1+1}, \ldots, g_{k-1}, x_{j+N}, g_{k+1}, \ldots, g_N, x_1, \ldots, x_N\}}{\{g_1, \ldots, g_N, x_1, \ldots, x_N\}},$$

for $j = 1, \ldots, K$.

In particular, if we take $g_j = x_{N+j} - p_j(x_1, \ldots, x_N) = 0$, where $p_j = p_j(x_1, \ldots, x_N)$ are suitable functions for $j = 1, \ldots, N$, then from (4.7) we obtain

$$\dot{x}_j = \lambda_{N+j}, \quad \dot{x}_{N+j} = \Phi_j + \sum_{n=1}^{N} \lambda_{N+n} \frac{\partial p_j}{\partial x_n},$$

thus

$$\dot{x}_j = \lambda_{N+j}, \quad \frac{d}{dt}(x_{N+j} - p_j) = \Phi_j. \qquad (4.8)$$

Now take the arbitrary functions λ_{N+j} and Φ_j as

$$\lambda_{N+j} = \sum_{n=1}^{N} \tilde{G}_{jn} x_{N+n}, \quad \Phi_j = \frac{\partial L}{\partial x_j},$$

for $j = 1, \ldots, N$, where $\tilde{G}_{jn} = \tilde{G}_{jn}(x_1, \ldots, x_N)$ are elements of a symmetric definite positive matrix $\tilde{G}$, and

$$L = \frac{1}{2} \sum_{n,j=1}^{N} G_{jn}(\mathbf{x})(\dot{x}_j - v_j)(\dot{x}_n - v_n) = \frac{1}{2}||\dot{\mathbf{x}} - \mathbf{v}||^2 = \frac{1}{2}||\dot{\mathbf{x}}||^2 - \langle \mathbf{v}, \dot{\mathbf{x}} \rangle + \frac{1}{2}||\mathbf{v}||^2$$

$$= T - \langle \mathbf{v}, \dot{\mathbf{x}} \rangle + \frac{1}{2}||\mathbf{v}||^2,$$

where $G = (G_{jk})$ is the inverse matrix of $\tilde{G} = (\tilde{G}_{jk})$.

We can write g_j as

$$g_j = x_{j+N} - p_j = \sum_{n=1}^{N} G_{jn}(\dot{x}_n - v_n) = 0.$$

for $j = 1, \ldots, N$. Then, $g_j = 0$ if and only if $\dot{x}_1 - v_1 = \cdots = \dot{x}_N - v_N = 0$. Since

$$\Phi_j = \frac{\partial L}{\partial x_j} = -\left\langle \dot{\mathbf{x}} - \mathbf{v}, \frac{\partial \mathbf{v}}{\partial x_j} \right\rangle, \quad \text{we have } \Phi_j|_{g_j=0} = \frac{\partial L}{\partial x_j}\Big|_{\dot{\mathbf{x}}=\mathbf{v}} = 0,$$

for $j = 1, \ldots, N$.

On the other hand, in view of the relations

$$g_j = x_{j+N} - p_j = \sum_{n=1}^{N} G_{jn}(\dot{x}_n - v_n) = \frac{\partial L}{\partial \dot{x}_j},$$

we finally deduce that the equations (4.8) can be written as the *Lagrangian differential equations*

$$\frac{d}{dt}\left(\frac{\partial L}{\partial \dot{x}_j}\right) - \frac{\partial L}{\partial x_j} = 0, \quad \text{for} \quad j = 1, \ldots, N.$$

After some computation and recalling the constraints (4.3) we finally obtain the differential system (4.5). This completes the proof of the theorem.   □

In view of the second Newton law: *acceleration is equal to force* (see for instance [151]), the right-hand sides of the equations of motion (4.5) are the generalized forces acting on the mechanical system, which depend only on its position. Consequently, force field **F** with components

$$F_j = \frac{\partial}{\partial x_j}\left(\frac{1}{2}\|\mathbf{v}\|^2\right) + \sum_{n=1}^{N} v_n \left(\frac{\partial p_j}{\partial x_n} - \frac{\partial p_n}{\partial x_j}\right)$$

is the most general field of force depending only on the position of the *natural mechanical system* which is constrained to move on the N-dimensional subset of the phase space given by (4.3). Thus, the equations of motion (4.5) provide a complete answer to the inverse problem (i) when the constraints are given in the form (4.3).

Now we want to solve the inverse problem (i) for the classical constraints

$$\sum_{n=1}^{N} a_{jn}(\mathbf{x})\dot{x}_n = 0 \quad \text{for} \quad j = 1, \ldots, M. \tag{4.9}$$

We recall that the equations of motion of a constrained Lagrangian system with Lagrangian $\tilde{L} = \frac{1}{2}\|\dot{\mathbf{x}}\|^2 - U(\mathbf{x})$, and constraints given by (4.9), but with a force field $\tilde{\mathbf{F}} = (\tilde{F}_1, \ldots, \tilde{F}_N)$ depending on positions and velocities, are the

Lagrange differential equations with multipliers

$$\frac{d}{dt}\left(\frac{\partial T}{\partial \dot{x}_k}\right) - \frac{\partial T}{\partial x_k} = \tilde{F}_k(\mathbf{x}, \dot{\mathbf{x}}) = -\frac{\partial U}{\partial x_k} + \sum_{j=1}^{M} \mu_j a_{jk}, \quad \text{for} \quad k = 1, \ldots, N,$$

$$\sum_{n=1}^{N} a_{jn}(\mathbf{x})\dot{x}_n = 0, \qquad\qquad \text{for} \quad j = 1, \ldots, M,$$

(4.10)

where $\mu_j = \mu_j(\mathbf{x}, \dot{\mathbf{x}})$ are the *Lagrangian multipliers*. As we can observe, the forces $\tilde{\mathbf{F}}$ are composed of the potential forces with components $-\partial U/\partial x_k$ and the reactive forces generated by constraints with components $\sum_{j=1}^{M} \mu_j a_{jk}$ for $k = 1, \ldots, N$. For more details, see [4].

Thus, we have two equations of motions: the ones given in (4.4), or what is the same (4.5), for constraints of type (4.3), and the classical ones given in (4.10) for the constraints (4.9). In order to solve the problem (i) for the constraints (4.9) we establish the relationship between these two sets of equations. To this aim we shall choose conveniently the vector field $\mathbf{v}$ which appear in (4.3).

Since the constraints (4.3) are equivalent to the constraints $\dot{x}_j = v_j(\mathbf{x})$, $j = 1, \ldots, N$. On the other hand, from (4.9) we obtain that $\langle a_j, \mathbf{v} \rangle = \sum_{n=1}^{N} a_{jn} v_n = 0$, thus $\mathbf{v}$ must be orthogonal to the independent vectors $a_j = (a_{j1}, \ldots, a_{jN})$ for $j = 1, \ldots, M$. So we introduce N independent 1-forms: the first M 1-forms are associated to the M constraints (4.9), namely,

$$\Omega_j = \sum_{n=1}^{N} a_{jn}(\mathbf{x})dx_n \quad \text{for} \quad j = 1, \ldots, M,$$

(4.11)

and we choose the 1-forms Ω_j for $j = M+1, \ldots, N$ arbitrarily, but satisfying that the determinant $|\Upsilon|$ of the matrix

$$\Upsilon = \begin{pmatrix} \Omega_1(\partial_1) & \cdots & \Omega_1(\partial_N) \\ \vdots & \vdots & \vdots \\ \Omega_N(\partial_1) & \cdots & \Omega_N(\partial_N) \end{pmatrix} = \begin{pmatrix} a_{11} & \cdots & a_{N1} \\ \vdots & \vdots & \vdots \\ a_{N1} & \cdots & a_{NN} \end{pmatrix},$$

is nonzero. The ideal case would be when this determinant is constant; in other words the N 1-forms Ω_j for $j = 1, \ldots, N$ are independent. Now we define the vector field $\mathbf{v}$ as

$$\mathbf{v} = -\frac{1}{|\Upsilon|} \begin{vmatrix} \Omega_1(\partial_1) & \cdots & \Omega_1(\partial_N) & 0 \\ \vdots & \vdots & \vdots & \vdots \\ \Omega_M(\partial_1) & \cdots & \Omega_M(\partial_N) & 0 \\ \Omega_{M+1}(\partial_1) & \cdots & \Omega_{M+1}(\partial_N) & v_{M+1} \\ \vdots & \vdots & \vdots & \vdots \\ \Omega_N(\partial_1) & \cdots & \Omega_N(\partial_N) & v_N \\ \partial_1 & \cdots & \partial_N & 0 \end{vmatrix} = \langle \Upsilon^{-1}\mathbf{P}, \partial_{\mathbf{x}} \rangle,$$

(4.12)

where $\mathbf{P} = (0, \ldots, 0, \nu_{M+1}, \ldots, \nu_N)^T$, $\nu_j = \nu_j(\mathbf{x})$ are nonzero arbitrary functions due to the arbitrariness of Ω_j for $j = M + 1, \ldots, N$.

Proposition 4.2.2. *The vector field* (4.12) *is the most general vector field satisfying the constraints* (4.9), *i.e.,* $\Omega_j(\mathbf{v}) = 0$ *for* $j = 1, \ldots, M$, *where the* Ω_j *are given in* (4.11).

Proof. First we prove that the vector field (4.12) is such that

$$
\begin{aligned}
\sum_{n=1}^{N} \Omega_j(\partial_n)v_n = \Omega_j(\mathbf{v}) = 0 & \qquad \text{for} \quad j = 1, \ldots, M, \\
\sum_{n=1}^{N} \Omega_{M+k}(\partial_n)v_n = \Omega_{M+k}(\mathbf{v}) = \nu_{M+k} & \qquad \text{for} \quad k = M + 1, \ldots, N.
\end{aligned}
\tag{4.13}
$$

Indeed, from the relation $\mathbf{v}(\mathbf{x}) = \mathbf{S}^{-1}\mathbf{P}$ we get that

$$
\begin{aligned}
\Upsilon\,\mathbf{v}(\mathbf{x}) &= (\Omega_1(\mathbf{v}), \ldots, \Omega_M(\mathbf{v}), \Omega_{M+1}(\mathbf{v}), \ldots, \Omega_N(\mathbf{v}))^T \\
&= \mathbf{P} = (0, \ldots, 0, \nu_{M+1}, \ldots, \nu_N)^T.
\end{aligned}
$$

Thus we obtain (4.13). Consequently the vector field $\mathbf{v}$ satisfies the constraints.

Now we show that vector field $\mathbf{v}$ is the most general satisfying these constraints. Let $\tilde{\mathbf{v}} = (\tilde{v}_1, \ldots, \tilde{v}_N)$ be another vector field satisfying the constraints, i.e., $\sum_{n=1}^{N} \Omega_j(\partial_n)\tilde{v}_n = \Omega_j(\tilde{\mathbf{v}}) = 0$ for $j = 1, \ldots, M$. Taking the arbitrary functions $\nu_{M+1}, \ldots, \nu_N$ as $\nu_{M+k} = \sum_{n=1}^{N} \Omega_{M+j}(\partial_n)\tilde{v}_n$, we obtain from (4.12) the relations

$$
\mathbf{v} = -\frac{1}{|\Upsilon|}
\begin{vmatrix}
\Omega_1(\partial_1) & \cdots & \Omega_1(\partial_N) & \sum_{n=1}^{N}\Omega_1(\partial_n)\tilde{v}_n \\
\vdots & \cdots & \vdots & \vdots \\
\Omega_M(\partial_1) & \cdots & \Omega_M(\partial_N) & \sum_{n=1}^{N}\Omega_M(\partial_n)\tilde{v}_n \\
\Omega_{M+1}(\partial_1) & \cdots & \Omega_{M+1}(\partial_N) & \sum_{n=1}^{N}\Omega_{M+1}(\partial_n)\tilde{v}_n \\
\vdots & \cdots & \vdots & \vdots \\
\Omega_N(\partial_1) & \cdots & \Omega_N(\partial_N) & \sum_{n=1}^{N}\Omega_N(\partial_n)\tilde{v}_n \\
\partial_1 & \cdots & \partial_N & 0
\end{vmatrix}
$$

$$
= -\frac{1}{|\Upsilon|} \sum_{n=1}^{N} \tilde{v}_n
\begin{vmatrix}
\Omega_1(\partial_1) & \cdots & \Omega_1(\partial_N) & \Omega_1(\partial_n) \\
\vdots & \cdots & \vdots & \vdots \\
\Omega_M(\partial_1) & \cdots & \Omega_M(\partial_N) & \Omega_M(\partial_n) \\
\Omega_{M+1}(\partial_1) & \cdots & \Omega_{M+1}(\partial_N) & \Omega_{M+1}(\partial_n) \\
\vdots & \cdots & \vdots & \vdots \\
\Omega_N(\partial_1) & \cdots & \Omega_N(\partial_N) & \Omega_N(\partial_n) \\
\partial_1 & \cdots & \partial_N & 0
\end{vmatrix}.
$$

Thus

$$
\mathbf{v} = -\frac{1}{|\Upsilon|} \sum_{n=1}^{N} \tilde{v}_n
\begin{vmatrix}
\Omega_1(\partial_1) & \cdots & \Omega_1(\partial_N) & 0 \\
\vdots & \cdots & \vdots & \vdots \\
\Omega_M(\partial_1) & \cdots & \Omega_M(\partial_N) & 0 \\
\Omega_{M+1}(\partial_1) & \cdots & \Omega_{M+1}(\partial_N) & 0 \\
\vdots & \cdots & \vdots & \vdots \\
\Omega_N(\partial_1) & \cdots & \Omega_N(\partial_N) & 0 \\
\partial_1 & \cdots & \partial_N & -\partial_n
\end{vmatrix}
= \sum_{j,k=1}^{N} \tilde{v}_n \partial_n = \tilde{\mathbf{v}},
$$

and Proposition 4.2.2 is proved.   □

We define

$$
\Lambda = \Lambda(\mathbf{x}) = (\Lambda_1(\mathbf{x}), \ldots, \Lambda_N(\mathbf{x}))^T = (\Upsilon^T)^{-1} H \mathbf{v}(\mathbf{x}) = A\mathbf{P}, \tag{4.14}
$$

where $A = (A_{jk})$ is the antisymmetric $N \times N$ matrix given by

$$
A = (\Upsilon^T)^{-1} H \Upsilon^{-1}, \quad H = (H_{jn}) = \left(\frac{\partial p_n}{\partial x_j} - \frac{\partial p_j}{\partial x_n} \right). \tag{4.15}
$$

Theorem 4.2.3. *Let Σ be a constrained Lagrangian mechanical system with configuration space Q, kinetic energy T given in (4.2), and constraints given by (4.3) with $\mathbf{v} = (v_1, \ldots, v_N)^T$ given by (4.12).*

The equations of motion of Σ are

$$
\frac{d}{dt} \left(\frac{\partial T}{\partial \dot{x}_j} \right) - \frac{\partial T}{\partial x_j} = F_j(\mathbf{x}) = \frac{\partial}{\partial x_j} \left(\frac{1}{2} \|\mathbf{v}\|^2 \right) + \sum_{k=1}^{M} \Lambda_k a_{kj}, \tag{4.16}
$$

for $j = 1, \ldots, N$, where the Λ_k's are defined in (4.14) with

$$
\Lambda_k = \sum_{j=1}^{M} A_{kj} \nu_j = 0 \quad for \quad k = M+1, \ldots, N. \tag{4.17}
$$

Proof. Let σ be the 1-*form associated to the vector field* $\mathbf{v}$, i.e.,

$$\sigma = \langle \mathbf{v}, d\mathbf{x} \rangle = \sum_{j,k=1}^{N} G_{jk} v_j dx_k = \sum_{n=1}^{N} p_n \, dx_n. \tag{4.18}$$

Then the 2-form $d\sigma$ admit the decomposition

$$d\sigma = \sum_{n,j=1}^{N} \left(\frac{\partial p_n}{\partial x_j} - \frac{\partial p_j}{\partial x_n} \right) dx_j \wedge dx_n = \frac{1}{2} \sum_{n,j=1}^{N} A_{nj} \Omega_n \wedge \Omega_j. \tag{4.19}$$

Here we have used that the 1-forms $\Omega_1, \ldots, \Omega_N$ are independent, and consequently they form a basis of the space of 1-forms. Hence $\Omega_k \wedge \Omega_n$ for $k, n = 1, \ldots, N$ form a basis of the space of 2-forms. From (4.19) it follows that the inner product of vector field $\mathbf{v}$ and $d\sigma$, i.e., $\imath_{\mathbf{v}} d\sigma$, is such that

$$\imath_{\mathbf{v}} d\sigma = \sum_{n,j=1}^{N} v_n \left(\frac{\partial p_n}{\partial x_j} - \frac{\partial p_j}{\partial x_n} \right) dx_j = \langle H\mathbf{v}, d\mathbf{x} \rangle, \tag{4.20}$$

where H is the matrix with elements $\dfrac{\partial p_n}{\partial x_j} - \dfrac{\partial p_j}{\partial x_n}$.

Again from (4.19) we have that

$$\imath_{\mathbf{v}} d\sigma(\cdot) = d\sigma(\mathbf{v}, \cdot) = \frac{1}{2} \sum_{n,j=1}^{N} b_{nj} \Omega_n \wedge \Omega_j(\mathbf{v}, \cdot)$$

$$= \frac{1}{2} \sum_{n,j=1}^{N} A_{nj} \left(\Omega_n(\mathbf{v})\Omega_j(\cdot) - \Omega_j(\mathbf{v})\Omega_n(\cdot) \right)$$

$$= \frac{1}{2} \sum_{n,j=1}^{N} A_{nj} \Omega_n(\mathbf{v})\Omega_j(\cdot) - \frac{1}{2} \sum_{n,j=1}^{N} A_{jn} \Omega_n(\mathbf{v})\Omega_j(\cdot) \tag{4.21}$$

$$= \frac{1}{2} \sum_{n,j=1}^{N} (A_{nj} - A_{jn}) \, \Omega_n(\mathbf{v})\Omega_j(\cdot)$$

$$= \sum_{n,j=1}^{N} A_{nj}\Omega_n(\mathbf{v})\Omega_j(\cdot) = \sum_{n=1}^{N} \Lambda_n \Omega_n(\cdot).$$

Now from the last equality and (4.19) we have

$$\imath_{\mathbf{v}} d\sigma(\partial_j) = \sum_{n=1}^{N} \Lambda_n \Omega_n(\partial_j) = \sum_{n,j=1}^{N} v_n \left(\frac{\partial p_n}{\partial x_j} - \frac{\partial p_j}{\partial x_n} \right). \tag{4.22}$$

Clearly, from these relations it follows that $H\mathbf{v}(\mathbf{x}) = \Upsilon^T \Lambda$, hence

$$\Lambda = (\Upsilon^T)^{-1} H\mathbf{v}(\mathbf{x}) = (\Upsilon^T)^{-1} H\Upsilon^{-1}\mathbf{P} = A\mathbf{P},$$

because $\mathbf{v}(\mathbf{x}) = \Upsilon^{-1}\mathbf{P}$.

From (4.22) and (4.5) we obtain

$$\frac{d}{dt}\left(\frac{\partial T}{\partial \dot{x}_j}\right) - \frac{\partial T}{\partial x_j} = \frac{\partial}{\partial x_j}\left(\frac{1}{2}||\mathbf{v}||^2\right) + \sum_{j=1}^{N} \Lambda_j \Omega_j(\partial_k), \tag{4.23}$$

for $k = 1, \ldots, N$. Finally, (4.23), (4.11), and (4.17) yield (4.16). Theorem 4.2.3 is proved. $\qquad\square$

Remark 4.2.4. Equations (4.17) form a system of *first-order partial differential equations* with unknown functions $\nu_{M+1}, \ldots, \nu_N$ (see (4.12), (4.15), and (4.17)).

We observe that equations (4.17) can be rewritten as

$$\tilde{A}\mathbf{b} = \mathbf{0}, \quad \text{with} \quad \mathbf{b} = (\nu_{M+1}, \ldots, \nu_N)^T, \tag{4.24}$$

where $\tilde{A}$ is an antisymmetric $(N - M) \times (N - M)$ matrix. Thus if $N - M$ is even then, from (4.12), it follows that the vector $\mathbf{b}$ is nonzero, consequently the determinant of the matrix $|\tilde{A}| = \mu_{N,M}^2$ must be zero, i.e., $\mu_{N,M} = 0$. If $N - M$ is odd, then $|\tilde{A}|$ is necessarily zero. If in this case $\text{rank}(\tilde{A}) = r$, then without loss of generality we can assume that (4.17) takes the form

$$\sum_{j=M+1}^{N} A_{kj}\nu_j = 0 \quad \text{for} \quad k = M + 1, \ldots, M + r.$$

In particular, for $M = 1$, $N = 3$ and $M = 2$, $N = 4$ we obtain, respectively,

$$\begin{aligned}
\mu_{3,1} &= a_1 H_{23} + a_2 H_{31} + a_3 H_{12} = 0, \\
\mu_{4,2} &= (\alpha_{42}\alpha_{31} - \alpha_{32}\alpha_{41}) H_{12} + (\alpha_{41}\alpha_{22} - \alpha_{21}\alpha_{42}) H_{13} \\
&\quad + (\alpha_{21}\alpha_{32} - \alpha_{31}\alpha_{22}) H_{14} + (\alpha_{42}\alpha_{11} - \alpha_{12}\alpha_{41}) H_{23} \\
&\quad + (\alpha_{12}\alpha_{31} - \alpha_{32}\alpha_{11}) H_{24} + (\alpha_{22}\alpha_{11} - \alpha_{12}\alpha_{21}) H_{34} = 0.
\end{aligned} \tag{4.25}$$

Remark 4.2.5. Equations (4.16) can be interpreted as the equations of motion of the constrained Lagrangian system with Lagrangian $\tilde{L} = T + \frac{1}{2}||\mathbf{v}||^2$ and constraints (4.9). The force field with components

$$F_j(\mathbf{x}) = \frac{\partial}{\partial x_j}\left(\frac{1}{2}||\mathbf{v}||^2\right) + \sum_{k=1}^{M} \Lambda_k a_{kj}, \quad j = 1, \ldots, N, \tag{4.26}$$

has the same structure than the force field appearing in (4.10), but there are three important differences. First, the potential and reactive components in (4.26) are related through the vector field $\mathbf{v}$ (which itself is determined by the constraints), while in (4.10) the potential U is completely independent of the *reactive forces* with components $\sum_{k=1}^{M} \mu_k a_{kj}$. Second, the multipliers $\Lambda_1, \ldots, \Lambda_M$ in (4.26) depend only on the position of the mechanical system, while in (4.10) the *Lagrangian multipliers* μ_j depends on both the position and velocity, and finally, the system (4.16) was derived from the Lagrangian differential system (4.4), while the system (4.10) in general has no relations with the Lagrangian equations.

In the applications of Theorem 4.2.3 we will determine the functions ν_{M+1}, ..., ν_N as solutions of (4.17) together with the condition

$$U = -\frac{1}{2}||\mathbf{v}||^2 + h, \tag{4.27}$$

where h is a constant. Under the potential (4.27), the only difference between the force fields $\tilde{\mathbf{F}}$ given in (4.10) and $\mathbf{F}$ given in (4.26) consists in the coefficients which determine the reactive forces.

The following two questions arise: Do there exist solutions of equations (4.17) and (4.27) such that the solutions of the differential system

$$\frac{d}{dt}\left(\frac{\partial T}{\partial \dot{x}_j}\right) - \frac{\partial T}{\partial x_j} = -\frac{\partial U}{\partial x_j} + \sum_{k=1}^{M} \Lambda_k a_{kj}, \quad j = 1, \ldots, N, \tag{4.28}$$

where the Λ_k's are defined in (4.14), coincide with the solutions of (4.10)?

If the answer to the previous question is always positive, then there are equations of motion with force fields depending only on the positions (4.16) equivalent to the Lagrangian equations of motions with constraints (4.10). In other words, we would have a new model to describe the behavior of the mechanical systems with constraints linear in the velocity.

The second question is: What is the mechanical meaning of the differential equations generated by the vector field (4.12), i.e.,

$$\dot{\mathbf{x}} = \mathbf{v}(\mathbf{x}) = \Upsilon^{-1}\mathbf{P}, \tag{4.29}$$

under the conditions (4.17), and of the differential equations

$$\frac{d}{dt}\left(\frac{\partial T}{\partial \dot{x}_j}\right) - \frac{\partial T}{\partial x_j} = \frac{\partial \frac{1}{2}||\mathbf{v}||^2}{\partial x_j} + \sum_{k=1}^{M} \mu_k a_{kj} \ ?$$

Partial answer to theses questions are given in the examples of the next section.

4.3   Examples

In this section we illustrate in some particular cases the relations among three mathematical models:

(i) the classical model deduced from the *d'Alembert–Lagrange principle* (see (4.10));

(ii) the model deduced from the Lagrangian equations (4.4) (see (4.16));

(iii) the model obtained from the first-order differential equations (4.29) under the conditions (4.17).

Example 4.3.1. *Suslov problem.* In this example we study a classical problem of nonholonomic dynamics formulated by Suslov [147]. We consider the rotational motion of a *rigid body* around a fixed point and subject to the nonholonomic constraint $\langle \tilde{\mathbf{a}}, \omega \rangle = 0$, where $\omega = (\omega_1, \omega_2, \omega_3)$ is the *angular velocity* of the body, $\tilde{\mathbf{a}}$ is a constant vector, and $\langle\,,\,\rangle$ is the scalar product. Suppose that the body rotates in a force field with potential $U(\gamma) = U(\gamma_1, \gamma_2, \gamma_3)$. Applying the method of Lagrange multipliers, we write the equations of motion (4.10) in the form

$$I\dot{\omega} = I\omega \wedge \omega + \gamma \wedge \frac{\partial U}{\partial \gamma} + \mu\tilde{\mathbf{a}}, \qquad \dot{\gamma} = \gamma \wedge \omega, \qquad \langle \tilde{\mathbf{a}}, \omega \rangle = 0, \qquad (4.30)$$

where

$$\gamma = (\gamma_1, \gamma_2, \gamma_3) = (\sin z \sin x, \sin z \cos x, \cos z), \qquad (4.31)$$

$(x, y, z) = (\varphi, \psi, \theta)$ are the Euler angles, and I is the inertia tensor.

We observe that the equations $\dot{\gamma} = \gamma \wedge \omega$ are called the *Poisson equations*.

Using the constraint equation $\langle \tilde{\mathbf{a}}, \omega \rangle = 0$, the Lagrange multiplier μ can be expressed as a function of ω and γ as follows:

$$\mu = -\frac{\left\langle \tilde{\mathbf{a}}, I\omega \wedge \omega + \gamma \wedge \dfrac{\partial U}{\partial \gamma} \right\rangle}{\langle \tilde{\mathbf{a}}, I^{-1}\tilde{\mathbf{a}} \rangle}.$$

We shall assume that the vector $\mathbf{a} = (0, 0, 1)$, consequently the constraint takes the form

$$\omega_3 = \dot{x} + \dot{y} \cos z = 0.$$

To determine the vector field $\mathbf{v}$ we take for the manifold Q the special orthogonal group of rotations of $\mathbb{R}^3$, i.e., $Q = SO(3)$, with the Riemann metric G given by

$$\begin{pmatrix} I_3 & I_3 \cos z & 0 \\ I_3 \cos z & (I_1 \sin^2 x + I_2 \cos^2 x) \sin^2 z + I_3 \cos^2 z & (I_1 - I_2) \sin x \cos x \sin z \\ 0 & (I_1 - I_2) \sin x \cos x \sin z & I_1 \cos^2 x + I_2 \sin^2 x \end{pmatrix},$$

with determinant $|G| = I_1 I_2 I_3 \sin^2 z$.

By choosing the 1-form Ω_j for $j = 1, 2, 3$ as $\Omega_1 = dx + \cos z \, dy$, $\Omega_2 = dy$, $\Omega_3 = dz$, we obtain $|\Upsilon| = 1$. Thus the vector field $\mathbf{v}$ takes the form

$$\mathbf{v} = \nu_2 \cos z \frac{\partial}{\partial x} - \nu_2 \frac{\partial}{\partial y} + \nu_3 \frac{\partial}{\partial z}.$$

Accordingly, the differential system (4.29) can be written as

$$\dot{x} = \nu_2 \cos z, \quad \dot{y} = -\nu_2, \quad \dot{z} = -\nu_3 \tag{4.32}$$

Using (4.6) we compute

$$p_1 = 0,$$
$$p_2 = (I_1 \sin^2 x + I_2 \cos^2 x)\nu_2 \sin^2 z + (I_2 - I_1)\nu_3 \cos x \sin x \sin z,$$
$$p_3 = -\nu_3(I_2 \sin^2 x + I_1 \cos^2 x) + (I_2 - I_1)\nu_2 \sin x \cos x \sin z.$$

Replacing ν_1 and ν_2 by μ_1 and μ_2, where

$$\mu_1 = I_2(\nu_3 \sin x - \nu_2 \sin z \cos x), \quad \mu_2 = I_1(\nu_3 \cos x + \nu_2 \sin z \sin x),$$

we obtain

$$p_1 = 0, \quad p_2 = \mu_1 \sin z \cos x - \mu_2 \sin z \sin x, \quad p_3 = \sin x \mu_1 + \cos x \mu_2.$$

Now the first of condition (4.25) takes the form

$$\mu_{3,1} = a_1 H_{23} + a_2 H_{31} + a_3 H_{12} = \partial_z p_2 - \partial_y p_3 + \cos z \partial_x p_3 = 0. \tag{4.33}$$

After the change $\gamma_1 = \sin z \sin x$, $\gamma_2 = \sin z \cos x$, $\gamma_3 = \cos z$, the system (4.32) with the constraints and condition (4.33) accounted for can be written as

$$\dot{\gamma}_1 = \frac{1}{I_2} \mu_1 \gamma_3, \quad \dot{\gamma}_2 = \frac{1}{I_1} \mu_2 \gamma_3, \quad \dot{\gamma}_3 = -\frac{1}{I_1 I_2} (I_1 \mu_1 \gamma_1 + I_2 \mu_2 \gamma_2), \tag{4.34}$$

$$\sin z \left(\gamma_3 \left(\frac{\partial \mu_1}{\partial \gamma_2} - \frac{\partial \mu_2}{\partial \gamma_1} \right) - \gamma_2 \frac{\partial \mu_1}{\partial \gamma_3} + \gamma_1 \frac{\partial \mu_2}{\partial \gamma_3} \right) - \cos x \, \partial_y \mu_2 - \sin x \, \partial_y \mu_1 = 0, \tag{4.35}$$

respectively.

Clearly, if $\mu_j = \mu_j(x, z, K_1, K_4)$ for $j = 1, 2$, where K_1 and K_4 are arbitrary constants, then the equation (4.35) takes the form

$$\gamma_3 \left(\frac{\partial \mu_1}{\partial \gamma_2} - \frac{\partial \mu_2}{\partial \gamma_1} \right) - \gamma_2 \frac{\partial \mu_1}{\partial \gamma_3} + \gamma_1 \frac{\partial \mu_2}{\partial \gamma_3} = 0. \tag{4.36}$$

By comparing the Poisson differential system under the condition $\omega_3 = 0$ with (4.34) we obtain that

$$\omega_1 = -\frac{\mu_2}{I_1}, \quad \omega_2 = \frac{\mu_1}{I_2}. \tag{4.37}$$

Example 4.3.2. The *Veselova problem describes the motion of a rigid body which rotates around a fixed point and is subject to the nonholonomic constraint*

$$\langle \gamma, \omega \rangle = \dot{y} + \dot{x} \cos z = 0,$$

To determinate the vector field **v**, *we choose the 1-form* Ω_1, Ω_2, *and* Ω_3 *as follows:*

$$\Omega_1 = dy + \cos z\, dx, \quad \Omega_2 = dx, \quad \Omega_3 = dz.$$

Hence, the differential system (6.3) *for the Veselova problem is*

$$\dot{x} = \nu_2, \quad \dot{y} = -\nu_2 \cos z, \quad \dot{z} = \nu_3, \tag{4.38}$$

Now the first of conditions (4.25) *takes the form*

$$\mu_{3,1} = a_1 H_{23} + a_2 H_{31} + a_3 H_{12} = \frac{\partial p_3}{\partial x} - \frac{\partial p_1}{\partial z} + \cos z \Big(\frac{\partial p_2}{\partial z} - \frac{\partial p_3}{\partial y} \Big) = 0, \tag{4.39}$$

From (4.6) *we compute*

$$p_1 = I_3 \sin^2 z \nu_2,$$

$$p_2 = \nu_2 \cos z \left(I_3 - (I_1 \sin^2 x + I_2 \cos^2 x) \right) \sin^2 z + (I_2 - I_1)\nu_3 \cos x \sin x \sin z,$$

$$p_3 = \nu_3(I_2 \sin^2 x + I_1 \cos^2 x) - (I_1 - I_2)\nu_2 \sin x \cos x \sin z \cos z. \tag{4.40}$$

We want to determine the functions ν_2 *and* ν_3 *that appear in* (4.38) *solving the partial differential equations* (4.39). *We assume that* $\nu_j = \nu_j(x, z)$ *for* $j = 1, 2.$

It then follows from (4.40) *that*

$$p_2 = \frac{\cos z \left(I_3(I_1 \cos^2 x + I_2 \sin^2 x) - I_2 I_1 \right)}{I_3(I_1 \cos^2 x + I_2 \sin^2 x)} p_1 + \frac{(I_1 - I_2) \sin z \cos x \sin x}{I_1 \cos^2 x + I_2 \sin^2 x} p_3.$$

Inserting this expression into (4.39) *we obtain the differential equation*

$$\Theta := (I_1 - I_2) \sin z \cos z \sin x \cos x \frac{\partial \Psi_3}{\partial z}$$
$$+ \left(I_1 \cos^2 x + I_2 \sin^2 x \right) \frac{\partial \Psi_3}{\partial x} - \sin z\, p \frac{\partial \Psi_1}{\partial z} = 0, \tag{4.41}$$

where

$$\Psi_3 = p_3 \sin z, \quad \Psi_1 = \frac{p}{I_3} p_1, \quad p = \sqrt{I_1 I_2 I_3 \left(\frac{\gamma_1^2}{I_1} + \frac{\gamma_2^2}{I_2} + \frac{\gamma_3^2}{I_3} \right)}. \tag{4.42}$$

When $I_1 \neq I_2$, *particular solutions of* (4.41) *are*

$$\Psi_3 = \Psi_3(\tan z \sqrt{I_1 \cos^2 x + I_2 \sin^2 x}), \quad \Psi_1 = \Psi_1(x),$$

where Ψ_1 and Ψ_3 are arbitrary functions of their arguments. If $I_1 = I_2$, then from (4.41) we obtain the equation

$$\frac{\partial}{\partial x}\left(\frac{\Psi_3}{\sin z\sqrt{I_3\left((\sin^2 z/I_1) + (\cos^2 z/I_3)\right)}}\right) - \frac{\partial\Psi_1}{\partial z} = 0.$$

Thus we get that

$$\frac{\Psi_3}{\sin z\sqrt{I_3\left(\dfrac{\sin^2 z}{I_1} + \dfrac{\cos^2 z}{I_3}\right)}} = \frac{\partial S}{\partial z}, \quad \Psi_1 = \frac{\partial S}{\partial x}. \tag{4.43}$$

Hence, from (4.42) we have

$$p_1 = \frac{I_3\dfrac{\partial S}{\partial x}}{\sqrt{I_3\left(\dfrac{\sin^2 z}{I_1} + \dfrac{\cos^2 z}{I_3}\right)}}, \quad p_3 = \frac{\sqrt{I_3\left(\dfrac{\sin^2 z}{I_1} + \dfrac{\cos^2 z}{I_3}\right)}\dfrac{\partial S}{\partial z}}{I_1},$$

where $S = S(z,x)$ is an arbitrary smooth function.
By considering that (see [91])

$$\omega_1 = \dot{y}\sin z\sin x + \dot{z}\cos x,$$
$$\omega_2 = \dot{y}\sin z\cos x - \dot{z}\sin x,$$
$$\omega_3 = \dot{y}\cos z + \dot{x},$$

we obtain that on the solutions of (4.38)

$$\omega_1 = \gamma_2\frac{\nu_3}{\sin z} - \gamma_1\gamma_3\nu_2,$$
$$\omega_2 = -\gamma_1\frac{\nu_3}{\sin z} - \gamma_2\gamma_3\nu_2, \tag{4.44}$$
$$\omega_3 = \sin^2 z\nu_2.$$

By solving (4.40) with respect to ν_2 and ν_3 and using (4.43) we obtain

$$\nu_2 = \frac{\Psi_1}{\sin^2 z\,p}, \quad \nu_3 = \frac{\Psi_3 p + (I_1 - I_2)\Psi_1\cos x\sin x\cos z}{p\sin z(I_1\cos^2 x + I_2\sin^2 x)}. \tag{4.45}$$

Hence, the system (4.38) takes the form

$$\dot{x} = \frac{\Psi_1}{\sin^2 z\,p}, \quad \dot{y} = -\frac{\cos z\Psi_1}{\sin^2 z\,p}, \quad \dot{z} = \frac{\Psi_3 p + (I_1 - I_2)\Psi_1\cos x\sin x\cos z}{p\sin z(I_1\cos^2 x + I_2\sin^2 x)}. \tag{4.46}$$

Inserting (4.45) into (4.44) we get

$$\omega_1 = \frac{p\Psi_3\gamma_2 - I_2\,\Psi_1\gamma_1\gamma_3}{p(I_1\gamma_2^2 + I_2\gamma_1^2)},$$

$$\omega_2 = -\frac{p\Psi_3\gamma_1 + I_1\,\Psi_1\gamma_2\gamma_3}{p(I_1\gamma_2^2 + I_2\gamma_1^2)}, \tag{4.47}$$

$$\omega_3 = \frac{\Psi_1}{p}.$$

Remark 4.3.3. The importance of equations (4.34), (4.37), (4.36), (4.41), (4.47) and (4.42) will be demonstrated in Chapter 2, devoted to the *integrability of the constrained rigid body problem*.

Example 4.3.4. Nonholonomic Chaplygin systems.

We illustrated Theorem 4.2.3 in the case of nonholonomic Chaplygin systems.

In many nonholonomic systems the generalized coordinates $x_1, \ldots, x_N$ can be chosen in such a way that the equations of the non-integrable constraints, can be written in the form

$$\dot{x}_j = \sum_{k=M+1}^{N} \hat{a}_{jk}(x_{M+1}, \ldots, x_N)\dot{x}_k, \quad j = 1, 2, \ldots, M. \tag{4.48}$$

A *constrained Chaplygin mechanical system* is a mechanical system with Lagrangian $\tilde{L} = \tilde{L}(x_{M+1}, \ldots, x_N, \dot{x}_1, \ldots, \dot{x}_N)$, subject to M linear nonholonomic constraints (4.48).

We shall solve the inverse problem for this constrained system when the Lagrangian function is the following

$$\tilde{L} = T = \frac{1}{2} \sum_{n,j=1}^{N} G_{jn}(x_{M+1}, \ldots, x_N)\dot{x}_j\dot{x}_n \tag{4.49}$$

In this section we determine the vector field (4.12) and the differential system (4.16) for the constrained Chaplygin-Lagrangian mechanical system with Lagrangian (4.49).

First we determine the 1-forms Ω_j for $j = 1, \ldots, N$. Taking

$$\Omega_j = dx_j - \sum_{k=M+1}^{N} \hat{a}_{jk}(x_{M+1}, \ldots, x_N)dx_k, \quad \text{for} \quad j = 1, 2, \ldots, M,$$

$$\Omega_k = dx_k, \qquad\qquad\qquad\qquad \text{for} \quad k = M+1, \ldots, N,$$

we obtain that

$$
\Upsilon = \begin{pmatrix}
1 & 0 & \cdots & 0 & 0 & -\hat{a}_{1\,M+1} & \cdots & -\hat{a}_{1\,N} \\
0 & 1 & \cdots & 0 & 0 & -\hat{a}_{2\,M+1} & \cdots & -\hat{a}_{2\,N} \\
\vdots & \vdots & \vdots & \vdots & \vdots & \vdots & & \vdots \\
0 & 0 & 0 & \cdots & 1 & -\hat{a}_{M\,M+1} & \cdots & -\hat{a}_{M\,N} \\
0 & 0 & 0 & \cdots & 0 & 1 & 0 & 0 \\
\vdots & \vdots & \vdots & \cdots & \vdots & \vdots & \vdots & \vdots \\
0 & 0 & 0 & \cdots & 0 & 0 & 0 & 1
\end{pmatrix}.
\tag{4.50}
$$

Thus $|\Upsilon| = 1$ and consequently

$$
\Upsilon^{-1} = \begin{pmatrix}
1 & 0 & \cdots & 0 & 0 & \hat{a}_{1\,M+1} & \cdots & \hat{a}_{1\,N} \\
0 & 1 & \cdots & 0 & 0 & \hat{a}_{2\,M+1} & \cdots & \hat{a}_{2\,N} \\
\vdots & \vdots & \vdots & \vdots & \vdots & \vdots & & \vdots \\
0 & 0 & 0 & \cdots & 1 & \hat{a}_{M\,M+1} & \cdots & \hat{a}_{M\,N} \\
0 & 0 & 0 & \cdots & 0 & 1 & 0 & 0 \\
\vdots & \vdots & \vdots & \cdots & \vdots & \vdots & \vdots & \vdots \\
0 & 0 & 0 & \cdots & 0 & 0 & 0 & 1
\end{pmatrix}.
$$

Thus the vector field (4.12) in this case generates the following differential equations:

$$
\dot{x}_j = \sum_{n=M+1}^{N} \hat{a}_{jn}\nu_n, \quad \dot{x}_k = \nu_k \quad \text{for} \quad j = 1,\ldots,M, \quad k = M+1,\ldots,N.
$$

In this case the differential system (4.16) admits the representation

$$
\begin{aligned}
\frac{d}{dt}\left(\frac{\partial T}{\partial \dot{x}_k}\right) &= \frac{\partial}{\partial x_k}\left(\frac{1}{2}\|\mathbf{v}\|^2\right) + \Lambda_k, \\
\frac{d}{dt}\left(\frac{\partial T}{\partial \dot{x}_j}\right) - \frac{\partial T}{\partial x_j} &= \frac{\partial}{\partial x_j}\left(\frac{1}{2}\|\mathbf{v}\|^2\right) - \sum_{k=1}^{M}\Lambda_k\hat{a}_{kj},
\end{aligned}
\tag{4.51}
$$

for $j = M+1,\ldots,N$, $k = 1,\ldots,M$, where $\Lambda_1,\ldots,\Lambda_M$ are determined by the formulas (4.14), (4.15), and (4.17).

Note that the system (4.51) coincides with the *Chaplygin system*. Indeed, excluding Λ_k from the first of the equations of (4.51) and denoting by L^* the expression in which the velocities $\dot{x}_1,\ldots,\dot{x}_M$ have been eliminated by means of the constraints equations (4.48), i.e.,

$$
L^* = L\bigg|_{\dot{x}_j = \sum_{k=M+1}^{N}\hat{a}_{jk}\dot{x}_k} = \left(T + \frac{1}{2}\|\mathbf{v}\|^2\right)\bigg|_{\dot{x}_j = \sum_{k=M+1}^{N}\hat{a}_{jk}\dot{x}_k},
$$

we obtain

$$\frac{\partial L^*}{\partial \dot{x}_j} = \frac{\partial L}{\partial \dot{x}_j} + \sum_{\alpha=1}^{M} \frac{\partial L}{\partial \dot{x}_\alpha} \hat{a}_{\alpha j}, \quad \frac{\partial L^*}{\partial x_j} = \frac{\partial L}{\partial x_j} + \sum_{\alpha=1}^{M} \sum_{m=M+1}^{N} \frac{\partial L}{\partial \dot{x}_\alpha} \dot{x}_m \frac{\partial \hat{a}_{\alpha m}}{\partial x_j},$$

for $j = M + 1, \ldots, N$.

From these relations, we obtain

$$\frac{d}{dt}\left(\frac{\partial L^*}{\partial \dot{x}_j}\right) - \frac{\partial L^*}{\partial x_j} = \sum_{m=M+1}^{N} \sum_{l=1}^{M} \left(\frac{\partial \hat{a}_{lj}}{\partial x_m} - \frac{\partial \hat{a}_{lm}}{\partial x_j}\right) \dot{x}_m \frac{\partial L}{\partial \dot{x}_l},$$

for $j = M + 1, \ldots, N, \ k = 1, \ldots, M$. These are the equations that Chaplygin published in the Proceeding of the Society of the Friends of Natural Science in 1897.

Example 4.3.5. The Chaplygin–Carathéodory sleigh.

We shall now analyze one of the classical nonholonomic systems, the *Chaplygin–Carathéodory sleigh* (for more details see [120]). This is described by the constrained Lagrangian system with the configuration space $Q = \mathbb{S}^1 \times \mathbb{R}^2$, with the Lagrangian function

$$\tilde{L} = \frac{m}{2}\left(\dot{y}^2 + \dot{z}^2 + \frac{J_C}{2}\dot{x}^2\right) - U(x, y, z),$$

and with the constraint $\varepsilon \dot{x} + \sin x \dot{y} - \cos x \dot{z} = 0$, where m, J_C, and ε are parameters related to the sleigh. We note that the *Chaplygin skate* is a particular case of this mechanical system, namely, it is obtained when $\varepsilon = 0$.

To determine the vector field (4.12) in this case we choose the 1-forms Ω_j for $j = 1, 2, 3$ as (see [139]) $\Omega_1 = \varepsilon dx + \sin x \, dy - \cos x \, dz$, $\Omega_2 = \cos x \, dy + \sin x \, dz$, $\Omega_3 = dx$, hence $|\Upsilon| = 1$.

The differential equations (4.29) and the first condition of (4.25) take the respective forms

$$\dot{x} = \nu_3, \quad \dot{y} = \nu_2 \cos x - \varepsilon \lambda_3 \sin x, \quad \dot{z} = \nu_2 \sin x + \varepsilon \nu_3 \cos x, \tag{4.52}$$

where $\nu_j = \nu_j(x, y, z, \varepsilon)$ for $j = 2, 3$ are solutions of the partial differential equation

$$\begin{aligned} 0 &= \mu_{3,1} = a_1 H_{23} + a_2 H_{31} + a_3 H_{12} \\ &= \sin x(J\partial_z \nu_3 + \varepsilon m \partial_y \nu_2) + \cos x(J\partial_y \nu_3 - \varepsilon m \partial_z \nu_2) - m(\partial_x \nu_2 - \varepsilon \nu_3), \end{aligned} \tag{4.53}$$

with $J = J_C + \varepsilon^2 m$.

For the Chaplygin skate ($\varepsilon = 0$) we have

$$\dot{x} = \nu_3, \quad \dot{y} = \nu_2 \cos x, \quad \dot{z} = \nu_2 \sin x, \quad \dot{y} \cos x - \dot{x} \cos x = 0, \tag{4.54}$$

$$J_C(\sin x \partial_z \nu_3 + \cos x \partial_y \nu_3) - m \partial_x \nu_2 = 0, \tag{4.55}$$

where $\nu_j = \nu_j(x, y, z, 0)$ for $j = 2, 3$. Now we study the behavior of the Chaplygin skate by using the differential equations generated by the vector field $\mathbf{v}$ with ν_2 and ν_3 satisfying the partial differential equation (4.55).

Proposition 4.3.6. *All the trajectories of the Chaplygin skate ($\varepsilon = 0$) under the action of the potential force field with potential $U = mgy$ can be obtained from the differential system (4.54), where ν_2 and ν_3 are solutions of (4.55).*

Proof. Indeed, taking $\varepsilon = 0$ the equation of motions of the Chaplygin skate obtained from (4.10) read

$$\ddot{x} = 0, \quad \ddot{y} = mg + \sin x \mu, \quad \ddot{z} = -\cos x \mu, \quad \sin x \ddot{y} - \cos x \ddot{z} = 0.$$

Hence, we obtain $\dfrac{d}{dt}\left(\dfrac{\dot{z}}{\sin x}\right) = g \cos x$. We study only the case when $\dot{x}|_{t=t_0} = C_0 \neq 0$. Then

$$\dot{x} = C_0, \quad \dot{y} = \left(\frac{g \sin x}{C_0} + C_1\right) \cos x, \quad \dot{z} = \left(\frac{g \sin x}{C_0} + C_1\right) \sin x. \tag{4.56}$$

The solutions of these equations coincide with the solutions of (4.54) and (4.55) under the condition $\|\mathbf{v}\|^2 = J_C \nu_3^2 + m \nu_2^2 = 2(-mgy + h)$. Indeed, taking

$$\nu_3 = C_0, \quad \nu_2 = \sqrt{\frac{2(-mgy + h) - J_C C_0^2}{m}},$$

where C_0 is an arbitrary constant, we obtain the differential system

$$\dot{x} = C_0, \ \dot{y} = \sqrt{\frac{2(-mgy + h) - J_C C_0^2}{m}} \cos x, \ \dot{z} = \sqrt{\frac{2(-mgy + h) - J_C C_0^2}{m}} \sin x.$$

The solutions of this system coincide with the solutions of (4.56), so the proposition is proved. □

In what follows we study the motion of the Chaplygin–Carathéodory sleigh without the action of active forces.

Proposition 4.3.7. *All the trajectories of the Chaplygin–Carathéodory sleigh in the absence of active forces can be obtained from (4.52) with the condition (4.53).*

Proof. Indeed, taking in (4.53) $\nu_j = \nu_j(x, \varepsilon)$, $j = 1, 2$, such that $\partial_x \nu_2 = \varepsilon \nu_3$, all the trajectories of equation (4.52) are given by

$$y = y_0 + \int \frac{(\nu_2 \cos x - \varepsilon \nu_3 \sin x) dx}{\nu_3},$$

$$z = z_0 - \int \frac{(\nu_2 \sin x - \varepsilon \nu_3 \cos x) dx}{\nu_3},$$

$$t = t_0 + \int \frac{dx}{\lambda_3(x, \varepsilon)}.$$

On the other hand, for the Chaplygin–Carathéodory sleigh in absence of active forces we obtain from (4.10)

$$J_C\ddot{x} = \varepsilon\mu, \quad m\ddot{y} = \sin x\mu, \quad m\ddot{z} = -\cos x\mu, \quad \varepsilon\dot{x} + \sin x\dot{y} - \cos x\dot{z} = 0.$$

Hence, after integration we obtain the system

$$\dot{x} = qC_0\cos\theta, \quad \dot{y} = C_0(\sin\theta\cos x - q\varepsilon\cos\theta\sin x), \quad \dot{z} = C_0(\sin\theta\sin x + q\varepsilon\cos\theta\cos x),$$

where $\theta = q\varepsilon x + C$ and $q^2 = \dfrac{m}{J_C + m\varepsilon^2}$, which is a particular case of the system (4.52) with $\nu_2 = C_0\sin\theta$, $\nu_3 = C_0 q\cos\theta$. Clearly in this case

$$2\|\mathbf{v}\|^2 = (J_C + m\varepsilon^2)\nu_3^2(x,\varepsilon) + m\nu_2^2(x,\varepsilon) = mC_0^2 = 2(-U + h),$$

and the equation $\partial_x\nu_2 = \varepsilon\nu_3$ holds. Thus the proposition follows. □

Example 4.3.8. We shall illustrate this case by the following system, which we call *Gantmacher's system* (for more details see [62]).

Two particles m_1 and m_2 with equal masses are linked by a metal rod with fixed length l and small mass. The system can move only in the vertical plane and so the speed of the midpoint of the rod is directed along the rod. It is necessary to determine the trajectories of the particles m_1 and m_2.

Let (q_1, r_1) and (q_2, r_2) be the coordinates of the points m_1 and m_2. Making the change of coordinates: $x_1 = (q_2 - q_1)/2$, $x_2 = (r_1 - r_2)/2$, $x_3 = (r_2 + r_1)/2$, $x_4 = (q_1 + q_2)/2$, we obtain the mechanical system with configuration space $Q = \mathbb{R}^4$, the Lagrangian function $L = \dfrac{1}{2}\sum_{j=1}^{4}\dot{x}_j^2 - gx_3$, and constraints

$$x_1\dot{x}_1 + x_2\dot{x}_2 = 0, \quad x_1\dot{x}_3 - x_2\dot{x}_4 = 0.$$

The equations of motion (4.10) obtained from the d'Alembert–Lagrange principle are

$$\ddot{x}_1 = \mu_1 x_1, \quad \ddot{x}_2 = \mu_1 x_2, \quad \ddot{x}_3 = -g + \mu_2 x_1, \quad \ddot{x}_4 = -\mu_2 x_2, \tag{4.57}$$

where μ_1, μ_2 are the Lagrangian multipliers, which we determine as follows

$$\mu_1 = -\frac{\dot{x}_1^2 + \dot{x}_2^2}{x_1^2 + x_2^2}, \quad \mu_2 = \frac{\dot{x}_2\dot{x}_4 - \dot{x}_1\dot{x}_3 + gu_1}{x_1^2 + x_2^2}. \tag{4.58}$$

After the integration of (4.57) we obtain (for more details see [62])

$$\dot{x}_1 = -\dot{\varphi}x_2, \quad \dot{x}_2 = \dot{\varphi}x_1, \quad \dot{x}_3 = \frac{f}{r}x_2, \quad \dot{x}_4 = \frac{f}{r}x_1, \tag{4.59}$$

where (φ, r) are the polar coordinates: $x_1 = r\cos\varphi$, $x_2 = r\sin\varphi$, and f is a solution of the equation $\dot{f} = -\dfrac{2g}{r}x_2$.

To construct the differential systems (4.29) and (4.12) we introduce the 1-forms Ω_j for $j = 1, 2, 3, 4$, as follows (see [139])

$$\Omega_1 = x_1 dx_1 + x_2 dx_2, \qquad \Omega_2 = x_1 dx_3 - x_2 dx_4,$$
$$\Omega_3 = -x_1 dx_2 + x_2 dx_1, \qquad \Omega_4 = x_2 dx_3 + x_1 dx_4.$$

Here Ω_1 and Ω_2 are given by the constraints, and Ω_3 and Ω_4 are chosen in order that the determinant $|\Upsilon|$ becomes nonzero, and if it can be chosen constant one would be in the ideal situation. Hence we obtain that $|\Upsilon| = -(x_1^2 + x_2^2)^2 = -l^2/4 \neq 0$. By considering that in this case $N = 4$ and $M = 2$, (4.25) yields

$$\mu_{4,2} = x_2 \partial_{x_3} \nu_3 - x_1 \partial_{x_4} \nu_3 + x_2 \partial_{x_1} \nu_4 + x_1 \partial_{x_2} \nu_4 = 0. \tag{4.60}$$

The differential equations (4.29) take the form

$$\dot{x}_1 = -\nu_3 x_2, \quad \dot{x}_2 = \nu_3 x_1, \quad \dot{x}_3 = \nu_4 x_2, \quad \dot{x}_4 = \nu_4 x_1. \tag{4.61}$$

It is easy to show that the functions ν_3 and ν_4 given by

$$\nu_3 = g_3(x_1^2 + x_2^2), \qquad \nu_4 = \sqrt{\frac{2(-gx_3 + h)}{(x_1^2 + x_2^2)} - g_3^2(x_1^2 + x_2^2)}, \tag{4.62}$$

where g and h are constants, and g_3 is an arbitrary function of the variable $x_1^2 + x_2^2$, are solutions of (4.60). Consequently, from the relation (4.27) yields

$$2||\mathbf{v}||^2 = (x_1^2 + x_2^2)(\nu_3^2 + \nu_4^2) = 2(-g\,x_3 + h) = 2(-U + h).$$

The solutions of (4.61) with ν_3 and ν_4 given in (4.62) are

$$x_1 = r\cos\alpha, \quad x_2 = r\sin\alpha, \quad \alpha = \alpha_0 + g_3(r)t,$$
$$x_3 = u_3^0 + \frac{g}{2g_3(r)}t - \frac{g}{4g_3^2(r)}\sin 2\alpha - \frac{\sqrt{2g}C}{g_3(r)}\cos\alpha, \tag{4.63}$$
$$x_4 = -h + \frac{r^2 g_3^2(r)}{2g} + \left(\frac{\sqrt{g}}{\sqrt{2g_3(r)}}\sin\alpha + C\right)^2,$$

where C, r, α_0, u_3^0, h, are arbitrary constants, g_3 is an arbitrary function of r.

To compare these solutions with the solutions obtained from (4.59) we observe that they coincide. We note that we have obtained the trajectories of the particles m_1 and m_2 solving the first-order differential equations (4.61) with the functions (4.62).

Finally, we observe that for the *Gantmacher system* the system (4.16) takes the form

$$\ddot{x}_1 = \Lambda_1 x_1, \quad \ddot{x}_2 = \Lambda_1 x_2, \quad \ddot{x}_3 = -g + \Lambda_2 x_1, \quad \ddot{x}_4 = \Lambda_2 x_2, \tag{4.64}$$

and admits as solutions the ones given in (4.63) (see Remark 4.2.5).

Remark 4.3.9. Using these examples we provide a partial answer to the questions stated in Remark 4.2.5. The differential equations generated by the vector field (4.12) under the conditions (4.24) can be applied to study the behavior of the nonholonomic systems with constraints linear in the velocity (at least for certain class of such system). Is it possible to apply this mathematical model to describe the behavior of the nonholonomic systems with linear constraints with respect to velocity in general? For the moment we have no answer to this question.

4.4 Inverse problem in dynamics. Generalized Dainelli inverse problems

Now we consider a mechanical system with configuration space Q of dimension N and with kinetic energy T given by (4.2). The problem of determining the most general force field depending only on the position of the system, for which the curves defined by

$$f_j = f_j(\mathbf{x}) = c_j \in \mathbb{R} \quad \text{for} \quad j = 1, \ldots, N-1, \tag{4.65}$$

are formed by orbits of the mechanical system, is called the *generalized Dainelli inverse problem in dynamics.* If we assume that the given family of curves (4.65) admits the family of orthogonal hypersurfaces $S = S(\mathbf{x}) = c_N$, then this problem is called the *generalized Dainelli–Joukovsky inverse problem.*

If the force field is potential in the generalized Dainelli inverse problems, then such problems coincide with the *Suslov inverse problem,* or the *inverse problem in Celestial Mechanics,* and the generalized Dainelli–Joukovsky inverse problem coincides with the *Joukovsky problem* (for more details see [139]).

The solutions of the generalized Dainelli problem for $N = 2$, and of the Joukovsky problems for $N = 2, 3$ can be found in [166, 42, 78, 60]. A complete solution of the Suslov problem can be found in [149], but this solution in general is not easy to implement.

The following result provides a solution of these inverse problems.

Theorem 4.4.1. *Under the assumptions of Theorem 4.2.3 if the given $M = N - 1$ 1-forms (4.11) are closed, i.e., $\Omega_j = df_j$ for $j = 1, \ldots, N-1$, then the following statements hold.*

(a) *System (4.16) takes the form*

$$\frac{d}{dt}\left(\frac{\partial T}{\partial \dot{x}_j}\right) - \frac{\partial T}{\partial x_j} = \frac{\partial}{\partial x_j}\left(\frac{1}{2}||\mathbf{v}||^2\right) + \nu_N \sum_{k=1}^{N-1} A_{Nk}\frac{\partial f_k}{\partial x_j} =: F_j, \tag{4.66}$$

for $j = 1, \ldots, N$, where $\nu_N = \nu_N(\mathbf{x})$ is an arbitrary function. Clearly F_j are the components of the most general force field that depends only on the position under which a given $(N-1)$-parameter family of curves (4.65) can be described as orbits of the mechanical system.

(b) *If*

$$\nu_N \sum_{k=1}^{N-1} A_{Nk} \frac{\partial f_k}{\partial x_j} = -\frac{\partial h}{\partial x_j} \tag{4.67}$$

for $j = 1, \ldots, N-1$, *where* $h = h(f_1, \ldots, f_{N-1})$, *then the family of curves* (4.65) *can be freely described by a mechanical system under the influence of forces derived from the potential function* $V = -U = \frac{1}{2}||\mathbf{v}||^2 - h(f_1, \ldots, f_{N-1})$.

(c) *If one assumes that the given family of curves* (4.65) *admits the family of orthogonal hypersurface* $S = S(\mathbf{x}) = c_N$ *defined by*

$$\left\langle \frac{\partial S}{\partial \mathbf{x}}, \frac{\partial f_j}{\partial \mathbf{x}} \right\rangle = 0, \quad j = 1, \ldots, N-1, \tag{4.68}$$

then the most general force field that depends only on the position of the system under which the given family of curves is formed by orbits of (4.66) *is*

$$\mathbf{F} = \frac{\partial}{\partial \mathbf{x}} \left(\frac{\nu}{\sqrt{2}} \left\| \frac{\partial S}{\partial \mathbf{x}} \right\| \right)^2 + \left\langle \frac{\partial}{\partial \mathbf{x}} \left(\frac{\nu^2}{2} \right), \frac{\partial S}{\partial \mathbf{x}} \right\rangle \frac{\partial S}{\partial \mathbf{x}} - \left\| \frac{\partial S}{\partial \mathbf{x}} \right\|^2 \frac{\partial}{\partial \mathbf{x}} \left(\frac{\nu^2}{2} \right), \tag{4.69}$$

where $\nu = \nu(\mathbf{x})$ *is an arbitrary function on* Q. *If we choose* ν *and* $h = h(f_1, \ldots, f_{N-1})$ *satisfying the first-order partial differential equation*

$$\left\langle \frac{\partial}{\partial \mathbf{x}} \left(\frac{\nu^2}{2} \right), \frac{\partial S}{\partial \mathbf{x}} \right\rangle \frac{\partial S}{\partial \mathbf{x}} - \left\| \frac{\partial S}{\partial \mathbf{x}} \right\|^2 \frac{\partial}{\partial \mathbf{x}} \left(\frac{\nu^2}{2} \right) = -\frac{\partial h}{\partial \mathbf{x}}, \tag{4.70}$$

then the force field $\mathbf{F}$ *is given by the potential*

$$V = \frac{\nu^2}{2} \left\| \frac{\partial S}{\partial \mathbf{x}} \right\|^2 - h(f_1, \ldots, f_{N-1}). \tag{4.71}$$

If (4.65) *is such that* $f_j = x_j = c_j$ *for* $j = 1, \ldots, N-1$, *then* (4.71) *takes the form*

$$V = \frac{\nu^2 |\tilde{G}|}{2\Delta} \left(\frac{\partial S}{\partial x_N} \right)^2 - h(x_1, \ldots, x_{N-1}), \tag{4.72}$$

where $\tilde{G} = (\tilde{G}_{nm})$ *is the inverse of the matrix* G *and*

$$\Delta = \begin{vmatrix} \tilde{G}_{11} & \cdots & \tilde{G}_{1,N-1} \\ \vdots & \cdots & \vdots \\ \tilde{G}_{1,N-1} & \cdots & \tilde{G}_{N-1,N-1} \end{vmatrix}.$$

Clearly, (4.70) *holds in particular if* $\nu = \nu(S)$ *and* h *is a constant.*

(d) *Under the assumption* (b) *we have that* $\displaystyle\int_{g_{\mathbf{v}}^t(\gamma)} \sigma = $ const, *where* $\sigma = \langle \mathbf{v}, d\mathbf{x} \rangle$ *is the 1-form associated to vector field* $\mathbf{v}$, $g_{\mathbf{v}}^t$ *is the flow of* $\mathbf{v}$, *and* γ *is an arbitrary closed curve on* Q.

We note that statement (a) of Theorem 4.4.1 provides the answer to the generalized Dainelli inverse problem, which before was solved only for $N = 2$ by Dainelli. Statement (b) of Theorem 4.4.1 gives a simpler solution to the Suslov inverse problem, already solved by Suslov himself. Statement (c) of Theorem 4.4.1 provides the answer to the generalized Dainelli–Joukovsky problem solved by Joukovsky for the case when the force field is potential and $N = 2, 3$. Finally, statement (d) of Theorem 4.4.1 is the well-known *Thomson Theorem* (see [92]) in our context.

Proof of Theorem 4.4.1. In this case we obtain that the vector field (4.12) is

$$
\mathbf{v} = -\frac{1}{|\Upsilon|}
\begin{vmatrix}
df_1(\partial_1) & \cdots & df_1(\partial_N) & 0 \\
\vdots & \cdots & \vdots & \vdots \\
df_{N-1}(\partial_1) & \cdots & df_{N-1}(\partial_N) & 0 \\
df_N(\partial_1) & \cdots & df_N(\partial_N) & \nu_N \\
\partial_1 & \cdots & \partial_N & 0
\end{vmatrix}
\tag{4.73}
$$

$$
= \frac{\nu_N}{|\Upsilon|}
\begin{vmatrix}
df_1(\partial_1) & \cdots & df_1(\partial_N) \\
\vdots & \cdots & \vdots \\
df_{N-1}(\partial_1) & \cdots & df_{N-1}(\partial_N) \\
\partial_1 & \cdots & \partial_N
\end{vmatrix}
= \tilde{\nu}\{f_1, \ldots, f_{N-1}, *\}.
$$

Condition (4.17) now takes the form $\Lambda_N = A_{NN}\nu_N = 0$. Since the matrix A is antisymmetric, $A_{NN} = 0$. On the other hand, since $\Lambda_j = A_{Nj}\nu_N$ for $j = 1, \ldots, N - 1$, we deduce that system (4.16) takes the form

$$
\frac{d}{dt}\frac{\partial T}{\partial \dot{x}_j} - \frac{\partial T}{\partial x_j} = F_j = \frac{\partial}{\partial x_j}\left(\frac{1}{2}||\mathbf{v}||^2\right) + \sum_{k=1}^{N-1} \Lambda_k df_k(\partial_j)
$$

$$
= \frac{\partial}{\partial x_j}\left(\frac{1}{2}||\mathbf{v}||^2\right) + \nu_N \sum_{k=1}^{N-1} A_{Nk} df_k(\partial_j).
$$

These relations yield statement (a) of the theorem.

Statement (b) follows trivially from the previous result.

Statement (c) follows by observing that under the assumption (4.68) we have

$$
\left\langle \frac{\partial S}{\partial \mathbf{x}}, \frac{\partial \Psi}{\partial \mathbf{x}} \right\rangle = \varrho
\begin{vmatrix}
df_1(\partial_1) & \cdots & df_1(\partial_N) \\
\vdots & \cdots & \vdots \\
df_{N-1}(\partial_1) & \cdots & df_{N-1}(\partial_N) \\
d\Psi(\partial_1) & \cdots & d\Psi(\partial_N)
\end{vmatrix}
= \varrho\{f_1, \ldots, f_{N-1}, \Psi\},
$$

where Ψ and $\varrho = \varrho(x_1, \ldots, x_N)$ are an arbitrary functions. Hence, the 1-form associated to the vector field $\mathbf{v}$ is $\sigma = \langle \mathbf{v}, d\mathbf{x} \rangle = \left\langle \nu\frac{\partial S}{\partial \mathbf{x}}, d\mathbf{x} \right\rangle = \nu\, dS$, where $\nu = \frac{\tilde{\nu}}{\varrho}$

(see (4.18)) Thus $d\sigma = d\nu \wedge dS$ and consequently from (4.20) we have

$$
\imath_{\mathbf{v}} d\sigma = \sum_{n,j=1}^{N} v_n \left(\frac{\partial p_n}{\partial x_j} - \frac{\partial p_j}{\partial x_n} \right) dx_j = d\nu(\mathbf{v}) dS - dS(\mathbf{v}) d\nu
$$

$$
= \mathbf{v}(\nu) dS - \mathbf{v}(S) d\nu = \left\langle \mathbf{v}(\mathbf{x}), \frac{\partial \nu}{\partial \mathbf{x}} \right\rangle dS - \left\langle \mathbf{v}(\mathbf{x}), \frac{\partial S}{\partial \mathbf{x}} \right\rangle d\nu
$$

$$
= \frac{1}{2} \left(\left\langle \frac{\partial \nu^2}{\partial \mathbf{x}}, \frac{\partial S}{\partial \mathbf{x}} \right\rangle dS - \left\| \frac{\partial S}{\partial \mathbf{x}} \right\|^2 d\nu^2 \right).
$$

After some computations, we deduce that in view of (4.22) the force field $\mathbf{F}$ admits the representation $F_j = \dfrac{\partial}{\partial x_j} \left(\dfrac{1}{2} \|\mathbf{v}\|^2 \right) + \imath_{\mathbf{v}} d\sigma(\partial_j)$. Hence we obtain (4.69).

If the curve is given by intersection of the hyperplane $f_j = x_j$ for $j = 1, \ldots, N-1$, then condition (4.68) takes the form

$$
\sum_{k=1}^{N} \tilde{G}_{\alpha k} \frac{\partial S}{\partial x_k} = 0, \quad \alpha = 1, \ldots, N-1, \tag{4.74}
$$

where $\tilde{G}$ is the inverse of the matrix G.

By solving these equations with respect to $\dfrac{\partial S}{\partial x_k}$ for $k = 1, \ldots, N-1$ we obtain that $\partial S/\partial x_k$ is equal to

$$
\frac{1}{\Delta} \frac{\partial S}{\partial x_N}
\begin{vmatrix}
\tilde{G}_{11} & \cdots & \tilde{G}_{1,k-1} & -\tilde{G}_{1N} & \tilde{G}_{1,k+1} & \cdots & \tilde{G}_{1,N-1} \\
\vdots & \cdots & \vdots & \vdots & \vdots & \cdots & \vdots \\
\tilde{G}_{1,N-1} & \cdots & \tilde{G}_{N-1,k-1} & -\tilde{G}_{N-1,N} & \tilde{G}_{N-1,k+1} & \cdots & \tilde{G}_{N-1,N-1}
\end{vmatrix}
:= L_k \frac{\partial S}{\partial x_N}.
$$

Using these relations and (4.74), and the fact that $\sum\limits_{n=1}^{N} L_n \tilde{G}_{Nn} = |\tilde{G}|$ we deduce after some computations that

$$
\left\langle \frac{\partial S}{\partial \mathbf{x}}, \frac{\partial F}{\partial \mathbf{x}} \right\rangle := \sum_{j,k=1}^{N} \tilde{G}_{jk} \frac{\partial S}{\partial x_k} \frac{\partial F}{\partial x_j} = \sum_{j=1}^{N} \tilde{G}_{Nk} \frac{\partial S}{\partial x_k} \frac{\partial F}{\partial x_N} = \frac{|\tilde{G}|}{\Delta} \frac{\partial S}{\partial x_N} \frac{\partial F}{\partial x_N}. \tag{4.75}
$$

Consequently, the equations (4.70) are recast as

$$
-\frac{\partial h}{\partial \mathbf{x}} = \left\langle \frac{\partial}{\partial \mathbf{x}} \left(\frac{\nu^2}{2} \right), \frac{\partial S}{\partial \mathbf{x}} \right\rangle \frac{\partial S}{\partial \mathbf{x}} - \left\| \frac{\partial S}{\partial \mathbf{x}} \right\|^2 \frac{\partial}{\partial \mathbf{x}} \left(\frac{\nu^2}{2} \right)
$$

$$
= \frac{|\tilde{G}|}{\Delta} \frac{\partial S}{\partial x_N} \left(\frac{\partial}{\partial x_N} \left(\frac{\nu^2}{2} \right) \frac{\partial S}{\partial \mathbf{x}} - \left(\frac{\partial S}{\partial x_N} \right) \frac{\partial}{\partial \mathbf{x}} \left(\frac{\nu^2}{2} \right) \right). \tag{4.76}
$$

In view of (4.75) we obtain that the potential function V takes the form

$$V = \frac{\nu^2}{2}\left\|\frac{\partial S}{\partial \mathbf{x}}\right\|^2 - h(f_1, \ldots, f_{N-1}) = \frac{\nu^2}{2}\frac{|\tilde{G}|}{\Delta}\left(\frac{\partial S}{\partial x_N}\right)^2 - h(x_1, \ldots, x_{N-1}).$$

We observe that if $\tilde{G}_{\alpha N} = 0$ for $\alpha = 1, \ldots, N-1$, then $|\tilde{G}| = \Delta\tilde{G}_{NN}$ and $S_N = x_N = c_N$ is a orthogonal family of hypersurfaces orthogonal to $f_j = x_j = c_j$ for $j = 1, \ldots, N-1$. After integrating (4.76) we obtain that

$$V = \frac{1}{2}\tilde{G}_{NN}\nu^2 - h = \left(g(x_N) - \sum_{j=1}^{N-1}\int h(x_1, \ldots, x_{N-1})\frac{\partial}{\partial x_j}\left(\frac{1}{\tilde{G}_{NN}}\right)dx_j\right)\tilde{G}_{NN},$$

where $g = g(x_N)$ and $h = h(x_1, \ldots, x_{N-1})$ are arbitrary functions.

Clearly, if $\nu = \nu(S)$, then $\sigma = d\Phi(S)$ where $\Phi = \int \nu(S)dS$. Therefore $d\sigma = 0$. So $\iota_\mathbf{v}d\sigma = 0$. This yields statement (c).

Now we prove statement (d). We use the *homotopy formula* $\mathbf{L}_\mathbf{v} = \iota_\mathbf{v}d + d\iota_\mathbf{v}$, see [67]. In view of (4.21), condition (4.67) is equivalent to

$$\iota_\mathbf{v}d\sigma = \sum_{j=1}^{N-1}\Lambda_j df_j = \nu_N\sum_{j=1}^{N-1}A_{Nj}df_j = -dh.$$

Thus $\mathbf{L}_\mathbf{v}\sigma = \iota_\mathbf{v}d\sigma + d\iota_\mathbf{v}\sigma = -dh + d\sigma(\mathbf{v}) = -dh + d\|\mathbf{v}\|^2 = d\left(\|\mathbf{v}\|^2 - h\right)$, where we used the relation $\sigma(\mathbf{v}) = \langle\mathbf{v}, \mathbf{v}\rangle = \|\mathbf{v}\|^2$. Hence, if $g_\mathbf{v}^t$ is the flow of $\mathbf{v}$ and γ is a closed curve on Q, then the integral $I = \int_{g_\mathbf{v}^t(\gamma)}\sigma$ is a function on t. By the well-known formula (see [92]) $\dot{I} = \int_{g_\mathbf{v}^t(\gamma)}\mathbf{L}_\mathbf{v}\sigma$, we obtain that $\dot{I} = 0$. Theorem 4.4.1 is proved. $\qquad\square$

In the two following subsections we illustrate the statements (b) and (c) of Theorem 4.4.1.

4.5   Generalized inverse Bertrand problem

For a particle with kinetic energy $T = \frac{1}{2}(\dot{x}^2 + \dot{y}^2)$, we determine the most general force field $\mathbf{F} = (F_x, F_y)$ that generates the family of planar orbits $f(x, y) = \text{const}$. From (4.66) we obtain for $N = 2$ the equation

$$\begin{aligned}
\mathbf{F} &= \frac{\partial}{\partial\mathbf{x}}\left(\frac{1}{2}\|\mathbf{v}\|^2\right) + \nu a_{21}\frac{\partial f}{\partial\mathbf{x}} \\
&= \frac{\partial}{\partial\mathbf{x}}\left(\frac{1}{2}\nu^2\left((\partial_x f)^2 + (\partial_y f)^2\right)\right) - \nu\left(\partial_x(\nu\partial_x f) + \partial_y(\nu\partial_y f)\right)\frac{\partial f}{\partial\mathbf{x}}.
\end{aligned}$$

This force field coincides with the solutions of Dainelli's problem given in [166].
Clearly, if the arbitrary function ν is chosen as a solution of the equation

$$\nu \left(\partial_x (\nu \partial_x f) + \partial_y (\nu \partial_y f) \right) = \frac{\partial h(f)}{\partial f}, \tag{4.77}$$

then the vector field $\mathbf{F}$ is potential with

$$U = \frac{1}{2} \nu^2 \left((\partial_x f)^2 + (\partial_y f)^2 \right) - h(f)$$

In order to apply this result, we prove that the potential energy function U capable of generating a one-parameter family of conics $f = r + bx = c$, where $r = \sqrt{x^2 + y^2}$, is given by

$$U = a_{-1} H_{-1}(\tau) + K_{-1} \log \left(r(1 + b\tau) \right)$$
$$+ \sum_{j \in \mathbb{Z} \setminus \{-1\}} a_j r^{j+1} \left(H_j(\tau) + K_j \frac{(1 + b\tau)^{j+1}}{j+1} \right),$$

if $b \neq 0$.

Here a_j and K_j are real constants and H_j, $j \in \mathbb{Z}$, are functions given by

$$H_j(\tau) = \mathbb{M}_j(\tau) \left(C_j - \frac{2K_j}{b} \int \frac{(1 + b\tau)^j}{(1 - \tau^2) \mathbb{M}_j(\tau)} d\tau \right),$$
$$\mathbb{M}_j(\tau) = (1 - \tau)^{\frac{j+1}{2} + \frac{j+3}{2b}} (\tau + 1)^{\frac{j+1}{2} - \frac{j+3}{2b}}, \tag{4.78}$$

where C_j are arbitrary constants $\tau = \cos \theta$, $' = \dfrac{d}{d\tau}$. If $b = 0$, then

$$U = \frac{\Psi(\tau)}{r^2} - \frac{2}{r^2} \int h(r) dr,$$

where $\Psi = \Psi(\tau)$ and $h = h(r)$ are arbitrary functions (see [139]).
Indeed, from (4.77) it follows that the sought-for potential force field exists if and only if

$$\left(\frac{x}{\sqrt{x^2 + y^2}} + b \right) \frac{\partial \nu^2}{\partial x} + \frac{y}{\sqrt{x^2 + y^2}} \frac{\partial \nu^2}{\partial y} + \frac{2\nu^2}{r} = 2 \frac{\partial h}{\partial f}.$$

In polar coordinates $x = r \cos \theta$, $y = r \sin \theta$, this equation reads

$$(1 + b \cos \theta) \frac{\partial \nu^2}{\partial r} - \frac{b \sin \theta}{r} \frac{\partial \nu^2}{\partial \theta} + \frac{2\nu^2}{r} = 2 \frac{\partial h}{\partial f},$$

or, equivalently,

$$(1 + b\tau)\frac{\partial \nu^2}{\partial r} + \frac{b(1 - \tau^2)}{r}\frac{\partial \nu^2}{\partial \tau} + \frac{2\nu^2}{r} = 2\frac{\partial h}{\partial f}, \tag{4.79}$$

where $f = r(1 + b\tau)$, $\tau = \cos\theta$.

Now we shall study the case when $b \neq 0$ and h is such that

$$h(f) = -a_{-1}K_{-1}\ln|f| - \sum_{\substack{j \in \mathbb{Z} \\ j \neq -1}} a_j K_j \frac{f^{j+1}}{j+1}, \tag{4.80}$$

where ν_j, $j \in \mathbb{Z}$, are real constants, and λ is determined in such a way that

$$\nu^2 = \sum_{j \in \mathbb{Z}} a_j r^{j+1} H_j(\tau). \tag{4.81}$$

It is clear that (4.80) and (4.81) are *formal series*.

By inserting (4.80) and (4.81) into (4.79) we obtain

$$b(1 - \tau^2)H_j'(\tau) + \big((j+1)b\tau + j + 3\big)H_j(\tau) + 2K_j(1 + b\tau)^j = 0,$$

for $j \in \mathbb{Z}$,

The general solutions of these equations are the functions (4.78).

Consequently, the sought-for potential function U has the form

$$U(r, \tau) = \frac{1}{2}\lambda^2(1 + b^2 + 2b\tau) - h(f) = \sum_{j \in \mathbb{Z}} a_j U_j(r, \tau),$$

where

$$U_j(r, \tau) = \frac{1}{2}r^{j+1}H_j(\tau)(1 + b^2 + 2b\tau) + \frac{K_j}{j+1}f^{j+1}$$

for $j \neq -1$, and

$$U_{-1}(r, \tau) = \frac{1}{2}H_{-1}(\tau)(1 + b^2 + 2b\tau) + K_{-1}\ln|f|.$$

Let us examine the subcase $b = 1$ separately from the subcase $b \neq 1$.

If $b = 1$, then

$$U(r, \tau) = \lambda^2(1 + \tau) - h(f) = \sum_{j \in \mathbb{Z}} a_j U_j(r, \tau),$$

where

$$U_j(r, \tau) = r^{j+1}(1 - \tau)^{j+2}\left(C_j - 2K_j\int\frac{(1 + \tau)^j}{(1 - \tau)^{j+3}}d\tau\right) + \frac{K_j}{j+1}f^{j+1},$$

for $j \neq -1$ and

$$U_{-1}(r,\tau) = (1-\tau)\left(C_{-1} - 2K_{-1}\int \frac{d\tau}{(1-\tau)^2(1+\tau)}\right) + K_{-1}\ln|f|.$$

One easily verifies that

$$U_{-2} = \frac{C_{-2}}{r} - 2\frac{K_{-2}}{r}\left(\int \frac{d\tau}{(1+\tau)^2(1-\tau)} + \frac{1}{1+\tau}\right)$$

$$= \frac{C_{-2}}{r} + \frac{K_{-2}}{r}g(\tau),$$

where $g(\tau) = \ln\sqrt{\dfrac{1-\tau}{1+\tau}}$.

Therefore, if $b = 1$, then

$$U(r,\tau) = \frac{a_{-2}C_{-2}}{r} + \frac{a_{-2}K_{-2}g(\tau)}{r} + \sum_{\substack{j\in\mathbb{Z}\\j\neq-2}} a_j U_j(r,\tau).$$

If $b \neq 1$ and $b \neq 0$, it is easy to prove that

$$H_{-2}(\tau) = \frac{(1-\tau)^{\frac{1-b}{2b}}}{(1+\tau)^{\frac{1+b}{2b}}}C_{-2} - \frac{2K_{-2}}{(b\tau+1)(1-b^2)},$$

$$U_{-2}(r,\tau) = \frac{H_{-2}}{2r}(1+b^2+2b\tau) - \frac{K_{-2}}{r(b\tau+1)} = \frac{2K_{-2}}{r(b^2-1)} + \frac{C_{-2}}{r}G(\tau),$$

where

$$G(\tau) = \frac{1}{2}(1+b^2+2b\tau)\sqrt{\left(\frac{1-\tau}{1+\tau}\right)^{1/b}\frac{1}{1-\tau^2}}.$$

Under these conditions, the potential function U takes the form

$$U(r,\tau) = \frac{a_{-2}C_{-2}}{r}G(\tau) + \frac{2a_{-2}K_{-2}}{r(b^2-1)} + \sum_{\substack{j\in\mathbb{Z}\\j\neq-2}} a_j U_j(r,\tau).$$

Summarizing the above computations, we deduce that if $b \neq 0$ the function U is represented as

$$U(r,\tau) = \frac{\alpha}{r} + \frac{\beta(\tau)}{r} + \sum_{\substack{j\in\mathbb{Z}\\j\neq-2}} a_j U_j(r,\tau),$$

where α is a constant and $\beta = \beta(\tau)$ is a suitable function.

If $b = 0$, then $f = r$ and condition (4.79) takes the form

$$\partial_r \lambda^2 + 2\frac{\lambda^2}{r} = 2\partial_f h(f).$$

Therefore,

$$r^2 \lambda^2 = 2 \int r^2 \partial_r h(r) dr + 2\Psi(\tau),$$

which rearranged results in the expression:

$$\lambda^2 = \frac{2}{r^2} \int r^2 \partial_r h(r) dr + \frac{2\Psi(\tau)}{r^2} = 2h(r) - \frac{4}{r^2} \int h(r) r dr + \frac{2\Psi(\tau)}{r^2},$$

where Ψ is an arbitrary function.

Hence,

$$U(r, \tau) = \frac{\Psi(\tau)}{r^2} - \frac{2}{r^2} \int h(r) dr.$$

4.6   Inverse Stäckel problem

Let

$$f_j = f_j(\mathbf{x}) = \sum_{k=1}^{n} \int \frac{\varphi_{kj}(x_k)}{\sqrt{K_k(x_k)}} dx_k = c_j, \quad j = 1, \ldots, N-1, \tag{4.82}$$

be a given $(N-1)$-parameter *family of orbits* in the configuration space Q of the mechanical system with N degrees of freedom and kinetic energy

$$T = \frac{1}{2} \sum_{j=1}^{N} \frac{\dot{x}_j^2}{A_j}, \tag{4.83}$$

where

$$K_k(x_k) = 2\Psi_k(x_k) + 2\sum_{j=1}^{N} \alpha_j \varphi_{kj}(x_k),$$

α_k for $k = 1, \ldots, N$ are constants, $\Psi_k = \Psi_k(x_k)$ are arbitrary functions, and $A_j = A_j(\mathbf{x})$ are such that

$$\frac{\{\varphi_1, \ldots, \varphi_{N-1}, x_j\}}{\{\varphi_1, \ldots, \varphi_{N-1}, \varphi_N\}} = A_j, \tag{4.84}$$

for $j = 1, \ldots, N$. Here

$$d\varphi_\alpha = \sum_{k=1}^{N} \varphi_{k\alpha}(x_k) dx_k, \quad \varphi_{k\alpha} = \varphi_{k\alpha}(x_k),$$

for $k = 1, \ldots, N, \alpha = 1, \ldots, N$ are arbitrary functions.

From (4.83) it follows that the metric G is diagonal with $G_{jj} = \dfrac{1}{A_j}$.

The *inverse Stäckel problem* is the problem of determining the potential force field under which any curve of the family (4.82) is a trajectory of the mechanical system. The solution is as follows (see [139]).

Proposition 4.6.1. *For a mechanical system with configuration space* Q *and kinetic energy* (4.83), *the potential force field* $\mathbf{F} = \dfrac{\partial V}{\partial \mathbf{x}}$, *for which the family of curves* (4.82) *are trajectories is*

$$V = -U = \nu^2(S)\left(\frac{\{\varphi_1, \ldots, \varphi_{N-1}, \Psi\}}{\{\varphi_1, \ldots, \varphi_{N-1}, \varphi_N\}} + \alpha_N\right) - h_0, \qquad (4.85)$$

where

$$S = \int \sum_{j=1}^{N} \sqrt{\Psi_k(x_k) + \sum_{k=1}^{N} \alpha_j \varphi_{kj}(x_k)}\, dx_k = \int \sum_{k=1}^{N} \frac{dx_k}{q_k(x_k)}$$

is a function such that the hypersurface $S = c_N$ *is orthogonal to the given hypersurfaces* $f_j = c_j$.

We observe that from (4.84) and (4.85) it follows that the metric G and the potential function U can be determined from the given functions (4.82).

Proof. After some tedious computations we get the equality

$$\frac{\{f_1, \ldots, f_{N-1}, *\}}{\{f_1, \ldots, f_{N-1}, f_N\}} = \frac{\begin{vmatrix} q_1 d\varphi_1(\partial_1) & \cdots & q_N d\varphi_1(\partial_N) \\ \vdots & & \vdots \\ q_1 d\varphi_{N-1}(\partial_1) & \cdots & q_N d\varphi_{N-1}(\partial_N) \\ \partial_1 & \cdots & \partial_N \end{vmatrix}}{\displaystyle\prod_{j=1}^{N} q_j \{\varphi_1, \ldots, \varphi_N\}}$$

$$= \sum_{j=1}^{N}\left(\frac{A_j}{q_j}\partial_j\right) = \sum_{j=1}^{N}\left(A_j \frac{\partial S}{\partial x_j}\partial_j\right),$$

By (4.73), $\mathbf{v}(\mathbf{x}) = \nu G^{-1}\dfrac{\partial S}{\partial \mathbf{x}}$, hence in view of the identity (viii) for the Nambu bracket we obtain

$$\left\langle \frac{\partial S}{\partial \mathbf{x}}, \frac{\partial f_j}{\partial \mathbf{x}} \right\rangle = \sum_{k=1}^{N} A_k \varphi_{kj} = \sum_{k=1}^{N} A_k \frac{\partial \varphi_j}{\partial x_k} = \frac{\{\varphi_1, \ldots, \varphi_{N-1}, \varphi_j\}}{\{\varphi_1, \ldots, \varphi_{N-1}, \varphi_N\}} = 0,$$

for $j = 1, \ldots, N-1$, which establishes the orthogonality of the surfaces.

On the other hand, we have

$$\|\mathbf{v}\|^2 = \nu^2 \sum_{k=1}^{N} A^k (K_k(x_k))^2 = \nu^2 \sum_{k=1}^{N} A_k \left(2\Psi_k(x_k) + 2 \sum_{j=1}^{N} \alpha_j \varphi_{kj}(x_k) \right)$$

$$= 2\nu^2 \sum_{k=1}^{N} A^k \Psi_k(x_k) + 2\nu^2 \sum_{j=1}^{N} \alpha_j \sum_{k=1}^{N} A_k \varphi_{kj}(x_k)$$

$$= 2\nu^2 \left(\frac{\{\varphi_1, \ldots, \varphi_{N-1}, \Psi\}}{\{\varphi_1, \ldots, \varphi_{N-1}, \varphi_N\}} + \sum_{j=1}^{N} \alpha_j \frac{\{\varphi_1, \ldots, \varphi_{N-1}, \varphi_j\}}{\{\varphi_1, \ldots, \varphi_{N-1}, \varphi_N\}} \right)$$

$$= 2\nu^2 \left(\frac{\{\varphi_1, \ldots, \varphi_{N-1}, \Psi\}}{\{\varphi_1, \ldots, \varphi_{N-1}, \varphi_N\}} + \alpha_N \right),$$

where we used the identity (iv) for the Nambu bracket, and $d\Psi = \sum_{j=1}^{N} \Psi_k(x_k) dx_k$.

We observe that if we take $\nu = \nu(S)$, then from (4.70) we obtain that the force field which generates the given family of orbits (4.82) is potential, with potential function given by (4.85). In particular, if $\nu = 1$ and $h_0 = \alpha_N$, then we obtain the classical *Stäckel potential* (see [23]).   □

Example 4.6.2. The example below is a particular case of a previously studied one. We will call it the *inverse problem of two fixed centers* (for more details, see [139]).

Let P be a particle of infinitesimal mass which is attracted by two fixed centers C_0 and C_1 of mass m_1 and m_2, respectively. Selecting coordinates so that the origin coincides with the center of mass and the abscissa passing through the points C_0 and C_1, and designating by r_0, r_1, and $2c$ the distances between $C_0(x_0, 0, 0)$, $P(x, y, z)$, $C_1(x_1, 0, 0)$, $P(x, y, z)$ and $C_0(x_0, 0, 0)$, $C_1(x_1, 0, 0)$, respectively, we have that

$$r_0 = \sqrt{(x - x_0)^2 + y^2 + z^2}, \quad r = \sqrt{(x - x_1)^2 + y^2 + z^2}, \quad 2c = |x_1 - x_0|.$$

Thus, we are dealing with a particle with configuration space $\mathbb{R}^3$ and the Lagrangian function

$$L = \frac{1}{2} \left(\dot{x}^2 + \dot{y}^2 + \dot{z}^2 \right) - \left(\frac{m_0}{r_0} + \frac{m_1}{r_1} \right) f,$$

where f is the attraction constant (see [48]).

After the change of coordinates

$$x = \frac{m_0 - m_1}{m_1 + m_0} c + c \lambda \mu,$$
$$y = c \sqrt{(\lambda^2 - 1)(1 - \mu^2)} \cos w,$$
$$z = c \sqrt{(\lambda^2 - 1)(1 - \mu^2)} \sin w,$$

we obtain

$$L = \frac{c^2(\lambda^2 - \mu^2)}{4(\lambda^2 - 1)}\dot{\lambda}^2 - \frac{c^2(\lambda^2 - \mu^2)}{4(1 - \mu^2)}\dot{\mu}^2 + \frac{c^2(\lambda^2 - 1)(1 - \mu^2)}{2}\dot{w}^2$$
$$- f\frac{(m_0 + m_1)\lambda + (m_1 - m_0)\mu}{c(\lambda^2 - \mu^2)},$$

and $r_0 = c(\lambda + \mu)$, $r_1 = c(\lambda - \mu)$, where $1 \le \lambda < +\infty$, $-1 \le \mu \le 14$, and $0 \le w \le 2\pi$.

Clearly, in this case the matrix $\tilde{G}$ is

$$\tilde{G} = \begin{pmatrix} \dfrac{2(\lambda^2 - 1)}{c^2(\lambda^2 - \mu^2)} & 0 & 0 \\ 0 & \dfrac{2(1 - \mu^2)}{c^2(\lambda^2 - \mu^2)} & 0 \\ 0 & 0 & \dfrac{1}{c^2(\lambda^2 - 1)(1 - \mu^2)} \end{pmatrix}.$$

The inverse problem of two fixed centers requires the construction of the potential force field for which the given family of curves

$$f_1(\lambda, \mu, w) = \int \frac{d\lambda}{\sqrt{R_2(\lambda)}} - \int \frac{d\mu}{\sqrt{R_1(\mu)}} = c_1,$$

$$f_2(\lambda, \mu, w)w - \frac{A}{2}\left(\int \frac{d\lambda}{(\lambda^2 - 1)\sqrt{R_2(\lambda)}} + \int \frac{d\mu}{(1 - \mu^2)\sqrt{R_1(\mu)}}\right) = c_2,$$

is formed by trajectories of the equations of motion, where R_1 and R_2 are the functions

$$R_1(\mu) = h_0 c^2 \mu^4 + fc(m_0 - m_1)\mu^3 + (a_2 - h_0 c^2)\mu^2 - fc(m_0 - m_1)\mu - \frac{A^2}{2} - a_2$$

$$R_2(\lambda) = h_0 c^2 \lambda^4 + fc(m_0 + m_1)\lambda^3 + (a_2 - h_0 c^2)\lambda^2 - fc(m_0 + m_1)\lambda - \frac{A^2}{2} - a_2,$$

with C, h_0, f, A, a_2 real constants.

After some computations we get that

$$\{f_1, f_2, F\} = -\frac{\partial_\lambda F}{\sqrt{R_1(\mu)}} - \frac{\partial_\mu F}{\sqrt{R_2(\lambda)}} - \frac{A(\lambda^2 - \mu^2)\partial w F}{2\sqrt{R_1(\mu)R_2(\lambda)}(\lambda^2 - 1)(1 - \mu^2)}$$

$$= \frac{c^2(\mu^2 - \lambda^2)}{2\sqrt{R_1(\mu)R_2(\lambda)}}\left(\frac{2(\lambda^2 - 1)\sqrt{R_2(\lambda)}\partial_\lambda F}{c^2(\lambda^2 - \mu^2)(\lambda^2 - 1)} + \frac{2(1 - \mu^2)\sqrt{R_1(\mu)}\partial_\mu F}{c^2(\lambda^2 - \mu^2)(1 - \mu^2)}\right.$$

$$\left. + \frac{A\partial_w F}{c^2(\lambda^2 - 1)(1 - \mu^2)}\right) := \varrho\left\langle \frac{\partial S}{\partial \mathbf{x}}, \frac{\partial F}{\partial \mathbf{x}}\right\rangle$$

$$= \varrho(\tilde{G}_{11}\partial_\lambda S\partial_\lambda F + \tilde{G}_{22}\partial_\mu S\partial_\mu F + \tilde{G}_{33}\partial_w S\partial_w F),$$

where F is an arbitrary function, and

$$\varrho = -\frac{c^2(\lambda^2 - \mu^2)}{2\sqrt{R_1(\mu)R_2(\lambda)}}, \quad S(\lambda, \mu, w) = \int \frac{\sqrt{R_1(\mu)}}{(1 - \mu^2)}d\mu + \int \frac{\sqrt{R_2(\lambda)}}{(\lambda^2 - 1)}d\lambda + A\,w.$$

Hence, from (4.85) it follows that

$$V = \frac{1}{2}\nu^2(S)\left\|\frac{\partial S}{\partial \mathbf{x}}\right\|^2 - h_0$$

$$= \frac{\nu^2}{c^2}\left(\frac{R_1(\mu)}{(1 - \mu^2)(\lambda^2 - \mu^2)} + \frac{R_2(\lambda)}{(\lambda^2 - 1)(\lambda^2 - \mu^2)} + \frac{A^2}{(\lambda^2 - 1)(1 - \mu^2)}\right) - h_0.$$

Since

$$\frac{R_1(\mu)}{1 - \mu^2} = -h_0 c^2 \mu^2 + (m_1 - m_0)cf\mu - a_2 - \frac{A^2}{2(1 - \mu^2)},$$

$$\frac{R_2(\lambda)}{\lambda^2 - 1} = h_0 c^2 \lambda^2 + (m_0 + m_1)cf\lambda + a_2 - \frac{A^2}{2(\lambda^2 - 1)},$$

we finally obtain that

$$U = \nu^2\left(h_0 + \frac{(m_0 + m_1)\lambda + (m_1 - m_0)\mu}{c(\lambda^2 - \mu^2)}f\right) - h_0.$$

If we take $\nu = 1$, then $U = f\dfrac{(m_0 + m_1)\lambda + (m_1 - m_0)\mu}{c(\lambda^2 - \mu^2)}$, which coincides with the well-known potential (see [48, 23].)

Example 4.6.3 (Joukovsky's example.). We shall study a mechanical systems with three degrees of freedom. Denote $x_1 = p$, $x_2 = q$, $x_3 = r$. Then we consider the mechanical system with kinetic energy

$$T = \frac{1}{2r^2}\left(\dot{p}^2 - 2p\,\dot{p}\,\dot{r} + \dot{q}^2 - 2q\,\dot{q}\,\dot{r} + \left(\frac{p^2 + q^2}{r^2} + r^2\right)\dot{r}^2\right).$$

The matrix $\tilde{G}$ is

$$\tilde{G} = \begin{pmatrix} \dfrac{p^2 + r^4}{r^2} & \dfrac{pq}{r^2} & \dfrac{p}{r} \\[3mm] \dfrac{pq}{r^2} & \dfrac{q^2 + r^4}{r^2} & \dfrac{q}{r} \\[3mm] \dfrac{p}{r} & \dfrac{q}{r} & 1 \end{pmatrix}.$$

Then we get $|\tilde{G}| = r^4$, $\quad \Delta = p^2 + q^2 + r^4$. We determine the force field derived from the potential-energy function (4.72) in such a way that the family of curves $p = c_1$, $q = c_2$ can be freely traced by a particle with kinetic energy T.

In this case equations (4.74) are

$$\tilde{g}_{11}\frac{\partial S}{\partial p} + \tilde{g}_{12}\frac{\partial S}{\partial q} + \tilde{g}_{13}\frac{\partial S}{\partial r} = \frac{p^2 + r^4}{r^2}\frac{\partial S}{\partial p} + \frac{pq}{r^2}\frac{\partial S}{\partial q} + \frac{p}{r}\frac{\partial S}{\partial r} = 0,$$

$$\tilde{g}_{21}\frac{\partial S}{\partial p} + \tilde{g}_{22}\frac{\partial S}{\partial q} + \tilde{g}_{23}\frac{\partial S}{\partial r} = \frac{q^2 + r^4}{r^2}\frac{\partial S}{\partial q} + \frac{pq}{r^2}\frac{\partial S}{\partial p} + \frac{q}{r}\frac{\partial S}{\partial r} = 0.$$

The solutions of these partial differential equations are $S = S\left(\dfrac{p^2 + q^2}{r^2} - r^2\right)$

where S is an arbitrary function of the variable $\dfrac{p^2 + q^2}{r^2} - r^2$.

Without loss of generality we assume that $S = \dfrac{p^2 + q^2}{r^2} - r^2$. After some computations we obtain that conditions (4.76) take the form

$$\frac{\partial h}{\partial p} = \frac{2p}{r}\frac{\partial \nu^2}{\partial r} + \frac{(p^2 + q^2 + r^4)}{r^2}\frac{\partial \nu^2}{\partial p},$$

$$\frac{\partial h}{\partial q} = \frac{2q}{r}\frac{\partial \nu^2}{\partial r} + \frac{(p^2 + q^2 + r^4)}{r^2}\frac{\partial \nu^2}{\partial q}. \tag{4.86}$$

From the compatibility conditions of these equations we obtain that $h = h(p^2 + q^2)$, $\nu = \nu(p^2 + q^2, r)$. In the coordinates $\xi = p^2 + q^2$, $r = r$, the conditions (4.86) read

$$\frac{\partial h}{\partial \xi} = \frac{1}{r^2}\left(r\frac{\partial \nu^2}{\partial r} + 2(\xi + r^4)\frac{\partial \nu^2}{\partial \xi}\right). \tag{4.87}$$

Thus, from (4.72), the potential function takes the form

$$V = \frac{1}{2}\nu^2(\xi, r)\left(\frac{\xi}{r^2} + r^2\right) - h(\xi), \tag{4.88}$$

where $\nu = \nu(\xi, r)$ and $h = h(\xi)$ are solutions of (4.87).

We shall look for the solution $h = h(\xi)$ of (4.87) when the function ν^2 is given by

$$\nu^2 = \Psi\left(\frac{\xi}{r^2} - r^2\right) + \sum_{j=-\infty}^{+\infty} a_j(\xi)r^j.$$

where the series is a *formal Laurent series*, and $\Psi = \Phi\left(\dfrac{\xi}{r^2} - r^2\right)$ is an arbitrary function.

Inserting this expression for ν^2 in (4.87) we obtain

$$\sum_{j=-\infty}^{+\infty}\left(ja_j + 2\xi\frac{da_j}{d\xi} + 2\frac{da_{j-4}}{d\xi}\right)r^j = \frac{r^2}{2}\frac{dh}{d\xi}.$$

We choose the coefficients a_j so that they satisfy

$$ja_j + 2\xi\frac{da_j}{d\xi} + 2\frac{da_{j-4}}{d\xi} = 0 \quad \Longleftrightarrow \quad (j-2)a_j + \frac{d}{d\xi}\left(2\xi a_j + 2a_{j-4}\right), \text{ for } j \neq 2,$$

$$2a_2 + 2\xi\frac{da_2}{d\xi} + 2\frac{da_{-2}}{d\xi} = \frac{dh}{2d\xi} \quad \Longleftrightarrow \quad \frac{d}{d\xi}\left(2\xi a_2 + 2a_{-2} - \frac{h}{2}\right) = 0.$$

Then the potential function (4.88) takes the form

$$V = 4\left(\Psi\left(\frac{\xi}{r^2} - r^2\right) + \sum_{j=-\infty}^{+\infty} a_j(\xi)r^j\right)\left(\frac{\xi}{r^2} + r^2\right) - 4\xi a_2 - 4a_{-2} - h_0.$$

If we change variables by $p = xz$, $q = yz$, $r = z$, where x, y, z are the cartesian coordinates, then the kinetic and potential function take the respective forms

$$T = \frac{1}{2}\left(\dot{x}^2 + \dot{y}^2 + \dot{z}^2\right),$$

$$V = 4\left(\Psi\left(x^2 + y^2 - z^2\right) + \sum_{j=-\infty}^{+\infty} a_j(z^2(x^2 + y^2))z^j\right)\left(x^2 + y^2 + z^2\right)$$
$$- 4z^2(x^2 + y^2)a_2(z^2(x^2 + y^2)) - 4a_{-2}(z^2(x^2 + y^2)) - h_0.$$

Clearly, if $a_j = 0$ for $j \in \mathbb{Z}$, then we obtain the potential

$$V = \Psi\left(x^2 + y^2 - z^2\right)\left(x^2 + y^2 + z^2\right) - h_0$$

obtained by *Joukovsky* in [78]. On the other hand, if $\Psi = 0$, $a_j = 0$ for $j \in \mathbb{Z}\setminus\{2\}$, and $4a_2 = a$, then we obtain the potential $V = az^4 - h_0$ given in [139].

Chapter 5

Inverse Problem for Constrained Hamiltonian Systems

5.1 Introduction

Constrained Hamiltonian systems arise in many fields, for instance in multi-body dynamics or in molecular dynamics. The theory of such systems goes back to by P.A.M. Dirac (see for instance [44]).

Dirac in [45] writes that the route one should take to get the most general relativistic quantum field theory is to start with an action principle, find the Lagrangian, find the Hamiltonian, and then quantize the Hamiltonian to get a first approximation to Quantum Field Theory.

When one writes down the most general Lagrangians which may not be quadratic in the velocity, however, one immediately runs into complications in the Hamiltonian formulation. Dirac generalizes the Hamiltonian and Poisson brackets to handle general Lagrangians, possibly with constraints.

Following Dirac, we give the following definition. Two functions on the phase space, f and F, are *weakly equal* if they are equal when the constraints are satisfied, but not everywhere in the phase space; this is denoted $f \approx F$.

Dirac argues that we should generalize the Hamiltonian (somewhat analogously to the method of Lagrange multipliers) to

$$H^* = H + \sum_j c_j g_j \approx H,$$

where the c_j are functions of the coordinates and momenta. Since this new Hamiltonian is the most general function of coordinates and momenta weakly equal to the naive Hamiltonian, H^* is the broadest generalization of the possible Hamiltonian. Then the equations of motion become

$$\dot{y}_j = -\frac{\partial H}{\partial x_j} - \sum_k u_k \frac{\partial g_k}{\partial x_j} = \{H, x_j\}^* + \sum_k u_k\{g_k, x_j\}^*,$$

$$\dot{x}_j = \frac{\partial H}{\partial y_j} + \sum_k u_k \frac{\partial g_k}{\partial y_j} = \{H, y_j\}^* + \sum_k u_k\{g_k, y_j\}^*,$$

$$g_j(x, y) = 0,$$

where the u_k are functions of coordinates and velocities that can be determined, in principle, from the second equation of motion above. These equations are called *Dirac's equations*. More information on the Dirac ideas can found in [40, 45, 76, 143]. For an introduction to constrained dynamics see [148].

The *inverse problem for constrained Hamiltonian systems* can be stated as follows: for a given submanifold $\mathcal{M}$ of a *symplectic manifold* $\mathbb{M}$ we must determine the differential systems whose flows leave $\mathcal{M}$ invariant.

We find the equations of motion of a constrained Hamiltonian system in the following cases:

(i) The given constraints are l first integrals with $\dim \mathbb{M}/2 \le l < \dim \mathbb{M}$. In particular, the differential equations obtained solving this inverse problem are Hamiltonian only if the first integrals are in involution.

(ii) The given constraints are $M < \dim \mathbb{M}/2$ partial integrals. We deduce the differential equations which can be interpreted as a normal form of the equations of motion of a nonholonomic system with constraints nonlinear in the momenta.

We note that these two statements of the inverse problem for constrained Hamiltonian systems are new.

Let $\mathbb{M}$ be a $2N$-dimensional smooth manifold with local coordinates $(\mathbf{x}, \mathbf{y}) = (x_1, \ldots, x_N, y_1, \ldots, y_N)$. Let Ω^2 be a closed non-degenerate 2-form on $\mathbb{M}$. Then $(\mathbb{M}, \Omega^2)$ is a *symplectic manifold*. Let $H : \mathbb{M} \to \mathbb{R}$ be a smooth function, and let $\mathcal{M}$ be a submanifold of $\mathbb{M}$.

The 4-tuple $(\mathbb{M}, \Omega^2, \mathcal{M}, H)$ is called a *constrained Hamiltonian system* (see [4]). We essentially study two inverse problems for the constrained Hamiltonian systems: in the first the submanifolds $\mathcal{M}$ are obtained by fixing the values of given first integrals, while in the second these submanifolds are defined by the hypersurfaces given by partial integrals.

Now we can formulate the inverse problem for constrained Hamiltonian systems: determine the vector fields $\mathbf{W}$ with components $(W_1, \ldots, W_{2N})$, where $W_j = W_j(\mathbf{x}, \mathbf{y})$, such that the submanifold $\mathcal{M}$ is invariant under the flow of the differential system

$$\begin{aligned} \dot{x}_k &= \{H, x_k\}^* + W_k, \\ \dot{y}_k &= \{H, y_k\}^* + W_{N+k}, \end{aligned} \qquad k = 1, \ldots, N, \qquad (5.1)$$

where

$$\{H, G\}^* = \sum_{k=1}^{N} \left(\frac{\partial H}{\partial y_k} \frac{\partial G}{\partial x_k} - \frac{\partial H}{\partial x_k} \frac{\partial G}{\partial y_k} \right) \tag{5.2}$$

is the Poisson bracket. In this chapter we solve this inverse problem.

We note that if $W_k = 0$ for $k = 1, \ldots, N$, then the equations (5.1) are the standard Hamiltonian equations for a mechanical system subject to the action of an external force with components $W_{N+1}, \ldots, W_{2N}$.

5.2 Hamiltonian system with given first integrals

We have the following result.

Theorem 5.2.1. *Let* $(\mathbb{M}, \Omega^2, \mathcal{M}_1, H)$ *be a constrained Hamiltonian system and let* $f_j = f_j(\mathbf{x}, \mathbf{y})$ *for* $j = 1, \ldots, N$ *be a given set of independent functions defined on* $\mathbb{M}$ *and such that the submanifold* $\mathcal{M}_1$ *is given by* $\mathcal{M}_1 = \{(\mathbf{x}, \mathbf{y}) \in \mathbb{M} : f_j(\mathbf{x}, \mathbf{y}) = c_j \in \mathbb{R} \text{ for } j = 1, \ldots, N\}$, *where* c_j, $j = 1, \ldots, N$ *are some constants.*
  (i) *Assume that*
$$\{f_1, \ldots, f_N, x_1, \ldots, x_N\} \neq 0, \quad in \quad \mathbb{M}.$$

Then the manifold $\mathbb{M}_1$ *is invariant under the flow of the differential system*

$$\dot{x}_k = \{H, x_k\}^*,$$

$$\dot{y}_k = \{H, y_k\}^* - \sum_{j=1}^{N} \frac{\{H, f_j\}^* \{f_1, \ldots, f_{j-1}, y_k, f_{j+1}, \ldots f_N, x_1, \ldots, x_N\}}{\{f_1, \ldots, f_N, x_1, \ldots, x_N\}}$$

$$= \{H, y_k\}^* + W_{k+N}, \quad k = 1, \ldots, N. \tag{5.3}$$

(ii) *Assume that*
$$\begin{aligned} \{f_1, \ldots, f_N, x_1, \ldots, x_N\} &= 0 \quad and \\ \{f_1, \ldots, f_N, x_1, \ldots, x_{N-1}, y_1\} &\neq 0. \end{aligned} \tag{5.4}$$

Then the submanifold $\mathcal{M}_1$ *is invariant under the flow of the differential system*

$$\dot{x}_k = \{H, x_k\}^*, \quad for \quad k = 1, \ldots, N - 1,$$
$$\dot{x}_N = \{H, x_N\}^*$$

$$- \sum_{j=1}^{N} \frac{\{H, f_j\}^* \{f_1, \ldots, f_{j-1}, x_N, f_{j+1}, \ldots f_N, x_1, \ldots, x_{N-1}, y_1\}}{\{f_1, \ldots, f_N, x_1, \ldots, x_{N-1}, y_1\}}$$

$$= \{H, x_N\}^* + W_N,$$
$$\dot{y}_1 = \{H, y_1\}^* + \lambda \{f_1, \ldots, f_N, x_1, \ldots, x_{N-1}, y_1\}$$
$$= \{H, y_1\}^* + W_{1+N},$$

$$\dot{y}_k = \{H, y_k\}^*$$

$$-\sum_{j=1}^{N} \frac{\{H, f_j\}^*\{f_1,\ldots,f_{j-1},y_k,f_{j+1},\ldots,f_N,x_1,\ldots,x_{N-1},y_1\}}{\{f_1,\ldots,f_N,x_1,\ldots,x_{N-1},y_1\}}$$

$$+\lambda\{f_1,\ldots,f_N,x_1,\ldots,x_{N-1},y_k\}$$

$$= \{H, y_k\}^* + W_{k+N}, \quad k = 2,\ldots,N, \tag{5.5}$$

where $\lambda = \lambda(\mathbf{x}, \mathbf{y})$ is an arbitrary function.

We observe that the solution (5.3) of the inverse problem for constrained Hamiltonian systems in the case when the first integrals are *pairwise in involution*, and $H = H(f_1,\ldots,f_N)$ becomes into the Hamiltonian system $\dot{x}_k = \{H, x_k\}^*$, $\dot{y}_k = \{H, y_k\}^*$. Moreover, when the first integrals are pairwise in involution and (5.4) holds and $H = H(f_1,\ldots,f_N)$, then (5.5) becomes the differential system

$$\dot{x}_k = \{H, x_k\}^*, \quad \dot{y}_k = \{H, y_k\}^* + \lambda\{f_1,\ldots,f_N,x_1,\ldots,x_{N-1},y_k\}, \quad k = 1,\ldots,N.$$

These are the equations of motion of the mechanical system with the constraints $\{f_1,\ldots,f_N,x_1,\ldots,x_N\} = 0$.

Proof of Theorem 5.2.1. Under the assumptions of Corollary 1.4.2, replacing the N in the corollary by $2N$, introducing the notations $y_j = x_{N+j}$, and choosing $g_{N+j} = x_j$ for $j = 1,\ldots,N$, the differential systems (1.20) takes the form

$$\dot{x}_j = \lambda_{N+j}, \quad \dot{y}_j = \sum_{k=1}^{N} \lambda_{N+k} \frac{\{f_1,\ldots,f_N,x_1,\ldots,x_{k-1},y_j,x_{k+1},\ldots,x_N\}}{\{f_1,\ldots,f_N,x_1,\ldots,x_N\}}, \tag{5.6}$$

$j = 1, 2,\ldots,N$. These equations are the most general differential equations that admit N independent first integrals and satisfy the condition

$$\{f_1,\ldots,f_N,x_1,\ldots,x_N\} \neq 0.$$

The proof of Theorem 5.2.1 is done by choosing the arbitrary functions λ_{N+j} as $\lambda_{N+j} = \{H, x_j\}^*$, $j = 1,\ldots,N$, where H is the Hamiltonian function. From the identity (xi) for the Nambu bracket with $G = y_k$, $f_{N+j} = x_j$ for $j = 1,\ldots,N$, we obtain that the differential system (5.6) can be rewritten as

$$\dot{x}_j = \{H, x_j\}^* + W_j,$$

$$\dot{y}_j = \sum_{k=1}^{N} \{H, x_k\}^* \frac{\{f_1,\ldots,f_N,x_1,\ldots,x_{k-1},y_j,x_{k+1},\ldots,x_N\}}{\{f_1,\ldots,f_N,x_1,\ldots,x_N\}}$$

$$= \{H, y_j\}^* - \sum_{k=1}^{N} \{H, f_k\}^* \frac{\{f_1,\ldots,f_{k-1},y_j,f_{k+1},\ldots,f_N,x_1,\ldots,,\ldots,x_N\}}{\{f_1,\ldots,f_N,x_1,\ldots,x_N\}}$$

$$+\sum_{k=1}^{N} W_j \frac{\{f_1,\ldots,f_N,x_1,\ldots,x_{k-1},y_j,x_{k+1},\ldots,x_N\}}{\{f_1,\ldots,f_N,x_1,\ldots,x_N\}}.$$

Clearly, if the first integrals are in involution and $W_j = 0$, then we conclude that the Hamiltonian system with Hamiltonian $H = H(f_1, \ldots, f_N)$ is *integrable by quadratures*.

Next we derive the equations (5.5). Since $\{f_1, \ldots, f_N, x_1, \ldots, x_N\} = 0$ and $\{f_1, \ldots, f_N, x_1, \ldots, x_{N-1}, y_1\} \neq 0$, if we take $W_j = 0$ for $j = 1, \ldots, N - 1$ and $\lambda_{N+j} = \dfrac{\partial H}{\partial y_j} = \{H, x_j\}^*$, for $j = 1, \ldots, N - 1$, where H is the Hamiltonian function, and use the identity (xi) with $G = x_N$, $f_{N+j} = x_j$ for $j = 1, \ldots, N - 1$, $f_{2N} = y_1$, and $G = y_j$, $f_{N+j} = x_j$ for $j = 1, \ldots, N - 1$, $f_{2N} = y_1$, we obtain that differential system (5.6) can be rewritten as

$$\dot{x}_j = \{H, x_j\}^*, \quad j = 1, \ldots, N - 1,$$

$$\dot{x}_N = \sum_{k=1}^{N-1} \{H, x_k\}^* \frac{\{f_1, \ldots, f_N, x_1, \ldots, x_{k-1}, x_N, x_{k+1}, \ldots, y_1\}}{\{f_1, \ldots, f_N, x_1, \ldots, x_{N-1}, y_1\}}$$

$$+ \lambda_{2N} \frac{\{f_1, \ldots, f_N, x_1, \ldots, x_{N-1}, x_N\}}{\{f_1, \ldots, f_N, x_1, \ldots, x_{N-1}, y_1\}}$$

$$= \{H, x_N\}^* - \sum_{k=1}^{N} \{H, f_j\}^* \frac{\{f_1, \ldots, f_{k-1}, x_N, f_{k+1}, \ldots, f_N, x_1, \ldots, y_1\}}{\{f_1, \ldots, f_N, x_1, \ldots, x_{N-1}, y_1\}}$$

$$+ (\lambda_{2N} - \{H, y_1\}^*) \frac{\{f_1, \ldots, f_N, x_1, \ldots, x_{N-1}, x_N\}}{\{f_1, \ldots, f_N, x_1, \ldots, x_{N-1}, y_1\}},$$

$$\dot{y}_1 = \lambda_{2N},$$

$$\dot{y}_j = \sum_{k=1}^{N-1} \{H, x_k\}^* \frac{\{f_1, \ldots, f_N, x_1, \ldots, x_{k-1}, y_j, x_{k+1}, \ldots, x_N\}}{\{f_1, \ldots, f_N, x_1, \ldots, x_{N-1}, y_1\}}$$

$$+ \lambda_{2N} \frac{\{f_1, \ldots, f_N, x_1, \ldots, x_{N-1}, y_j\}}{\{f_1, \ldots, f_N, x_1, \ldots, x_{N-1}, y_1\}}$$

$$= \{H, y_j\} - \sum_{k=1}^{N} \{H, f_j\}^* \frac{\{f_1, \ldots, f_{k-1}, y_j, f_{k+1}, \ldots, f_N, x_1, \ldots, x_N\}}{\{f_1, \ldots, f_N, x_1, \ldots, x_{N-1}, y_1\}}$$

$$+ (\lambda_{2N} - \{H, y_1\}^*) \frac{\{f_1, \ldots, f_N, x_1, \ldots, x_{N-1}, y_k\}}{\{f_1, \ldots, f_N, x_1, \ldots, x_{N-1}, y_1\}}.$$

Now choosing λ_{2N} as $\lambda_{2N} = \{H, y_1\}^* + \lambda\{f_1, \ldots, f_N, x_1, \ldots, x_{N-1}, y_1\}$, we get the differential system (5.5).

In view of the identity (vii) with $G = f_j$, (5.5) yields the relations

$$\dot{f}_k = \sum_{j=1}^{N} \frac{\partial f_k}{\partial y_j} \{f_1, \ldots, f_N, x_1, \ldots, y_j\} = \frac{\partial f_k}{\partial x_N} \{f_1, \ldots, f_N, x_1, \ldots, x_N\} = 0.$$

When $\{H, f_j\} = 0$ for $j = 1, \ldots, N$, the system (5.5) is the standard Hamiltonian system with the constraints $\{f_1, \ldots, f_N, x_1, \ldots, x_N\} = 0$. $\qquad \square$

Example 5.2.2. Neumann–Moser integrable system We shall illustrate the last theorem in the case of the *Neumann–Moser integrable system*.

We consider systems with N independent *involutive first integrals* of the form

$$f_\nu = (Ax_\nu + By_\nu)^2 + C \sum_{j\neq\nu}^N \frac{(x_\nu y_j - x_j y_\nu)^2}{a_\nu - a_j}, \quad \nu = 1, \ldots, N, \qquad (5.7)$$

where A, B, and C are constants such that $C(A^2 + B^2) \neq 0$, and study the constrained Hamiltonian system $\left(\mathbb{R}^{2N}, \, \Omega^2, \mathbb{M}, , H \right)$.

The cases when $A = 0$, $B = 1$, $C = 1$ and $A = 1$, $B = 0$, $C = 1$ were investigated in particular in [117]. The case when $AB \neq 0$ was introduced in [136]. In particular, if $C = (A + B)^2$, then from (5.7) we obtain that $f_\nu = A^2 f_\nu^{(1)} + B^2 f_\nu^{(2)} + 2AB f_\nu^{(3)}$, where

$$f_\nu^{(1)} = x_\nu^2 + \sum_{j\neq\nu}^N \frac{(x_\nu y_j - x_j y_\nu)^2}{a_\nu - a_j},$$

$$f_\nu^{(2)} = y_\nu^2 + \sum_{j\neq\nu}^N \frac{(x_\nu y_j - x_j y_\nu)^2}{a_\nu - a_j},$$

$$f_\nu^{(3)} = x_\nu y_\nu + \sum_{j\neq\nu}^N \frac{(x_\nu y_j - x_j y_\nu)^2}{a_\nu - a_j}.$$

It is easy to show that $\{f_k^{(\alpha)}, f_m^{(\alpha)}\}^* = 0$ for $\alpha = 1, 2, 3$ and $m, k = 1, \ldots, N$, i.e., the first integrals are in involution.

After some computations we obtain that $\{f_1, \ldots, f_N, x_1, \ldots, x_N\} \neq 0$ if $B \neq 0$. Then taking in (5.3) $H = H(f_1, \ldots, f_N)$, and $W_j = 0$ for $j = 1, \ldots, N$, we obtain a completely integrable Hamiltonian system $\dot{x}_j = \{H, x_j\}^*$, $\dot{y}_j = \{H, y_j\}^*$.

If $B = 0$ then $\{f_1, \ldots, f_N, x_1, \ldots, x_N\} = 0$ Then taking in (5.5) $H = H(f_1, \ldots, f_N)$, $W_j = 0$ for $j = 1, \ldots, N$ and using the relations

$$\{f_1, \ldots, f_N, x_1, \ldots, x_{N-1}, y_j\} = \varrho(x)\, x_j \quad j = 1, \ldots, N,$$

with a suitable function $\varrho = \varrho(\mathbf{x})$, we obtain the differential system

$$\dot{\mathbf{x}} = \{H, \mathbf{x}\}^*, \quad \dot{\mathbf{y}} = \{H, \mathbf{y}\}^* + \tilde{\lambda}\mathbf{x}, \qquad (5.8)$$

where $\tilde{\lambda} = \varrho\lambda$. In particular for $N = 3$, we deduced

$$\{f_1, f_2, f_3, x_1, x_2, x_3\} = 0, \qquad \{f_1, f_2, f_3, x_1, x_2, y_1\} = \frac{K}{\Delta} x_3 x_1,$$

$$\{f_1, f_2, f_3, , x_1, x_2, y_2\} = \frac{K}{\Delta} x_3 x_2, \quad \{f_1, f_2, f_3, x_1, x_2, y_3\} = \frac{K}{\Delta} x_3 x_3,$$

where $\Delta = (a_1 - a_2)(a_2 - a_3)(a_1 - a_3)$, and K is a suitable function. Thus the differential system (5.8) with $\varrho = Kx_3/\Delta$ describes the behavior of a particle with Hamiltonian $H = H(f_1, f_2, f_3)$ and constrained to move on the sphere $x_1^2 + x_2^2 + x_3^2 = 1$.

If we take

$$H = \frac{1}{2}(a_1 f_1 + a_2 f_2 + a_3 f_3) = \frac{1}{2}\left(||\mathbf{x}||^2 ||\mathbf{y}||^2 - \langle \mathbf{x}, \mathbf{y}\rangle^2 + a_1 x_1^2 + a_2 x_2^2 + a_3 x_3^2\right)$$

and $\lambda = \Psi(x_1^2 + x_1^2 + x_1^2)$, then from equations (5.8) we deduce the equations of motion of a particle on a 3-dimensional sphere, with an anisotropic harmonic potential (*Neumann's problem*). This system is one of the best understood integrable systems of classical mechanics.

Theorem 5.2.3. *Let* $(\mathbb{M}, \Omega^2, \tilde{\mathcal{M}}_1, H)$ *be a constrained Hamiltonian system and let* $f_j = f_j(\mathbf{x}, \mathbf{y})$ *for* $j = 1, \ldots, N + r$, *with* $r < N$ *be a given set of independent functions defined in* $\mathbb{M}$ *and such that* $\{f_1, \ldots, f_{N+r}, x_1, \ldots, x_{N-r}\} \neq 0$ *and the manifold* $\tilde{\mathcal{M}}_1$ *is given by*

$$\tilde{\mathcal{M}}_1 = \{(\mathbf{x}, \mathbf{y}) \in \mathbb{M} : f_j(\mathbf{x}, \mathbf{y}) = c_j \in \mathbb{R} \quad for \quad j = 1, \ldots, N + r\},$$

where c_j *are arbitrary constants. Then* $\tilde{\mathcal{M}}_1$ *is invariant under the flow of the differential system*

$$\dot{x}_k = \{H, x_k\}^*, \tag{5.9}$$

$$\dot{x}_n = \{H, x_n\}^* - \sum_{j=1}^{N+r} \frac{\{H, f_j\}^* \{f_1, \ldots, f_{j-1}, x_n, f_{j+1}, \ldots, f_{N+r}, x_1, \ldots, x_{N-r}\}}{\{f_1, \ldots, f_{N+r}, x_1, \ldots, x_{N-r}\}}$$

$$= \{H, x_n\}^* + W_n,$$

$$\dot{y}_m = \{H, y_m\}^* - \sum_{j=1}^{N+r} \frac{\{H, f_j\}^* \{f_1, \ldots, f_{j-1}, y_m, f_{j+1}, \ldots, f_{N+r}, x_1, \ldots, x_{N-r}\}}{\{f_1, \ldots, f_{N+r}, x_1, \ldots, x_{N-r}\}}$$

$$= \{H, y_m\}^* + W_{m+N}, \quad k = 1, \ldots, N - r, \ n = N - r + 1, \ldots, N, \ m = 1, \ldots, N.$$

Proof of Theorem 5.2.3. Under the assumptions of the theorem the differential systems (1.20) take the form

$$\dot{x}_j = \lambda_{N+j}, \quad \text{for} \quad j = 1, 2, \ldots, N - r,$$

$$\dot{x}_n = \sum_{k=N+1}^{2N} \lambda_k \frac{\{f_1, \ldots, f_N, x_1, \ldots, x_{k-1}, x_n, x_{k+1}, \ldots, x_N\}}{\{f_1, \ldots, f_N, x_1, \ldots, x_N\}},$$

$$\text{for} \quad n = N - r + 1, \ldots, N, \tag{5.10}$$

$$\dot{y}_m = \sum_{k=N+1}^{2N} \lambda_k \frac{\{f_1, \ldots, f_N, x_1, \ldots, x_{k-1}, y_m, x_{k+1}, \ldots, x_N\}}{\{f_1, \ldots, f_N, x_1, \ldots, x_N\}},,$$

$$\text{for} \quad m = 1, 2, \ldots, N.$$

This is the most general system of differential equations which admits $N + r$ first integrals satisfying the condition $\{f_1, \ldots, f_{N+r}, x_1, \ldots, x_{N-r}\} \neq 0$.

By choosing in (5.1) the arbitrary functions $W_j = 0$ and $\lambda_{N+j} = \{H, x_j\}^*$ for $j = 1, \ldots, N - r$, where H is the Hamiltonian, and using the identity (xi) for the Nambu bracket with $G = x_k$, $f_{N+r+j} = x_j$ for $j = 1, \ldots, N - r$, and $G = y_k$, $f_{N+r+j} = x_j$ for $j = 1, \ldots, N - r$, we obtain that differential system (5.10) can be rewritten as

$$\dot{x}_j = \{H, x_j\}^* \quad \text{for} \quad j = 1, 2, \ldots, N - r,$$

$$\dot{x}_k = \sum_{j=1}^{N-r} \{H, x_j\}^* \frac{\{f_1, \ldots, f_{N+r}, x_1, \ldots, x_{j-1}, x_k, x_{j+1}, \ldots, x_{N-r}\}}{\{f_1, \ldots, f_{N+r}, x_1, \ldots, x_{N-r}\}}$$

$$= \{H, x_k\}^* - \sum_{j=1}^{N+r} \{H, f_j\}^* \frac{\{f_1, \ldots, f_{j-1}, x_k, f_{j+1}, \ldots, f_{N+r}, x_1, \ldots, \ldots, x_{N-r}\}}{\{f_1, \ldots, f_{N+r}, x_1, \ldots, x_{N-r}\}},$$

$$\text{for} \quad k = N - r + 1, \ldots, N,$$

$$\dot{y}_j = \sum_{k=1}^{N-r} \{H, x_k\}^* \frac{\{f_1, \ldots, f_{N+r}, x_1, \ldots, x_{k-1}, y_j, x_{k+1}, \ldots, x_N\}}{\{f_1, \ldots, f_{N+r}, x_1, \ldots, x_{N-r}\}}$$

$$= \{H, y_j\}^* - \sum_{k=1}^{N+r} \{H, f_k\}^* \frac{\{f_1, \ldots, f_{k-1}, y_j, f_{k+1}, \ldots, f_{N+r}, x_1, \ldots, \ldots, x_{N-r}\}}{\{f_1, \ldots, f_{N+r}, x_1, \ldots, x_{N-r}\}},$$

$j = 1, 2, \ldots, N$. Hence, we get the differential system (5.9). □

Remark 5.2.4. Concerning Theorems 5.2.1 and 5.2.3 we observe the following. If we assume that $\{f_1, \ldots, f_N, x_1, \ldots, x_N\} \neq 0$ in $\mathbb{M}$, and $H = H(f_1, \ldots, f_N)$, then the system of equations $f_j(\mathbf{x}, \mathbf{y}) = c_j$, $j = 1, \ldots, N$ can be solved locally with respect to $\mathbf{y}$ momenta, i.e., $y_j = u_j(\mathbf{x}, \mathbf{c})$, for $j = 1, \ldots, N$ where $\mathbf{c} = (c_1, \ldots, c_N)$. If the given first integrals are pairwise in involution, i.e., $\{f_j, f_k\} = 0$, then $\sum_{j=1}^{N} u_j(\mathbf{x}, \mathbf{c}) dx_j = dS(\mathbf{x})$. Let us recall *Liouville's theorem*:

Theorem 5.2.5. *If a Hamiltonian system has N independent first integrals in involution, which satisfy a certain nondegeneracy condition, then its motion can be obtained with quadratures, that is, the equation of motion can be solved simply by evaluating integrals.*

Then applying Theorem 5.2.5 we obtain that the Hamiltonian system

$$\dot{x}_k = \{H, x_k\}^*, \quad \dot{y}_k = \{H, y_k\}^*, \quad k = 1, \ldots, N,$$

is integrable by quadratures (for more details see [92]).

In general, the given set of first integrals is not in involution. The solution of the inverse problem for constrained Hamiltonian system shows that in the noninvolution case the differential equations which admit as invariant the submanifold $\mathcal{M}_1$ are in general not Hamiltonian. The origin of the theory of noncommutative integration is the *Nekhoroshev Theorem* (see [121]). The following result holds (see [92]).

Theorem 5.2.6. *If a Hamiltonian system with N degrees of freedom has $N + r$ independent first integrals f_j for $j = 1,\ldots, N + r$, such that the first integrals $f_1,\ldots, f_{N-r}$ are in involution with all integrals $f_1,\ldots, f_{N+r}$, then the Hamiltonian system is integrable by quadratures.*

Thus, if $f_1,\ldots, f_{N-r}$ are the first integrals which are in involution with all the first integrals and $H = H(f_1,\ldots, f_{N-r})$, then the differential system (5.9) is Hamiltonian and is integrable by quadratures.

5.3 Hamiltonian system with given partial integrals

In this section we prove the following theorem.

Theorem 5.3.1. *Let $(\mathbb{M}, \Omega^2, \mathcal{M}_2, H)$ be a constrained Hamiltonian system and let $g_j : \mathbb{M} \to \mathbb{R}$ for $j = 1,\ldots, M < N$ be given independent functions in $\mathbb{M}$ such that*

$$\mathcal{M}_2 = \{(\mathbf{x}, \mathbf{y}) \in \mathbb{M} : g_j(\mathbf{x}, \mathbf{y}) = 0 \quad for \quad j = 1,\ldots, M < N\}.$$

Choose arbitrary functions g_m for $m = M + 1,\ldots, 2N$ in such a way that the determinant $\{g_1,\ldots, g_M, g_{M+1},\ldots, g_{2N}\} \neq 0$ in $\mathbb{M}$.

Assume that $\{g_1,\ldots, g_M, g_{M+1},\ldots, g_N, x_1,\ldots, x_N\} \neq 0$. Then the submanifold $\mathcal{M}_2$ is a manifold invariant under the flow of the differential system

$$\dot{x}_k = \{H, x_k\}^*,$$

$$\dot{y}_k = \{H, y_k\}^*$$

$$+ \sum_{j=1}^{M} \frac{(\Phi_j - \{H, g_j\}^*)\{g_1,\ldots, g_{j-1}, y_k, g_{j+1}, g_{2M}, g_{2M+1},\ldots, g_N, x_1,\ldots, x_N\}}{\{g_1,\ldots, g_N, x_1,\ldots, x_N\}}$$

$$+ \sum_{j=M+1}^{N} \frac{(\lambda_j - \{H, g_j\}^*)\{g_1,\ldots, g_{2M+1},\ldots, g_{j-1}, y_k, g_{j+1},\ldots g_N, x_1,\ldots, x_N\}}{\{g_1,\ldots, g_N, x_1,\ldots, x_N\}}$$

$$= \{H, y_k\}^* + W_{k+N}, \quad k = 1,\ldots, N, \tag{5.11}$$

where λ_j for $j = M+1,\ldots, N$, and Φ_j are arbitrary functions satisfying $\Phi_j|_{g_j=0} = 0$ for $j = 1,\ldots, M$.

We observe that when the arbitrary functions λ_k are $\lambda_k = \{H, g_k\}^*$ equations (5.11) on the submanifold $\mathcal{M}_2$ become

$$\dot{x}_j = \{H, x_j\}^*, \tag{5.12}$$

$$\dot{y}_j = \{H, y_j\}^* - \sum_{k=1}^{M} \{H, g_k\}^* \frac{\{g_1, \ldots, g_{k-1}, y_j, g_{k+1}, \ldots, g_{N_1}, \ldots, g_N, x_1, \ldots, x_N\}}{\{g_1, \ldots, g_N, x_1, \ldots, x_N\}},$$

for $j = 1, \ldots, N$. This system can be interpreted as the equations of motion of the constrained mechanical system with Hamiltonian H under the action of the external forces with components

$$W_{j+N} = -\sum_{k=1}^{M} \{H, g_k\}^* \frac{\{g_1, \ldots, g_{k-1}, y_j, g_{k+1}, \ldots, g_{N_1}, \ldots, g_N, x_1, \ldots, x_N\}}{\{g_1, \ldots, g_N, x_1, \ldots, x_N\}},$$

generated by the constraints $g_j = 0$, $j = 1, \ldots, M$.

Proof of Theorem 5.3.1. Analogously to the proof of Theorem 1.6.1 from formula (4.7), denoting by $(\partial_1, \ldots, \partial_{2N}) = (\partial_{x_1}, \ldots, \partial_{x_N}, \partial_{y_1}, \ldots, \partial_{y_N})$, and taking the arbitrary functions $\lambda_{N+j} = \{\tilde{H}, x_j\}^*$ for $j = 1, \ldots, N$, where $\tilde{H}$ is the Hamiltonian function, from identity (xi) with $f_j = g_j$, $f_{N+j} = x_j$, $G = y_j$, for $j = 1, \ldots, N$, we obtain the differential system (5.11). This completes the proof of Theorem 5.3.1 □

Example 5.3.2 (Gantmacher's system). Let us illustrate Theorem 5.3.1 for the nonholonomic system studied in Subsection 9.4. Thus we shall study the constrained Hamiltonian system $(\mathbb{R}^8, \Omega^2, \mathcal{M}_2, H)$ with $\mathcal{M}_2 = \{g_1 = x_1 y_1 + x_2 y_2 = 0, g_2 = x_1 y_3 - x_2 y_4 = 0\}$. We choose the arbitrary functions g_j for $j = 3, \ldots, 8$ as follows:

$$g_3 = x_1 y_2 - x_2 y_1, \quad g_4 = x_2 y_3 + x_1 y_4, \quad g_{j+4} = x_j, \quad j = 1, 2, 3, 4.$$

Now we apply Theorem 5.3.1. We have the relations

$$\{g_1, g_2, g_3, g_4, x_1, \ldots, x_4\} = -(x_1^2 + x_2^2)^2,$$
$$\{y_1, g_2, g_3, g_4, x_1, \ldots, x_4\} = -x_1(x_1^2 + x_2^2),$$
$$\{g_1, y_1, g_3, g_4, x_1, \ldots, x_4\} = 0,$$
$$\{g_1, g_2, y_1, g_4, x_1, \ldots, x_4\} = x_2(x_1^2 + x_2^2),$$
$$\{g_1, g_2, g_3, y_1, x_1, \ldots, x_4\} = 0,$$
$$\{g_1, g_2, g_3, g_4, y_1, x_2, x_3, x_4\} = (x_1 y_1 - x_2 y_2)(x_1^2 + x_2^2),$$
$$\{g_1, g_2, g_3, g_4, x_1, y_1, x_3, , x_4\} = (x_1 y_2 + x_2 y_1)(x_1^2 + x_2^2),$$
$$\{g_1, g_2, g_3, g_4, x_1, x_2, y_1, x_4\} = 0,$$
$$\{g_1, g_2, g_3, g_4, x_1, x_2, x_3, y_4\} = 0.$$

In a similar form we can obtain the remaining determinants. Thus system (5.11) takes the form

$$\dot{x}_j = \{\tilde{H}, x_j\}^*, \quad j = 1, 2, 3, 4,$$

$$\dot{y}_1 = \{\tilde{H}, y_1\}^* - \frac{x_1\{H, g_1\}^*}{x_1^2 + x_2^2} - (\lambda_3 - \{H, g_3\}^*)\frac{x_2}{x_1^2 + x_2^2},$$

$$\dot{y}_2 = \{\tilde{H}, y_2\}^* - \frac{x_2\{H, g_1\}^*}{x_1^2 + x_2^2} + (\lambda_3 - \{H, g_3\}^*)\frac{x_1}{x_1^2 + x_2^2},$$

$$\dot{y}_3 = \{\tilde{H}, y_3\}^* - \frac{x_1\{H, g_2\}^*}{x_1^2 + x_2^2} + (\lambda_4 - \{H, g_4\}^*)\frac{x_2}{x_1^2 + x_2^2},$$

$$\dot{y}_4 = \{\tilde{H}, y_4\}^* + \frac{x_2\{H, g_2\}^*}{x_1^2 + x_2^2} + (\lambda_4 - \{H, g_4\}^*)\frac{x_1}{x_1^2 + x_2^2}.$$

In particular, if we take

$$\lambda_3 = \{H, g_3\}^*, \quad \lambda_4 = \{H, g_4\}^*, \quad \text{and} \quad H = \frac{1}{2}\left(y_1^2 + y_2^2 + y_3^2 + y_4^2\right) - gx_3,$$

then (4.58) yields

$$\{H, g_1\}^* = y_1^2 + y_2^2 = -\mu_1(x_1^2 + x_2^2), \quad \{H, g_2\}^* = y_1y_3 - y_2y_4 + gx_1 = -\mu_2(x_1^2 + x_2^2).$$

Consequently, differential equations (5.12) take the form

$$\dot{x}_1 = y_1, \qquad \dot{x}_2 = y_2, \qquad \dot{x}_3 = y_3, \qquad \dot{x}_4 = y_4,$$
$$\dot{y}_1 = x_1\mu_1, \quad \dot{y}_2 = x_2\mu_1, \quad \dot{y}_3 = -g + x_1\mu_2, \quad \dot{y}_4 = -x_2\mu_2,$$

which coincides with the Hamiltonian form of equations (4.64).

Chapter 6

Integrability of the Constrained Rigid Body

6.1 Introduction

The integration theory of the differential equations which describe the motion of a mechanical system with constraints (constrained mechanical system) is not so complete as for mechanical systems without constraints (unconstrained mechanical systems). This can be due to several reasons. One of them is that the equations of motion of a constrained mechanical system in general have no invariant measure, in contrast to the unconstrained case, see for instance [91].

In this chapter we apply the results of Chapter 4 to study the integrability of the equations of motion of a rigid body around a fixed point. In the case where this mechanical system is free of constraints, its integrability is well known (see for instance [4]). In the presence of constraints, however, the theory is incomplete.

We shall study two classical the constrained rigid body problems: the Suslov problem and the Veselova problem. We present new cases of integrability for these two problems, which contain as particular cases previous known results on the integrability of these problems.

We also study the equations of motion of the constrained rigid body with a constraint linear in the velocity without resorting to a Lagrangian multiplier. By using these equations we provide a simple proof of the well-known *Veselova theorem* and improve Kozlov's result on the existence of an invariant measure. We give a new approach to solve the Suslov problem in absence of a force field and of an invariant measure.

6.2 Preliminaries and basic results

First we shall introduce the notations and definitions necessary for presenting our main results of this chapter.

Consider the differential system

$$\dot{\mathbf{x}} = \mathcal{X}(\mathbf{x}), \quad x = (x_1, \ldots, x_N) \in \mathbb{R}^N. \tag{6.1}$$

Let Ω be an open and dense subset of $\mathbb{R}^N$. A non-constant function $\Phi : \Omega \to \mathbb{R}$ such that Φ is constant on the solutions of system (6.1) contained in Ω is called a *first integral*. We say that the system (6.1) is *explicitly integrable* in Ω if it has $\Phi_k : \Omega \to \mathbb{R}$, $k = 1, \ldots, N - 1$, functionally independent first integrals, i.e., the rank of the $(N - 1) \times N$ Jacobian matrix

$$\frac{\partial(\Phi_1, \ldots, \Phi_{N-1})}{\partial(x_1, \ldots, x_N)},$$

is $N - 1$ in all the points $(x_1, \ldots, x_N)$ of Ω, except perhaps in a set of Lebesgue measure zero.

Let Σ be an open subset of $\mathbb{R}^M$ and let $F_j : \Omega \times \Sigma \to \mathbb{R}$, $j = 1, \ldots, M$, be a smooth map. The relation $F_j := F_j(\mathbf{x}, K_1, \ldots, K_M) = 0$, with $K_1, \ldots, K_M$ constants, is called a *general integral* of (6.1) if $\mathcal{X}F_j|_{F_1 = \cdots = F_M = 0} = 0$. The system (6.1) is *implicitly integrable* if it admits $M = N - 1$ general integrals $F_j = 0$, $j = 1, \ldots, N - 1$, such that the rank of the $(N - 1) \times (N - 1)$ Jacobian matrix

$$\frac{\partial(F_1, \ldots, F_{N-1})}{\partial(K_1, \ldots, K_{N-1})} \tag{6.2}$$

is $N - 1$ in all the points $(\mathbf{x}, K_1, \ldots, K_{N-1})$ of $\Omega \times \Sigma$, except perhaps in a set of Lebesgue measure zero. Indeed, under condition (6.2), the Implicit Function Theorem shows that we can extract from the set of $N - 1$ general integrals $N - 1$ local first integrals of the form $K_j = \Phi_j(\mathbf{x})$. Consequently, the system (6.1) is *locally explicitly integrable* in the sense of the previous definition.

The integration theory of the differential equations which describe the motion of mechanical systems with nonintegrable constraints (i.e., nonholonomic systems) is not so complete as for the unconstrained systems (i.e., holonomic systems). This is due to several reasons. One of them is that the equations of motion of nonholonomic systems in general have no invariant measure, in contrast to the case of *holonomic systems*, see for instance [91].

The existence of an invariant measure simplifies the integration of the differential equations. One has the well-known *Euler–Jacobi Theorem*: If the differential system (6.1) has $N - 2$ independent first integrals $\Phi_1, \ldots, \Phi_{N-2}$ and

$$\mathrm{div}\,(M(\mathbf{x})\mathcal{X}(\mathbf{x})) = \sum_{j=1}^{N} \frac{\partial(M(\mathbf{x})\mathcal{X}_j)}{\partial x_j} = 0$$

for some function $M(\mathbf{x}) > 0$, then the differential system is explicitly integrable. We observe that this condition involving the divergence is necessary and sufficient for the existence of an invariant measure with respect to the action of the vector field (6.1) (due to the Euler–Jacobi Theorem, see for instance [68, 91]).

Let $\bar{\Sigma}$ be an open subset of $\mathbb{R}^{N-2}$. System (6.1) is *quasi-implicitly integrable* in $\Omega \times \bar{\Sigma}$ if it has an invariant measure and admits $N-2$ general integrals $F_j(\mathbf{x}, K_1, \ldots, K_{N-2}) = 0$, $j = 1, \ldots, N-2$, such that the rank of $(N-2) \times (N-2)$ Jacobian matrix

$$\frac{\partial (F_1, \ldots, F_{N-2})}{\partial (K_1, \ldots, K_{N-2})}$$

is $N-2$, in all the points $(x_1, \ldots, x_N)$ of Ω, except perhaps in a set of Lebesgue measure zero and for arbitrary constants $K_1, \ldots, K_{N-2}$.

Now we study the integrability theory for the equations of motion of a rigid body around a fixed point. In the case where this mechanical system is free of constraints, its integrability is well known (see for instance [4]). But the integration of the equations of motion in the presence of constraints is incomplete. For example, the integrability is in general unknown when the constraint is of the form

$$\langle \nu, \omega \rangle = 0, \tag{6.3}$$

where $\nu = \nu(\gamma) = (\nu_1, \nu_2, \nu_3)$ is a vector of $\mathbb{R}^3$, γ is the unit vector of a spatially fixed axis in the coordinate system rigidly attached to the body and such that

$$\gamma = (\gamma_1, \gamma_2, \gamma_3) = (\sin z \sin x, \sin z \cos x, \cos z), \tag{6.4}$$

$(x, y, z) = (\varphi, \psi, \theta)$ are the Euler angles, and $\omega = (\omega_1, \omega_2, \omega_3)$ is the angular velocity.

Applying the method of Lagrange multipliers, we write the equations of motion of the rigid body around a fixed point with the constraint (6.3) as (for more details see for instance [4, 15])

$$I\dot{\omega} = I\omega \wedge \omega + \gamma \wedge \frac{\partial U}{\partial \gamma} + \mu \nu, \qquad \dot{\gamma} = \gamma \wedge \omega, \qquad \langle \nu, \omega \rangle = 0, \tag{6.5}$$

where I is the inertial tensor of the rigid body, i.e.,

$$I = \begin{pmatrix} I_{11} & I_{12} & I_{13} \\ I_{12} & I_{22} & I_{23} \\ I_{13} & I_{23} & I_{33} \end{pmatrix},$$

$U = U(\gamma_1, \gamma_2, \gamma_3)$ is the potential function, $\partial U(\gamma)/\partial \gamma$ is the gradient of $U(\gamma)$ with respect to γ, and $\wedge$ is the "wedge" product in $\mathbb{R}^3$.

The equations $\dot{\gamma} = \gamma \wedge \omega$ are known as the *Poisson differential equations*.

The system (6.5) always has three independent first integrals, namely

$$\Phi_1 = \langle \gamma, \gamma \rangle = \gamma_1^2 + \gamma_2^2 + \gamma_3^2, \quad \Phi_2 = \langle \nu, \omega \rangle, \quad \Phi_3 = \frac{1}{2}\langle I\omega, \omega \rangle + U(\gamma). \tag{6.6}$$

We shall study two particular cases with the constraint (6.3): the Suslov problem ($\nu = \mathbf{a}$) and the *Veselova problem* ($\nu = \gamma$). The main objective of this chapter is to present new cases of integrability for these two problems which contain as particular cases the previous results on integrability.

The Suslov problem (see Example 4.3.1) concerns the motion of a rigid body around a fixed point and subject to the nonholonomic constraint $\langle \mathbf{a}, \omega \rangle = 0$, where $\mathbf{a}$ is a constant vector (see [147]). Suppose that the body rotates in a force field with potential $U(\gamma) = U(\gamma_1, \gamma_2, \gamma_3)$. Applying the method of Lagrange multipliers, the equations of motion read

$$I\dot{\omega} = I\omega \wedge \omega + \gamma \wedge \frac{\partial U}{\partial \gamma} + \mu \mathbf{a}, \qquad \dot{\gamma} = \gamma \wedge \omega, \qquad \langle \mathbf{a}, \omega \rangle = 0, \qquad (6.7)$$

The system (6.7) has the three independent first integrals (6.6) with $\nu = \mathbf{a}$.

In order to have real motions we must take $\Phi_1 = 1$, $\Phi_2 = 0$ in (6.6). In this case using the first integrals Φ_3 we can reduce the problem of integration of (6.7) to the problem of existence of an invariant measure and a fourth independent first integral Φ_4. Under these assumptions, by the Euler–Jacobi Theorem (see for instance [68, 16]), the Suslov problem is integrable [90]. In general the system (6.7) has no invariant measure if the vector $\mathbf{a}$ is not an *eigenvector of the tensor of inertia*. The following result is well known, see [91].

Proposition 6.2.1. *If $\mathbf{a}$ is an eigenvector of the tensor of inertia I, i.e.,*

$$I\mathbf{a} = \kappa \mathbf{a} \qquad (6.8)$$

for some $\kappa \in \mathbb{R}$, then the flow of the system (6.7) preserves the Lebesgue measure in $\mathbb{R}^6 = \mathbb{R}^3\{\omega\} \times \mathbb{R}^3\{\gamma\}$.

We show in Section 6.5 that (6.8) is a necessary and sufficient condition for the existence of an invariant measure in the Suslov problem.

Suslov in [147] has considered the case when there are no *external forces*, i.e., $U = 0$. In this case the system

$$I\dot{\omega} = I\omega \wedge \omega + \mu \mathbf{a},$$

can be solved with respect to ω, i.e., it is integrable by quadratures. The analysis of theses quadratures shows that if (6.8) does not hold, then all trajectories $\omega = \omega(t)$ *approach asymptotically* as $t \to \pm\infty$ some fixed straight line in the plane $\langle \mathbf{a}, \omega \rangle = 0$. Consequently, equations (6.7) have no invariant measure. The question about the possibility of finding $\gamma = \gamma(t)$ by quadratures remains open in general (for more details, see Section 6.5 and [55]).

If (6.8) holds, then along the solutions of (6.7) the kinetic moment $\langle I\omega, I\omega \rangle$ is a first integral (see [91]).

From now on we assume that equality (6.8) is fulfilled.

Without loss of generality we can choose the vector $\mathbf{a}$ as the third axis, i.e., $\mathbf{a} = (0, 0, 1)$ and consequently the constraint becomes $\omega_3 = \dot{x} + \dot{y}\cos z = 0$. Then, the equations of motion are (for more details, see Section 6.5)

$$I_1\dot{\omega}_1 = \gamma_2\frac{\partial U}{\partial\gamma_3} - \gamma_3\frac{\partial U}{\partial\gamma_2}, \quad I_2\dot{\omega}_2 = \gamma_3\frac{\partial U}{\partial\gamma_1} - \gamma_1\frac{\partial U}{\partial\gamma_2},$$

$$\dot{\gamma}_1 = -\gamma_3\omega_2, \quad \dot{\gamma}_2 = \gamma_3\omega_1, \quad \dot{\gamma}_3 = \gamma_1\omega_2 - \gamma_2\omega_1,$$

$$(6.9)$$

where $I = \mathrm{diag}(I_1, I_2, I_3)$, and I_k are the principal moments of inertia of the body.

Kharlamova-Zabelina in [80] studied the case when the body rotates in the *homogenous force field* with the potential $U = \langle \mathbf{b}, \gamma \rangle$, where the vector $\mathbf{b}$ is orthogonal to the vector $\mathbf{a}$. Under these conditions the equations of motion have the first integral $\Phi_4 = \langle I\omega, \mathbf{b} \rangle$.

Kozlov considered the case when $\mathbf{b} = \lambda\mathbf{a}$, $\lambda \neq 0$. The integrability problem in this case was studied in [91, 114]. If $I_1 \neq I_2$, apparently the equations have no additional first integral independent of the energy integral. When $I_1 = I_2$ and $U = \lambda\langle \mathbf{a}, \gamma_3 \rangle$ there exists the fourth integral $\Phi_4 = \omega_1\gamma_1 + \omega_2\gamma_2$.

In the case when $U = \lambda|I|\langle I^{-1}\gamma, \gamma \rangle$, where $|I| = \det I$, system (6.7) has the Clebsch–Tisserand first integral $\Phi_4 = \frac{1}{2}\langle I\omega, I\omega \rangle - \frac{1}{2}\lambda|I|\langle I^{-1}\gamma, \gamma \rangle$ (see for instance [91]).

Okuneva in [124] proved the integrability of the Suslov problem for the potential $U = \alpha\gamma_1 + \beta\gamma_2 + \frac{\lambda}{2}\langle I^{-1}\gamma, \gamma \rangle$, where α, β and λ are constants. The first integral is $\Phi_4 = I_1\omega_1^2 - \lambda(I_2 - I_3)\gamma_2^2 - 2\beta\gamma_2$, or equivalently, $\Phi_4 = I_2\omega_2^2 - \lambda(I_3 - I_1)\gamma_1^2 - 2\alpha\gamma_1$.

Dragovic et al. in [47] considered the case when the potential is $U = c(\gamma_1, \gamma_2^2 + \gamma_3^2) - d(\gamma_2, \gamma_1^2 + \gamma_3^2)$ with arbitrary functions $c = c(\gamma_1, \gamma_2^2 + \gamma_3^2)$ and $d = d(\gamma_2, \gamma_1^2 + \gamma_3^2)$, proved that $\Phi_4 = \frac{1}{2}\langle I\omega, I\omega \rangle + I_2 c(\gamma_1, \gamma_2^2 + \gamma_3^2) - I_1 d(\gamma_2, \gamma_1^2 + \gamma_3^2)$ is a first integral of system (6.7).

The Veselova problem (see Example 4.3.2) describes the motion of a rigid body which rotates around a fixed point and is subject to the nonholonomic constraint

$$\langle \gamma, \omega \rangle = \dot{y} + \dot{x}\cos z = 0.$$

Thus, in the case of the Veselova constraint the projection of the angular velocity to a spatially fixed axis is zero.

Suppose that the body rotates in a force field with potential $U(\gamma_1, \gamma_2, \gamma_3)$. Applying the method of Lagrange multipliers we write the equations of motion in the form

$$I\dot{\omega} = I\omega \wedge \omega + \gamma \wedge \frac{\partial U}{\partial\gamma} + \mu\gamma, \quad \dot{\gamma} = \gamma \wedge \omega, \quad \langle \gamma, \omega \rangle = 0, \quad (6.10)$$

where $I = \mathrm{diag}(I_1, I_2, I_3)$. System (6.10) has always three independent integrals (6.6) with $\nu = \gamma$.

As proved in [155, 156], the system (6.10) has an invariant measure with density $\sqrt{\dfrac{\gamma_1^2}{I_1} + \dfrac{\gamma_2^2}{I_2} + \dfrac{\gamma_3^2}{I_3}}$ (for more details see Section 6.5). Thus, the Euler–Jacobi Theorem shows that if there exists a fourth first integral Φ_4 independent with Φ_1, Φ_2, Φ_3, then the Veselova problem is integrable. In order to have real motions we must take $\Phi_1 = 1$ and $\Phi_2 = 0$.

Remark 6.2.2. From (6.10) it follows that

$$
\begin{aligned}
\frac{d}{dt}\left(\gamma \wedge I\omega\right) &= \frac{d\gamma}{dt} \wedge I\omega + \gamma \wedge I\frac{d\omega}{dt} \\
&= (\gamma \wedge \omega) \wedge I\omega + \gamma \wedge \left(I\omega \wedge \omega + \gamma \wedge \frac{\partial U}{\partial \gamma} + \lambda\gamma\right) \\
&= (\gamma \wedge \omega) \wedge I\omega + (\omega \wedge I\omega) \wedge \gamma + \gamma \wedge \left(\gamma \wedge \frac{\partial U}{\partial \gamma}\right),
\end{aligned}
$$

and by considering the identities

$$
a \wedge (b \wedge c) = \langle a, c\rangle\, b - \langle a, b\rangle\, c, \quad a \wedge (b \wedge c) + b \wedge (c \wedge a) = -c \wedge (a \wedge b),
$$

we obtain

$$
\frac{d}{dt}\left(\gamma \wedge I\omega\right) = -\frac{\partial U}{\partial \gamma} + \gamma\left(\left\langle \gamma, \frac{\partial U}{\partial \gamma}\right\rangle - \langle I\omega, \omega\rangle\right). \tag{6.11}
$$

From this relation we deduce the equation

$$
\frac{d(p\,\omega)}{dt} = \frac{1}{p}\left(\gamma \wedge I\frac{\partial U}{\partial \gamma} + I\gamma \wedge \gamma\left(\left\langle \gamma, \frac{\partial U}{\partial \gamma}\right\rangle - \langle I\omega, \omega\rangle\right)\right), \tag{6.12}
$$

where $p = \sqrt{I_1 I_2 I_3\left(\dfrac{\gamma_1^2}{I_1} + \dfrac{\gamma_2^2}{I_2} + \dfrac{\gamma_3^2}{I_3}\right)}$.

6.3   Integrability of the Suslov problem

Our first main result is the following.

Theorem 6.3.1. *The motion of the rigid body in the Suslov problem under assumption* (6.8) *is described by the system* (6.9). *We assume that the rigid body rotates under the action of the force field defined by the potential*

$$
U = -\frac{1}{2I_1 I_2}(I_1\mu_1^2 + I_2\mu_2^2), \tag{6.13}
$$

where $\mu_1 = \mu_1(\gamma_1, \gamma_2, \gamma_3, K_3, K_4)$ and $\mu_2 = \mu_2(\gamma_1, \gamma_2, \gamma_3, K_3, K_4)$ with K_3 and K_4 constants. Let μ_1 and μ_2 be solutions of the first-order partial differential equation

$$\gamma_3 \left(\frac{\partial \mu_1}{\partial \gamma_2} - \frac{\partial \mu_2}{\partial \gamma_1} \right) - \gamma_2 \frac{\partial \mu_1}{\partial \gamma_3} + \gamma_1 \frac{\partial \mu_2}{\partial \gamma_3} = 0, \tag{6.14}$$

satisfying

$$\frac{\partial \mu_1}{\partial K_3} \frac{\partial \mu_2}{\partial K_4} - \frac{\partial \mu_2}{\partial K_3} \frac{\partial \mu_1}{\partial K_4} \neq 0 \quad \text{for all } (\gamma_1, \gamma_2, \gamma_3) \in \mathbb{R}^3. \tag{6.15}$$

Then the following statements hold.

(a) System (6.9) has the general integrals

$$\begin{aligned} F_1 &= I_1 \omega_1 - \mu_2(\gamma_1, \gamma_2, \gamma_3, K_3, K_4) = 0, \\ F_2 &= I_2 \omega_2 + \mu_1(\gamma_1, \gamma_2, \gamma_3, K_3, K_4) = 0. \end{aligned} \tag{6.16}$$

Moreover system (6.9) is quasi-implicitly integrable.

(b) Suslov, Kharlamova-Zabelina, Kozlov, Dragović–Gajić–Jovanović, Clebsch–Tisserand, Tisserand–Okuneva first integrals are particular cases of statement (a).

(c) Using (6.16), the Poisson equations for the Suslov problem take the form

$$\dot{\gamma}_1 = -\gamma_3 \frac{\mu_1}{I_2}, \quad \dot{\gamma}_2 = -\gamma_3 \frac{\mu_2}{I_1}, \quad \dot{\gamma}_3 = -\gamma_1 \frac{\mu_1}{I_2} - \gamma_2 \frac{\mu_2}{I_1}. \tag{6.17}$$

We provide the solution $\gamma(t) = (\gamma_1(t), \gamma_2(t), \gamma_3(t))$ for all the cases of statement (b).

Remark 6.3.2. It is easy to check that the functions

$$\begin{aligned} \mu_1 &= \frac{\partial \tilde{S}(\gamma_1, \gamma_2, \gamma_3, K_3, K_4)}{\partial \gamma_1} + \Psi_1(\gamma_2^2 + \gamma_3^2, \gamma_1, K_3, K_4), \\ \mu_2 &= \frac{\partial \tilde{S}(\gamma_1, \gamma_2, \gamma_3, K_3, K_4)}{\partial \gamma_2} + \Psi_2(\gamma_1^2 + \gamma_3^2, \gamma_2, K_3, K_4), \end{aligned} \tag{6.18}$$

are solutions of (6.14), where

$$\tilde{S}(\gamma_1, \gamma_2, \gamma_3, K_3, K_4) = S(\gamma_1, \gamma_2, K_3, K_4) + \int \Upsilon(\gamma_1^2 + \gamma_2^2, \gamma_3, K_3, K_4) d(\gamma_1^2 + \gamma_2^2),$$

and S, Ψ_1, Ψ_2, and Υ are arbitrary smooth functions for which (6.15) holds.

Remark 6.3.2 will be used for proving statement (b) above.

Proof of Theorem 6.3.1. After some calculations we obtain that the derivative of F_1 along the solutions of (6.9) is given by

$$\dot{F_1} = I_1 \dot{\omega}_1 - \dot{\mu}_2$$

$$= \gamma_2 \frac{\partial U}{\partial \gamma_3} - \gamma_3 \frac{\partial U}{\partial \gamma_2} + \frac{\partial \mu_2}{\partial \gamma_1} \gamma_3 \omega_2 - \frac{\partial \mu_2}{\partial \gamma_2} \gamma_3 \omega_1 - \frac{\partial \mu_2}{\partial \gamma_3} (\gamma_1 \omega_2 - \gamma_2 \omega_1)$$

$$- \frac{\partial \mu_2}{\partial K_1} \dot{K}_1 - \frac{\partial \mu_2}{\partial K_2} \dot{K}_2$$

$$= \gamma_2 \frac{\partial U}{\partial \gamma_3} - \gamma_3 \frac{\partial U}{\partial \gamma_2} + \omega_2 \left(\gamma_3 \frac{\partial \mu_2}{\partial \gamma_1} - \gamma_1 \frac{\partial \mu_2}{\partial \gamma_3} \right) + \omega_1 \left(\gamma_2 \frac{\partial \mu_2}{\partial \gamma_3} - \gamma_3 \frac{\partial \mu_2}{\partial \gamma_2} \right)$$

$$= \gamma_2 \frac{\partial U}{\partial \gamma_3} - \gamma_3 \frac{\partial U}{\partial \gamma_2} + \frac{F_2 - \mu_1}{I_2} \left(\gamma_3 \frac{\partial \mu_2}{\partial \gamma_1} - \gamma_1 \frac{\partial \mu_2}{\partial \gamma_3} \right)$$

$$+ \frac{F_1 + \mu_2}{I_1} \left(\gamma_2 \frac{\partial \mu_2}{\partial \gamma_3} - \gamma_3 \frac{\partial \mu_2}{\partial \gamma_2} \right)$$

$$= \gamma_2 \frac{\partial}{\partial \gamma_3} \left(U + \frac{1}{2 I_1 I_2} (I_1 \mu_1^2 + I_2 \mu_2^2) \right) - \gamma_3 \frac{\partial}{\partial \gamma_2} \left(U + \frac{1}{2 I_1 I_2} (I_1 \mu_1^2 + I_2 \mu_2^2) \right)$$

$$+ \frac{\mu_1}{I_2} \left(\gamma_3 \left(\frac{\partial \mu_1}{\partial \gamma_2} - \frac{\partial \mu_2}{\partial \gamma_1} \right) - \gamma_2 \frac{\partial \mu_1}{\partial \gamma_3} + \gamma_1 \frac{\partial \mu_2}{\partial \gamma_3} \right)$$

$$- \frac{F_2}{I_2} \left(\gamma_1 \frac{\partial \mu_2}{\partial \gamma_3} - \gamma_3 \frac{\partial \mu_2}{\partial \gamma_1} \right) - \frac{F_1}{I_1} \left(\gamma_3 \frac{\partial \mu_2}{\partial \gamma_2} - \gamma_2 \frac{\partial \mu_2}{\partial \gamma_3} \right),$$

where we used the fact that $\dot{K}_1 = \dot{K}_2 = 0$.

In view of (6.13) and (6.14) we obtain

$$\dot{F_1} = \frac{F_2}{I_2} \left(\gamma_1 \frac{\partial \mu_2}{\partial \gamma_3} - \gamma_3 \frac{\partial \mu_2}{\partial \gamma_1} \right) + \frac{F_1}{I_1} \left(\gamma_3 \frac{\partial \mu_2}{\partial \gamma_2} - \gamma_2 \frac{\partial \mu_2}{\partial \gamma_3} \right).$$

A similar relation can obtained for $\dot{F_2}$. Hence, by considering (6.16) we deduce that $\dot{F_j}|_{F_1 = F_2 = 0} = 0$ for $j = 1, 2$. Therefore, $F_1 = 0$ and $F_2 = 0$ are two general integrals. This yields two independent local first integrals

$$\Phi_1(\omega_1, \omega_2, \gamma_1, \gamma_2, \gamma_3) = K_3, \quad \Phi_4(\omega_1, \omega_2, \gamma_1, \gamma_2, \gamma_3) = K_4.$$

Thus system (6.9) is locally explicitly integrable. We have a third general integral

$$F_3(\gamma_1, \gamma_2, \gamma_3, K_3)|_{K_3 = 1} = \gamma_1^2 + \gamma_2^2 + \gamma_3^2 - 1 = 0.$$

On the other hand, system (6.9) has divergence zero due to the fact that its flow preserves the Lebesgue measure, see Proposition 6.2.1.

In short, applying the Euler–Jacobi Theorem it follows that system (6.9) is quasi-implicitly integrable, so statement (a) is proved.

In view of Remark 6.3.2, we consider first the functions

$$\mu_1 = \frac{\partial \tilde{S}(\gamma_1, \gamma_2, \gamma_3, K_1, K_4)}{\partial \gamma_1} = \frac{\partial \tilde{S}}{\partial \gamma_1}, \quad \mu_2 = \frac{\partial \tilde{S}(\gamma_1, \gamma_2, \gamma_3, K_1, K_4)}{\partial \gamma_2} = \frac{\partial \tilde{S}}{\partial \gamma_2}.$$

Then the equations (6.16) become

$$F_1 = I_1 \omega_1 - \frac{\partial \tilde{S}}{\partial \gamma_2} = 0, \quad F_2 = I_2 \omega_2 + \frac{\partial \tilde{S}}{\partial \gamma_1} = 0. \tag{6.19}$$

Now we show that Suslov's, Kharlamova-Zabelina's and Kozlov's first integrals can be obtained from (6.19).

For the *Suslov integrable case* (U =constant, and $I_1 \neq I_2$) we have that $\tilde{S} = C_1 \gamma_1 + C_2 \gamma_2$, where $C_1 = C_1(K_3, K_4)$ and $C_2 = C_2(K_3, K_4)$ are arbitrary constants. Since $\mu_1 = C_1$ and $\mu_2 = C_2$, taking

$$C_1 = \sqrt{\frac{I_2 (I_1 K_3 - K_4)}{I_1 - I_2}}, \quad C_2 = \sqrt{\frac{I_1 (K_4 - I_2 K_3)}{I_1 - I_2}},$$

(6.15) holds. Equations (6.16) read

$$F_1 = I_1 \omega_1 - C_2 = 0, \quad F_2 = I_2 \omega_2 + C_1 = 0.$$

By solving this system with respect to K_3 and K_4 we obtain

$$K_3 = I_1 \omega_1^2 + I_2 \omega_2^2 = \frac{C_2^2}{I_2} + \frac{C_1^2}{I_1}, \quad K_4 = I_1^2 \omega_1^2 + I_2^2 \omega_2^2 = C_2^2 + C_1^2.$$

Note that K_3 is the energy first integral, and K_4 is the kinetic moment (Suslov's first integral).

For the *Kharlamova-Zabelina integrable case* ($U = \langle \mathbf{b}, \gamma \rangle$) we take the function $\tilde{S}$ of the Remark 6.3.2 as

$$\tilde{S} = \frac{2/3}{\sqrt{I_1 b_1^2 + I_2 b_2^2}} (\tilde{h} + b_1 \gamma_1 + b_2 \gamma_2)^{3/2} - \frac{K_4}{b_1^2 I_1 + b_2^2 I_2} (b_2 I_2 \gamma_1 - b_1 I_1 \gamma_2),$$

where $\tilde{h} = I_1 I_2 \left(\frac{K_4^2 I_1 I_2}{b_1^2 I_1 + b_2^2 I_2} - K_3 \right)$, K_3 and K_4 are arbitrary constants. Then

$$\mu_1 = \frac{b_1}{\sqrt{I_1 b_1^2 + I_2 b_2^2}} \sqrt{\tilde{h} + b_1 \gamma_1 + b_2 \gamma_2} - \frac{K_4 b_2 I_2}{b_1^2 I_1 + b_2^2 I_2},$$

$$\mu_2 = \frac{b_2}{\sqrt{I_1 b_1^2 + I_2 b_2^2}} \sqrt{\tilde{h} + b_1 \gamma_1 + b_2 \gamma_2} + \frac{K_4 b_1 I_1}{b_1^2 I_1 + b_2^2 I_2}.$$

Therefore equations (6.16) take the form

$$F_1 = I_1\omega_1 - \left(\frac{b_2}{\sqrt{I_1 b_1^2 + I_2 b_2^2}}\sqrt{\tilde{h} + b_1\gamma_1 + b_2\gamma_2} + \frac{K_4 b_1 I_1}{b_1^2 I_1 + b_2^2 I_2}\right) = 0,$$

$$F_2 = I_2\omega_2 + \left(\frac{b_1}{\sqrt{I_1 b_1^2 + I_2 b_2^2}}\sqrt{\tilde{h} + b_1\gamma_1 + b_2\gamma_2} - \frac{K_4 b_2 I_2}{b_1^2 I_1 + b_2^2 I_2}\right) = 0.$$

Solving this system with respect to K_3 and K_4 we obtain

$$K_3 = I_1\omega_1^2 + I_2\omega_2^2 - \frac{1}{I_1 I_2}(b_1\gamma_1 + b_2\gamma_2), \quad K_4 = I_1\omega_1 b_1 + I_2\omega_2 b_2.$$

Again K_3 is the energy integral and K_4 is the well-known *Kharlamova-Zabelina first integral* [80].

For the *Kozlov integrable case* ($U = a\gamma_3$ and $I_1 = I_2$) we take

$$\tilde{S} = -K_4 \arctan\left(\frac{\gamma_1}{\gamma_2}\right) + \frac{1}{2}\int D(\gamma_1^2 + \gamma_2^2) d(\gamma_1^2 + \gamma_2^2),$$

where

$$D(u) = I_1 \sqrt{\frac{K_3 + a\sqrt{1-u}}{u} - \frac{K_4^2}{u^2}},$$

and a, K_3, and K_4 are constants. Hence

$$\mu_1 = -\frac{\gamma_2 K_4}{\gamma_1^2 + \gamma_2^2} + \gamma_1 D(\gamma_1^2 + \gamma_2^2), \quad \mu_2 = \frac{\gamma_1 K_4}{\gamma_1^2 + \gamma_2^2} + \gamma_2 D(\gamma_1^2 + \gamma_2^2).$$

Consequently, the functions F_1 and F_2 in (6.16) are

$$F_1 = \omega_1 - \left(\frac{\gamma_1 K_4}{\gamma_1^2 + \gamma_2^2} + \gamma_2 D(\gamma_1^2 + \gamma_2^2)\right) = 0,$$

$$F_2 = \omega_2 + \left(-\frac{\gamma_2 K_4}{\gamma_1^2 + \gamma_2^2} + \gamma_1 D(\gamma_1^2 + \gamma_2^2)\right) = 0.$$

Therefore, by solving this system with respect to K_3 and K_4 we obtain

$$K_3 = \omega_1^2 + \omega_2^2 - a\sqrt{1 - \gamma_1^2 - \gamma_2^2} = \omega_1^2 + \omega_2^2 - a\gamma_3, \quad K_4 = \omega_1\gamma_1 + \omega_2\gamma_2.$$

So K_3 is the energy integral and K_4 is the *Kozlov–Lagrange first integral*. In fact, this integral correspond to the well-known integrable "Lagrange case" of the Suslov problem [90].

Finally, we analyze the case when the functions μ_1 and μ_2 of (6.18) are given by the formulas

$$\mu_1 = \Psi_1(\gamma_1^2 + \gamma_3^2, \gamma_1, K_3, K_4), \quad \mu_2 = \Psi_2(\gamma_1^2 + \gamma_3^2, \gamma_2, K_3, K_4).$$

We observe that these solutions, obtained by integrating the system

$$\gamma_3\frac{\partial\mu_1}{\partial\gamma_2} - \gamma_2\frac{\partial\mu_1}{\partial\gamma_3} = 0, \quad \gamma_3\frac{\partial\mu_2}{\partial\gamma_1} - \gamma_1\frac{\partial\mu_2}{\partial\gamma_3} = 0,$$

are a particular case of equations (6.14). The potential function (6.13) in this case coincides with the potential obtained by Dragović, Gajić, and Jovanović in [47]. We call this case the *generalized Tisserand case*. In particular, if $I_1 \neq I_2$ and

$$\mu_1 = \sqrt{h_1 + (a_1 + a_3)(\gamma_3^2 + \gamma_2^2) + (b_1 + a_3)\gamma_1^2 + f_1(\gamma_1)},$$

$$\mu_2 = \sqrt{h_2 + (a_2 + a_4)(\gamma_3^2 + \gamma_1^2) + (b_2 + a_4)\gamma_2^2 + f_2(\gamma_2)},$$

where a_1, a_2, a_3, a_4, b_1, b_2 and

$$h_1 = \frac{I_2(I_1 K_3 - K_4)}{I_1 - I_2}, \quad h_2 = \frac{I_1(I_2 K_3 - K_4)}{I_1 - I_2},$$

are constants, and f_1 and f_2 are arbitrary functions, then the general integrals $F_1 = 0$ and $F_2 = 0$ take the form

$$F_1 = I_1\omega_1 - \sqrt{h_2 + (a_2 + a_4)(\gamma_3^2 + \gamma_1^2) + (b_2 + a_4)\gamma_2^2 + f_2(\gamma_2)} = 0,$$

$$F_2 = I_2\omega_2 + \sqrt{h_1 + (a_1 + a_3)(\gamma_3^2 + \gamma_2^2) + (b_1 + a_3)\gamma_1^2 + f_1(\gamma_1)} = 0.$$

The case when $f_j(\gamma_j) = \alpha_j\,\gamma_j$, $j = 1, 2$, where α_1 and α_2 are constants was studied by Okuneva in [124].

If $f_1 = f_2 = 0$, we obtain the *Tisserand case* [91]. By solving $F_j = 0$ for $j = 1, 2$ with respect to K_3 and K_4 we get that the first integrals in the *Clebsch–Tisserand case* are

$$K_3 = I_1\omega_1^2 + I_2\omega_2^2 - \left(\frac{b_1 + a_3}{I_2} + \frac{a_2 + a_4}{I_1}\right)\gamma_1^2$$

$$- \left(\frac{a_1 + a_3}{I_2} + \frac{b_2 + a_4}{I_1}\right)\gamma_2^2 - \left(\frac{a_1 + a_3}{I_2} + \frac{a_2 + a_4}{I_1}\right)\gamma_3^2,$$

$$K_4 = I_1^2\omega_1^2 + I_2^2\omega_2^2 - (b_1 + a_3 + a_2 + a_4)\,\gamma_1^2$$

$$- (a_1 + a_3 + b_2 + a_4)\,\gamma_2^2 - (a_1 + a_3 + a_2 + a_4)\,\gamma_3^2.$$

Statement (b) is proved.

Now we prove (c). First we introduce the following well-known definition. An elliptic integral is any integral which can be expressed in the form

$$\int R(x, \sqrt{P(x)})dx,$$

where R is a rational function of its two arguments, P is a polynomial of degree 3 or 4 with no repeated roots. In general, *elliptic integrals* cannot be expressed in terms of elementary functions. Exceptions to this general rule are when P has repeated roots, or when $R(x,y)$ contains no odd powers of y. However, with the appropriate reduction formula, every elliptic integral can be brought into a form that involves integrals of rational functions and the three Legendre canonical forms (i.e., the *elliptic integrals of the first, second, and third kind*).

We observe that from the general integrals (6.16) it follows that

$$\omega_1 = \frac{\mu_2}{I_1}, \quad \omega_2 = -\frac{\mu_1}{I_2},$$

thus by inserting in the Poisson equation (see formula (6.9)) we easily obtain (6.17).

Now we deal with the case $\mu_j = \dfrac{\partial \tilde{S}\left(\gamma_1, \gamma_2, \gamma_3, K_3, K_4\right)}{\partial \gamma_j}$ for $j = 1, 2$. Here equations (6.17) become

$$\dot{\gamma}_1 = -\frac{\gamma_3}{I_2}\frac{\partial \tilde{S}}{\partial \gamma_1}, \quad \dot{\gamma}_2 = -\frac{\gamma_3}{I_1}\frac{\partial \tilde{S}}{\partial \gamma_2}, \quad \dot{\gamma}_3 = \frac{\gamma_1}{I_2}\frac{\partial \tilde{S}}{\partial \gamma_1} + \frac{\gamma_2}{I_1}\frac{\partial \tilde{S}}{\partial \gamma_2}. \tag{6.20}$$

The vector $\gamma = \gamma(t)$ is determined by integrating system (6.20).

For the Suslov case the differential system (6.20) takes the form

$$\dot{\gamma}_1 = -\frac{\gamma_3 C_1}{I_2}, \quad \dot{\gamma}_2 = -\frac{\gamma_3 C_2}{I_1}, \quad \dot{\gamma}_3 = \frac{\gamma_1 C_1}{I_2} + \frac{\gamma_2 C_2}{I_1}.$$

After integration and taking into account that $\gamma_1^2 + \gamma_2^2 + \gamma_3^2 = 1$ we deduce that

$$\gamma_1(t) = \frac{C_1 I_1}{\sqrt{I_1^2 C_1^2 + I_2^2 C_2^2}} \sin\beta \sin\left(\frac{\sqrt{I_1^2 C_1^2 + I_2^2 C_2^2}}{I_1 I_2}t + \alpha\right) + \frac{I_2 C_2 \cos\beta}{\sqrt{I_1^2 C_1^2 + I_2^2 C_2^2}},$$

$$\gamma_2(t) = \frac{C_2 I_2}{\sqrt{I_1^2 C_1^2 + I_2^2 C_2^2}} \sin\beta \sin\left(\frac{\sqrt{I_1^2 C_1^2 + I_2^2 C_2^2}}{I_1 I_2}t + \alpha\right) - \frac{I_1 C_1 \cos\beta}{\sqrt{I_1^2 C_1^2 + I_2^2 C_2^2}},$$

$$\gamma_3(t) = \sin\beta \cos\left(\sqrt{\frac{I_1^2 C_1^2 + I_2^2 C_2^2}{I_1 I_2}}t + \alpha\right),$$

where α, β, C_1, C_2 are constants.

For the Kharlamova-Zabelina case the differential system (6.17) is

$$\dot{\gamma}_1 = \frac{\gamma_3}{I_2}\left(\frac{C_1}{\sqrt{I_1 C_1^2 + I_2 C_2^2}}\sqrt{\tilde{h} + C_1\gamma_1 + C_2\gamma_2} - \frac{K_4 C_2 I_2}{C_1^2 I_1 + C_2^2 I_2}\right),$$

$$\dot{\gamma}_2 = \frac{\gamma_3}{I_1} \left(\frac{C_2}{\sqrt{I_1 C_1^2 + I_2 C_2^2}} \sqrt{\tilde{h} + C_1 \gamma_1 + C_2 \gamma_2} + \frac{K_4 C_1 I_1}{C_1^2 I_1 + C_2^2 I_2} \right),$$

$$\dot{\gamma}_3 = -\frac{\gamma_1}{I_2} \left(\frac{C_1}{\sqrt{I_1 C_1^2 + I_2 C_2^2}} \sqrt{\tilde{h} + C_1 \gamma_1 + C_2 \gamma_2} - \frac{K_4 C_2 I_2}{C_1^2 I_1 + C_2^2 I_2} \right)$$

$$- \frac{\gamma_2}{I_1} \left(\frac{C_2}{\sqrt{I_1 C_1^2 + I_2 C_2^2}} \sqrt{\tilde{h} + C_1 \gamma_1 + C_2 \gamma_2} + \frac{K_4 C_1 I_1}{C_1^2 I_1 + C_2^2 I_2} \right),$$

and its solutions are

$$\gamma_1 = \frac{C_1}{I_2}(\tau - \tau_0)^2 - \frac{K_4}{C_1^2 I_1 + C_2^2 I_2}(\tau - \tau_1),$$

$$\gamma_2 = \frac{C_2}{I_1}(\tau - \tau_0)^2 - \frac{K_4}{C_1^2 I_1 + C_2^2 I_2}(\tau - \tau_2),$$

$$\gamma_3 = \sqrt{1 - \gamma_1^2 - \gamma_2^2} := \sqrt{P_4(\tau, \tau_0, \tau_1, \tau_2, K_4)},$$

$$t = t_0 + I_1 I_2 \int \frac{d\tau}{\sqrt{P_4(\tau, \tau_0, \tau_1, \tau_2, K_4)}},$$

where τ_0, τ_1, τ_2, K_4 are constants. Note that P_4 is a polynomial of degree four in the variable τ. Clearly, the equations of motion are integrable in elliptic functions of time.

The differential system (6.20) in the Kozlov case takes the form

$$\dot{\gamma}_1 = \gamma_3 \left(\gamma_1 D - \frac{\gamma_2 - K_4}{1 - \gamma_3^2} \right), \quad \dot{\gamma}_2 = \gamma_3 \left(\gamma_2 D + \frac{\gamma_1 - K_4}{1 - \gamma_3^2} \right), \quad \dot{\gamma}_3 = -(1 - \gamma_3^2)D,$$

where $D = D(1 - \gamma_3^2) = D(\sin^2 z)$, and K_4 is an arbitrary constant. Hence, by considering the constraint $\omega_3 = \dot{x} + \cos z \dot{y} = 0$ we deduce the differential equations

$$\dot{x} = -\frac{K_4 \cos z}{\sin^2 z}, \quad \dot{y} = \frac{K_4}{\sin^2 z}, \quad \dot{z} = D(\sin^2 z) \sin z := -\frac{\sqrt{P_3(\gamma_3, K_3, K_4, a)}}{\sin z},$$

which are easy to integrate. Note that P_3 is a polynomial of degree three in the variable γ_3. The solutions are

$$x = x_0 - K_4 \int \frac{\gamma_3 dz}{\sin^3 z D(\sin^2 z)} = x_0 + K_4 \int \frac{\gamma_3 d\gamma_3}{(1 - \gamma_3^2)\sqrt{P_3(\gamma_3, K_3, K_4)}},$$

$$y = y_0 + K_4 \int \frac{dz}{\sin^3 z D(\sin^2 z)} = y_0 - K_4 \int \frac{d\gamma_3}{(1 - \gamma_3^2)\sqrt{P_3(\gamma_3, K_3, K_4)}},$$

$$t = t_0 + I_1 I_2 \int \frac{d\gamma_3}{\sqrt{P_3(\gamma_3, K_3, K_4)}}.$$

As we can observe, in the *Kozlov case* as well as in Lagrange's classical problem of a heavy symmetric top x, y and t are expressed as elliptic integral of γ_3.

In the *generalized Tisserand case* the dependence $\gamma = \gamma(t)$ is determined as follows. Let Γ_1 and Γ_2 be the functions

$$\Gamma_1 = \Gamma_1(\gamma_1) = \Psi_1(\gamma_2^2 + \gamma_3^2, \gamma_1)|_{\gamma_2^2 + \gamma_3^2 = 1 - \gamma_1^2},$$
$$\Gamma_2 = \Gamma_2(\gamma_2) = \Psi_2(\gamma_1^2 + \gamma_3^2, \gamma_2)|_{\gamma_1^2 + \gamma_3^2 = 1 - \gamma_2^2}.$$

Then the *Poisson equations* (6.17) take the form

$$\dot{\gamma}_1 = \frac{\gamma_3}{I_2}\Gamma_1, \quad \dot{\gamma}_2 = \frac{\gamma_3}{I_1}\Gamma_2, \quad \dot{\gamma}_3 = -\frac{\gamma_1\Gamma_1}{I_2} - \frac{\gamma_2\Gamma_2}{I_2}. \tag{6.21}$$

The vector γ can be obtained as a function of time through the quadratures

$$\int \frac{d\gamma_1}{\Gamma_1(\gamma_1)} = I_1(\tau - \tau_0), \qquad \int \frac{d\gamma_2}{\Gamma_2(\gamma_2)} = I_2(\tau - \tau_0),$$
$$\sqrt{1 - \gamma_1^2(\tau) - \gamma_2^2(\tau)} = \gamma_3, \quad \int \frac{d\tau}{\sqrt{1 - \gamma_1^2(\tau) - \gamma_2^2(\tau)}} = \frac{t - t_0}{I_1 I_2}.$$

For the Tisserand case, if we suppose that $h_1 + a_1 + a_3 > 0$, $a_1 - b_1 > 0$ and $h_2 + a_2 + a_4 > 0$, $a_2 - b_2 > 0$, after the integration of equations (6.21) we obtain

$$\gamma_1 = \sqrt{\frac{h_1 + a_1 + a_3}{a_1 - b_1}} \sin\left(\sqrt{a_1 - b_1}I_1(\tau - \tau_1)\right) = \gamma_1(\tau),$$
$$\gamma_2 = \sqrt{\frac{h_2 + a_2 + a_4}{a_2 - b_2}} \sin\left(\sqrt{a_2 - b_2}I_2(\tau - \tau_2)\right) = \gamma_2(\tau),$$
$$\gamma_3 = \sqrt{1 - \gamma_1^2(\tau) - \gamma_2^2(\tau)},$$
$$t = t_0 + I_1 I_2 \int \frac{d\tau}{\sqrt{1 - \gamma_1^2(\tau) - \gamma_2^2(\tau)}}.$$

If

$$\sqrt{a_1 - b_1}I_1 = \sqrt{a_2 - b_2}I_2 = \alpha, \quad \tau_1 = \tau_2 = 0,$$
$$\frac{h_1 + h_2 + a_1 + a_2 + a_3 + a_4}{I_1 - I_2} = k^2 > 0,$$

then

$$t = t_0 + I_1 I_2 \int \frac{d\tau}{\sqrt{1 - k^2 \sin^2(\alpha\tau)}}.$$

In the most general case the analytical character of the solutions is considerably more complex. This completes the proof of statement (c), and hence of Theorem 6.3.1.

Remark 6.3.3.

(a) In all the known integrable cases studied before our work, condition (6.15) holds everywhere, and consequently we have local first integrals everywhere, but the known integrals are globally defined.

(b) In the previous results we determined γ as a function of time through quadratures. In general, these quadratures contain elliptic integrals, and so in order to obtain the explicit form of the time dependence one needs to first invert those integrals. To obtain the time dependence of the angular velocity vector we use the constraint $\omega_3 = 0$, and the general integrals $I_1\omega_1 - \mu_2 = 0$ and $I_2\omega_2 + \mu_1 = 0$.

Note that in Theorem 6.3.1 we are working with two general integrals because we are only using the two first integrals $\Phi_1 = \langle \gamma, \gamma \rangle = 1$ and $\Phi_2 = \langle \omega, \gamma \rangle = 0$. There are two main reasons for working with two general integrals instead of using also the energy integral $\Phi_3 = \langle I\omega, \omega \rangle + U(\gamma)$, which would allow to look for a unique general integral for determining the integrability of the Suslov problem. The first reason is that all the computations for studying the integrable cases are easier. The second is that working with two general integrals, which are linear in the angular velocity, the Poisson differential equations can be written in the form (6.17) which do not depend on the angular velocity, and consequently establishing their integrability is in general easier, mainly in the cases of statement (b).

Remark 6.3.4. The solutions of (6.14) can be represented as formal Laurent series

$$\mu_1 = \sum_{n,j,k\in\mathbb{Z}} a_{njk}\gamma_1^n\gamma_2^j\gamma_3^k, \quad \mu_2 = \sum_{n,j,k\in\mathbb{Z}} b_{njk}\gamma_1^n\gamma_2^j\gamma_3^k,$$

where $a_{njk} = a_{njk}(K_3, K_4)$ and $b_{njk} = b_{njk}(K_3, K_4)$ are coefficients which satisfy the relations

$$ja_{njk} - (n+1)b_{n+1,j-1,k} = (k+2)\left(a_{n,j-2,k+2} - b_{n-1,j-1,k+2}\right), \quad n, j, k \in \mathbb{Z}.$$

6.4 Integrability of the Veselova problem

Our second result on the Veselova problem is the following.

Theorem 6.4.1. *Consider the motion of the rigid body in the Veselova problem under the action of a force field with the potential*

$$U = -\frac{\Psi_1^2 + \Psi_2^2}{2(I_1\gamma_2^2 + I_2\gamma_1^2)},$$

where $\Psi_j = \Psi_j(x, z, K_3, K_4)$ *for* $j = 1, 2$ *and* K_3 *and* K_4 *are constants. Assume* *that Ψ_1 and Ψ_2 are solutions of the first-order partial differential equation*

$$\Theta := (I_1 - I_2) \sin z \cos z \sin x \cos x \frac{\partial \Psi_2}{\partial z}$$

$$+ \left(I_1 \cos^2 x + I_2 \sin^2 x\right) \frac{\partial \Psi_2}{\partial x} - p \sin z \frac{\partial \Psi_1}{\partial z} = 0, \tag{6.22}$$

satisfying

$$\frac{\partial \Psi_1}{\partial K_3} \frac{\partial \Psi_2}{\partial K_4} - \frac{\partial \Psi_2}{\partial K_3} \frac{\partial \Psi_1}{\partial K_4} \neq 0 \quad \text{for all } (\gamma_1, \gamma_2, \gamma_3) \in \mathbb{R}^3.$$

Then the following statements hold.

(a) *The system (6.10) has the general integrals*

$$F_1 = I_1 \omega_1 \gamma_2 - I_2 \omega_2 \gamma_1 - \Psi_2 = 0, \quad F_2 = p \omega_3 - \Psi_1 = 0. \tag{6.23}$$

Moreover, the system (6.10) is quasi-implicitly integrable.

(b) *If $\Psi_1^2 + \Psi_2^2 = 2\Psi(x)$, then the system (6.10) has the first integral*

$$I_1 \omega_1 \gamma_2 - I_2 \omega_2 \gamma_1 = K_4,$$

and consequently is explicitly integrable.

(c) *If $I_1 \neq I_2$ and $\Psi_1^2 + \Psi_2^2 = 2\Psi\left(\dfrac{I_1 \gamma_2^2 + I_2 \gamma_1^2}{\gamma_3^2}\right)$, then the system has the first integral*

$$\sqrt{I_1 I_2 I_3 \left(\frac{\gamma_1^2}{I_1} + \frac{\gamma_2^2}{I_2} + \frac{\gamma_3^2}{I_3}\right)} \, \omega_3 = K_4.$$

If $I_1 = I_2$ and $\Psi_1^2 + \Psi_3^2 = 2\Psi^2(z)$, then the system has the first integral

$$\sqrt{I_3 \left(\frac{\gamma_1^2 + \gamma_2^2}{I_1} + \frac{\gamma_3^2}{I_3}\right)} \, \omega_3 = K_4, \tag{6.24}$$

and consequently is explicitly integrable.

(d) *Using (6.23), the constraint $\gamma_1 \omega_1 + \gamma_2 \omega_2 + \gamma_3 \omega_3 = 0$, and the fact that $\gamma_1^2 + \gamma_2^2 + \gamma_3^2 = 1$, the Poisson equations for the Veselova problem take the form*

$$\dot{\gamma}_1 = -\frac{\gamma_2 \left(I_1 + (I_2 - I_1)\gamma_1^2\right) \Psi_1 + \gamma_3 \gamma_1 \Psi_2}{p(I_2 \gamma_1^2 + I_2 \gamma_2^2)},$$

$$\dot{\gamma}_2 = \frac{\gamma_1 \left(I_2 + (I_1 - I_2)\gamma_2^2)\right) \Psi_1 - \gamma_3 \gamma_2 \Psi_2}{p(I_2 \gamma_1^2 + I_2 \gamma_2^2)}, \tag{6.25}$$

$$\dot{\gamma}_3 = \frac{\gamma_1 \gamma_2 \gamma_3 (I_1 - I_2) \Psi_1 + p(\gamma_2^2 + \gamma_1^2)\Psi_2}{p(I_2 \gamma_1^2 + I_2 \gamma_2^2)}.$$

We observe that the integral (6.24) is well known (see for instance [47]).

We start with some preliminary computations that are necessary for proving Theorem 6.4.1 (see [135]).

First, in view of (6.23), using the constraint $\gamma_1\omega_1+\gamma_2\omega_2+\gamma_3\omega_3 = \dot{y}+\dot{x}\cos z = 0$ and the fact that $\gamma_1^2 + \gamma_2^2 + \gamma_3^2 = 1$ we deduced the relations

$$\omega_1 = \frac{p\Psi_2\gamma_2 - I_2\,\Psi_1\gamma_1\gamma_3}{p(I_1\gamma_2^2 + I_2\gamma_1^2)}, \quad \omega_2 = -\frac{p\Psi_2\gamma_1 + I_1\,\Psi_1\gamma_2\gamma_3}{p(I_1\gamma_2^2 + I_2\gamma_1^2)}, \quad \omega_3 = \frac{\Psi_1}{p}. \tag{6.26}$$

Consequently

$$\langle I\omega, \omega \rangle = \frac{\Psi_1^2 + \Psi_2^2}{I_1\gamma_2^2 + I_2\gamma_1^2}. \tag{6.27}$$

Using (6.26), the Poisson equations become (6.25), which in view of (4.31) yields

$$\dot{x} = \frac{\Psi_1}{\sin^2 z\, p}, \quad \dot{y} = -\frac{\cos z\,\Psi_1}{\sin^2 z\, p}, \quad \dot{z} = \frac{\Psi_2 p + (I_1 - I_2)\Psi_1 \cos x \sin x \cos z}{p\sin z(I_1 \cos^2 x + I_2 \sin^2 x)}. \tag{6.28}$$

From (6.25) and (6.28) we obtain the relation

$$\begin{aligned}
\frac{dG(\gamma)}{dt} &= (\gamma_2\omega_3 - \gamma_3\omega_2)\frac{\partial G}{\partial\gamma_1} + (\gamma_3\omega_1 - \gamma_1\omega_3)\frac{\partial G}{\partial\gamma_2} + (\gamma_1\omega_2 - \gamma_2\omega_1)\frac{\partial G}{\partial\gamma_3} \\
&= \frac{1}{p(I_1\gamma_2^2 + I_2\gamma_1^2)}\left(\left((I_2 - I_1)\left\langle\gamma, \frac{\partial G}{\partial\gamma}\right\rangle\gamma_1\gamma_2 + I_1\gamma_2\frac{\partial G}{\partial\gamma_1} - I_2\gamma_1\frac{\partial G}{\partial\gamma_2}\right)\Psi_1 \right.\\
&\qquad\left. + p\Psi_2\left(\gamma_3\left\langle\gamma, \frac{\partial G}{\partial\gamma}\right\rangle - \frac{\partial G}{\partial\gamma_3}\right)\right) \\
&= \frac{\Psi_1}{p(I_1\gamma_2^2 + I_2\gamma_1^2)}\left(I_1\cos^2 x + I_2\sin^2 x)\frac{\partial G}{\partial x}\right. \\
&\qquad\left. + (I_1 - I_2)\cos x\sin x\cos z\sin z\frac{\partial G}{\partial z}\right) \\
&\quad + \frac{p\Psi_2\sin z}{p(I_1\gamma_2^2 + I_2\gamma_1^2)}\frac{\partial G}{\partial z}.
\end{aligned}$$

Thus since Ψ_1 and Ψ_2 are arbitrary functions, we get that

$$\begin{aligned}
(I_2 - I_1)&\left\langle\gamma, \frac{\partial G}{\partial\gamma}\right\rangle\gamma_1\gamma_2 + I_1\gamma_2\frac{\partial G}{\partial\gamma_1} - I_2\gamma_1\frac{\partial G}{\partial\gamma_2} \\
&= (I_1\cos^2 x + I_2\sin^2 x)\frac{\partial G}{\partial x} + (I_1 - I_2)\cos x\sin x\cos z\sin z\frac{\partial G}{\partial z}, \tag{6.29}
\end{aligned}$$

$$\gamma_3\left\langle\gamma, \frac{\partial G}{\partial\gamma}\right\rangle - \frac{\partial G}{\partial\gamma_3} = \sin z\frac{\partial G}{\partial z}.$$

Let us calculate the derivative of $\Psi_1 = \Psi_j(x,z)$, $j = 1,2$, along the solutions of (6.28). Using (6.22) we obtain

$$
\begin{aligned}
\frac{d\Psi_2}{dt} &= \frac{\Psi_1}{p\left(I_1\gamma_2^2 + I_2\gamma_1^2\right)}(I_1\cos^2 x + I_2\sin^2 x)\frac{\partial\Psi_2}{\partial x} \\
&+ \frac{1}{(I_1\gamma_2^2 + I_2\gamma_1^2)}(I_1 - I_2)\cos x \sin x \cos z \sin z \frac{\partial\Psi_2}{\partial z} \\
&+ \frac{\sin z}{2(I_1\gamma_2^2 + I_2\gamma_1^2)}\frac{\partial\Psi_2^2}{\partial z},
\end{aligned}
$$

and so

$$
\begin{aligned}
\frac{d\Psi_2}{dt} &= \Theta + \frac{\sin z}{(I_1\gamma_2^2 + I_2\gamma_1^2)}\frac{\partial\Psi}{\partial z} \\
&= \Theta + \sin z \frac{\partial}{\partial z}\left(\frac{\Psi}{I_1\gamma_2^2 + I_2\gamma_1^2}\right) + 2\cos z \frac{\Psi}{I_1\gamma_2^2 + I_2\gamma_1^2}.
\end{aligned}
\tag{6.30}
$$

Similarly,

$$
\begin{aligned}
\frac{d}{dt}\Psi_1 &= \frac{\Psi_1}{p(I_1\gamma_2^2 + I_2\gamma_1^2)}\left((I_1\cos^2 x + I_2\sin^2 x)\frac{\partial\Psi_1}{\partial x}\right. \\
&+ \left.(I_1 - I_2)\cos x \sin x \cos z \sin z \frac{\partial\Psi_1}{\partial z}\right) \\
&+ \frac{p\Psi_2 \sin z}{p(I_1\gamma_2^2 + I_2\gamma_1^2)}\frac{\partial\Psi_1}{\partial z},
\end{aligned}
$$

and consequently

$$
\begin{aligned}
\frac{d\Psi_1}{dt} &= \Theta + \frac{1}{p\left(I_1\gamma_2^2 + I_2\gamma_1^2\right)}(I_1\cos^2 x + I_2\sin^2 x)\frac{\partial\Psi}{\partial x} \\
&+ \frac{1}{p\left(I_1\gamma_2^2 + I_2\gamma_1^2\right)}(I_1 - I_2)\cos x \sin x \cos z \sin z \frac{\partial\Psi}{\partial z} \\
&= \Theta + \frac{1}{(I_1\gamma_2^2 + I_2\gamma_1^2)}\left((I_2 - I_1)\left\langle\gamma,\frac{\partial\Psi}{\partial\gamma}\right\rangle\gamma_1\gamma_2 + I_1\gamma_2\frac{\partial\Psi}{\partial\gamma_1} - I_2\gamma_1\frac{\partial\Psi}{\partial\gamma_2}\right) \\
&= \Theta + (I_2 - I_1)\gamma_1\gamma_2\left(\left\langle\gamma,\frac{\partial}{\partial\gamma}\left(\frac{\Psi}{I_1\gamma_2^2 + I_2\gamma_1^2}\right)\right\rangle + 2\frac{\Psi}{I_1\gamma_2^2 + I_2\gamma_1^2}\right) \\
&+ I_1\gamma_2\frac{\partial}{\partial\gamma_1}\left(\frac{\Psi}{I_1\gamma_2^2 + I_2\gamma_1^2}\right) - I_2\gamma_1\frac{\partial}{\partial\gamma_2}\left(\frac{\Psi}{I_1\gamma_2^2 + I_2\gamma_1^2}\right)
\end{aligned}
\tag{6.31}
$$

where $\Psi = \dfrac{1}{2}\left(\Psi_1^2 + \Psi_3^2\right)$.

Proof of Theorem 6.4.1. After some tedious computations we deduce from (6.10), (6.26), (6.27), (6.28), (6.29), (6.30), and (6.31) the relations

$$\frac{dF_1}{dt} = \left(\frac{d}{dt} (I_1 \omega_1 \gamma_2 - I_2 \omega_2 \gamma_1) - \frac{d}{dt} \Psi_2 \right)$$

$$= -\Theta - \sin z \left(\frac{\partial}{\partial z} \left(U + \frac{\Psi}{I_1 \gamma_2^2 + I_2 \gamma_1^2} \right) \right) + \cos z \left(\langle I\omega, \omega \rangle - \frac{2\Psi}{I_1 \gamma_2^2 + I_2 \gamma_1^2} \right),$$

and so $\left. \dfrac{dF_1}{dt} \right|_{F_1 = F_2 = 0} = -\Theta$. Similarly,

$$\frac{dF_2}{dt} = \frac{d(p\omega)}{dt} - \frac{d\Psi_1}{dt}$$

$$= (I_1 - I_2) \gamma_1 \gamma_2 \left(\left\langle \gamma, \frac{\partial}{\partial \gamma} \left(U + \frac{\Psi}{I_1 \gamma_2^2 + I_2 \gamma_1^2} \right) \right\rangle + \langle I\omega, \omega \rangle - \frac{2\Psi}{I_1 \gamma_2^2 + I_2 \gamma_1^2} \right)$$

$$+ I_1 \gamma_2 \frac{\partial}{\partial \gamma_1} \left(U + \frac{\Psi}{I_1 \gamma_2^2 + I_2 \gamma_1^2} \right) - I_2 \gamma_1 \frac{\partial}{\partial \gamma_2} \left(U + \frac{\Psi}{I_1 \gamma_2^2 + I_2 \gamma_1^2} \right) - \Theta,$$

and so $\left. \dfrac{dF_2}{dt} \right|_{F_1 = F_2 = 0} = -\Theta$.

Here we apply the relation

$$\dot{\omega}_3 = \frac{1}{p} \left(\langle I\omega, \gamma \rangle (I_2 \omega_2 \gamma_1 - I_1 \omega_1 \gamma_2) + I_1 \gamma_2 \frac{\partial U}{\partial \gamma_1} - I_2 \gamma_1 \frac{\partial U}{\partial \gamma_2} \right.$$
$$\left. -(I_1 - I_2) \gamma_1 \gamma_2 \left\langle \gamma, \frac{\partial U}{\partial \gamma} \right\rangle \right),$$

which will be derived in Section 6.5. Moreover, we use the relation

$$\langle I\omega, \omega \rangle = 2(h - U) = \frac{\Psi_1^2 + \Psi_2^2}{I_1 \gamma_2^2 + I_2 \gamma_1^2}, \quad h = \text{const},$$

obtained from energy integral in view of (6.27) and

$$\frac{d(p\omega)}{dt} = (I_1 - I_2) \gamma_1 \gamma_2 \left(\left\langle \gamma, \frac{\partial U}{\partial \gamma} \right\rangle + \langle I\omega, \omega \rangle \right) + I_1 \gamma_2 \frac{\partial U}{\partial \gamma_1} - I_2 \gamma_1 \frac{\partial U}{\partial \gamma_2}$$

$$\frac{d}{dt} (I_1 \omega_1 \gamma_2 - I_2 \omega_2 \gamma_1) = -\gamma_3 \left\langle \gamma, \frac{\partial U}{\partial \gamma} \right\rangle + \frac{\partial U}{\partial \gamma_3} + \gamma_3 \langle I\omega, \omega \rangle,$$

which we deduce from (6.11) and (6.12). Consequently, in view of (6.22) we have $\left. \dfrac{dF_j}{dt} \right|_{F_1 = F_2 = 0} = 0$, $j = 1, 2$. Hence $F_1 = 0$ and $F_2 = 0$ are general integrals. This proves statement (a) of the theorem.

To prove statement (b), first we observe that if $\Psi_1^2 + \Psi_2^2 = 2\Psi(x)$, then from (6.31) we obtain that $\dfrac{d\Psi_2}{dt} = \Theta$. Thus if (6.22) holds, then $\Psi_2 = K_4 =$ arbitrary constant. Consequently, (a) yields the first integral $I_1\omega_1\gamma_2 - I_2\omega_2\gamma_1 = K_4$. By considering the energy integral, and applying the Euler–Jacobi Theorem we obtain (b).

Now let us prove statement (c). If

$$I_1 \neq I_2 \quad \text{and} \quad \Psi_1^2 + \Psi_2^2 = 2\Psi\left(\frac{I_1\gamma_2^2 + I_2\gamma_1^2}{\gamma_3^2}\right),$$

then from (6.31) it follows the proof of this statement by considering that the function

$$\Psi\left(\frac{I_1\gamma_2^2 + I_2\gamma_1^2}{\gamma_3^2}\right)$$

is a solution of the equation

$$0 = (I_2 - I_1)\left\langle \gamma, \frac{\partial\Psi}{\partial\gamma} \right\rangle \gamma_1\gamma_2 + I_1\gamma_2\frac{\partial\Psi}{\partial\gamma_1} - I_2\gamma_1\frac{\partial\Psi}{\partial\gamma_2}$$

$$= (I_1\cos^2 x + I_2\sin^2 x)\frac{\partial\Psi}{\partial x} + (I_1 - I_2)\cos x \sin x \cos z \sin z \frac{\partial\Psi}{\partial z}.$$

To prove (d), we first observe that the partial differential equation (6.22) for $I_1 = I_2$ takes the form

$$\frac{\partial\Psi_2}{\partial x} - \sin z \sqrt{\alpha + (1 - \alpha)\cos^2 z}\,\frac{\partial\Psi_1}{\partial z} = 0, \quad \alpha = \frac{I_3}{I_1}.$$

Hence

$$\frac{\partial}{\partial x}\left(\frac{\Psi_2}{\sin z \sqrt{\alpha + (1 - \alpha)\cos^2 z}}\right) - \frac{\partial\Psi_1}{\partial z} = 0.$$

Thus we get that

$$\Psi_2 = \sin z \sqrt{\alpha + (1 - \alpha)\cos^2 z}\,\frac{\partial S}{\partial z}, \quad \Psi_1 = \frac{\partial S}{\partial x}, \tag{6.32}$$

where $S = S(x, z)$ is an arbitrary function. If $\Psi_1^2 + \Psi_2^2 = 2\Psi(z, K_3, K_4)$, then from (6.22) we deduce that $\dfrac{d\Psi_1}{dt} = \Theta$, and so in view of (6.22) $\Psi_1 = K_4 =$ arbitrary constant.

In particular, if $S = S_1(z) + K_4 x$, then $\Psi_1 = K_4$, and so (6.32) yields the first integral $\sqrt{\alpha + (1 - \alpha)\cos^2 z}\,\omega_3 = K_4$. By considering the energy integral and applying the Euler–Jacobi Theorem we obtain (c).

Finally, we observe that when $\Psi_1 = K_4$ and $I_1 = I_2$ the differential system (6.28) takes the form

$$\dot{x} = \frac{K_4}{\sin^2 z \sqrt{\alpha + (1-\alpha)\cos^2 z}},$$

$$\dot{y} = \frac{-\cos z K_4}{\sin^2 z \sqrt{\alpha + (1-\alpha)\cos^2 z}},$$

$$\dot{z} = \frac{\sqrt{\alpha + (1-\alpha)\cos^2 z}}{I_1} S_1'.$$

The solutions of this system are easy to obtain because it has separable variables. This completes the proof of Theorem 6.4.1. $\qquad\square$

Corollary 6.4.2. *Equation* (6.22) *is equivalent to the equation*

$$(I_2 - I_1)\gamma_1\gamma_2 \left\langle \gamma, \frac{\partial \Psi_2}{\partial \gamma} \right\rangle + I_1\gamma_2 \frac{\partial \Psi_2}{\partial \gamma_1} - I_2\gamma_1 \frac{\partial \Psi_2}{\partial \gamma_2} - p \left(\gamma_3 \left\langle \gamma, \frac{\partial \Psi_1}{\partial \gamma} \right\rangle - \frac{\partial \Psi_1}{\partial \gamma_3} \right) = 0.$$

Proof. The proof follows from the relations (6.29). $\qquad\square$

Under the assumption $\Delta = (I_1 - I_2)(I_2 - I_3)(I_3 - I_1) \neq 0$ we introduce the coordinates (τ_1, τ_2, τ_3) by

$$\tau_1 = \gamma_1^2 + \gamma_2^2 + \gamma_3^2, \quad \tau_2 = I_1\gamma_1^2 + I_2\gamma_2^2 + I_3\gamma_3^2, \quad \tau_3 = \frac{\gamma_1^2}{I_1} + \frac{\gamma_2^2}{I_2} + \frac{\gamma_3^2}{I_3}. \quad (6.33)$$

Hence,

$$\gamma_1^2 = r_1 \left((I_2 + I_3)\tau_1 - \tau_2 - I_2 I_3 \tau_3 \right),$$
$$\gamma_2^2 = r_2 \left((I_1 + I_3)\tau_1 - \tau_2 - I_1 I_3 \tau_3 \right), \quad (6.34)$$
$$\gamma_3^2 = r_3 \left((I_2 + I_1)\tau_1 - \tau_2 - I_2 I_1 \tau_3 \right),$$

where

$$r_1 = \frac{(I_2 - I_3)I_1}{\Delta}, \quad r_2 = \frac{(I_3 - I_1)I_2}{\Delta}, \quad r_3 = \frac{(I_1 - I_2)I_3}{\Delta}.$$

Our last results on the Veselova problem are the following theorem and corollaries.

Theorem 6.4.3. *The following statements hold for the Veselova problem.*

(a) *The system* (6.10) *for* $\Delta \neq 0$ *and* $\mu \neq 0$ *(i.e., the system is constrained) admits the first integral*

$$\Phi_4 = \frac{1}{2}\|\gamma \wedge I\omega\|^2 - W(\tau_2, \tau_3) \quad (6.35)$$

if and only if the potential function U and the function W satisfy the first-order partial differential equations

$$\frac{\partial U}{\partial \tau_3} = \tilde{\nu}\tau_1, \quad \frac{\partial W}{\partial \tau_2} = \tilde{\nu}\tau_3, \quad |I|\frac{\partial U}{\partial \tau_2} + \frac{\partial W}{\partial \tau_3} = (-\tau_2 + (I_1 + I_2 + I_3)\tau_1)\tilde{\nu},$$

$$(6.36)$$

where $|I| = I_1 I_2 I_3$ and (τ_1, τ_2, τ_3) are the variables defined in (6.33) and $\tilde{\nu} = \tilde{\nu}(\tau_2, \tau_3)$ is an arbitrary function.

(b) The system (6.10) with $\Delta \neq 0$ and $\mu = 0$ (i.e., no constraints) admits the first integral (6.35) if and only if the potential function U and the function W satisfy the first-order partial differential equations

$$\frac{\partial U}{\partial \tau_3} = 0, \quad \frac{\partial W}{\partial \tau_2} = 0, \quad |I|\frac{\partial U}{\partial \tau_2} + \frac{\partial W}{\partial \tau_3} = 0. \tag{6.37}$$

Clearly for the real motions $\tau_1 = 1$.

Proof. From (6.11) and after an easy computation we obtain

$$\frac{d}{dt}\left(\|\gamma \wedge I\omega\|^2\right) = -2\left\langle \frac{\partial U}{\partial \gamma}, \gamma \wedge I\omega \right\rangle.$$

The function $\frac{1}{2}\|\gamma \wedge I\omega\|^2 - W(\gamma_1, \gamma_2, \gamma_3)$ is a first integral if and only if

$$\left\langle \frac{\partial U}{\partial \gamma}, \gamma \wedge I\omega \right\rangle = \left\langle \frac{\partial W}{\partial \gamma}, \gamma \wedge \omega \right\rangle,$$

consequently

$$\left\langle I\left(\frac{\partial U}{\partial \gamma} \wedge \gamma\right) - \frac{\partial W}{\partial \gamma} \wedge \gamma, \omega \right\rangle = 0.$$

Thus, by considering that $\langle \gamma, \omega \rangle = 0$, we have

$$I\left(\frac{\partial U}{\partial \gamma} \wedge \gamma\right) - \frac{\partial W}{\partial \gamma} \wedge \gamma = \lambda\gamma, \tag{6.38}$$

or equivalently

$$\frac{\partial U}{\partial \gamma} \wedge \gamma - I^{-1}\left(\frac{\partial W}{\partial \gamma} \wedge \gamma\right) = \lambda I^{-1}\gamma,$$

where $\lambda = \lambda(\gamma_1, \gamma_2, \gamma_3)$ is an arbitrary function.

To solve the partial differential equations (6.38) when $\Delta \neq 0$ we use the relations

$$\frac{\partial f}{\partial \gamma_j} = 2\gamma_j\left(\frac{\partial \bar{f}}{\partial \tau_1} + I_j\frac{\partial \bar{f}}{\partial \tau_2} + \frac{1}{I_j}\frac{\partial \bar{f}}{\partial \tau_3}\right),$$

for $j = 1, 2, 3$, where $f = f(\gamma_1, \gamma_2, \gamma_3) = \bar{f}(\tau_1, \tau_2, \tau_3)$. Hence

$$\frac{\partial f}{\partial \gamma} \wedge \gamma = v\left(\frac{(I_2 - I_3)}{\gamma_1}\left(\frac{\partial \bar{f}}{\partial \tau_2} - \frac{I_1}{|I|}\frac{\partial \bar{f}}{\partial \tau_3}\right), \frac{(I_3 - I_1)}{\gamma_2}\left(\frac{\partial \bar{f}}{\partial \tau_2} - \frac{I_2}{|I|}\frac{\partial \bar{f}}{\partial \tau_3}\right),\right.$$

$$\left.\frac{(I_1 - I_2)}{\gamma_3}\left(\frac{\partial \bar{f}}{\partial \tau_2} - \frac{I_3}{|I|}\frac{\partial \bar{f}}{\partial \tau_3}\right)\right),$$

where $v = 2\gamma_1\gamma_2\gamma_3$. After some computations we obtain

$$I\left(\frac{\partial U}{\partial\gamma}\wedge\gamma\right) - \frac{\partial W}{\partial\gamma}\wedge\gamma = v\left(\frac{(I_2-I_3)}{\gamma_1}\Phi_1, \frac{(I_3-I_1)}{\gamma_2}\Phi_2, \frac{(I_1-I_2)}{\gamma_3}\Phi_3\right) = \lambda\gamma,$$

$$\Phi_j = I_j\frac{\partial U}{\partial\tau_2} - \frac{I_j^2}{|I|}\frac{\partial U}{\partial\tau_3} - \frac{\partial W}{\partial\tau_2} + \frac{I_j}{|I|}\frac{\partial W}{\partial\tau_3},$$

for $j = 1, 2, 3$. Thus

$$(I_2-I_3)\Phi_1 = \frac{\lambda}{2\gamma_1\gamma_2\gamma_3}\gamma_1^2, \quad (I_3-I_1)\Phi_2 = \frac{\lambda}{2\gamma_1\gamma_2\gamma_3}\gamma_2^2, \quad (I_1-I_2)\Phi_3 = \frac{\lambda}{2\gamma_1\gamma_2\gamma_3}\gamma_3^2.$$

Hence, by using the relations

$$\frac{I_2-I_3}{I_1} + \frac{I_3-I_1}{I_2} + \frac{I_1-I_2}{I_3}$$
$$= \frac{I_1^2(I_3-I_2) + I_2^2(I_1-I_3) + I_3^2(I_2-I_1)}{I_1 I_2 I_3}$$
$$= \frac{(I_1-I_2)(I_2-I_3)(I_3-I_1)}{I_1 I_2 I_3} \neq 0,$$
$$I_1^3(I_2-I_3) + I_2^3(I_3-I_1) + I_3^3(I_1-I_2)$$
$$= (I_2-I_1)(I_2-I_3)(I_3-I_1)(I_1+I_2+I_3),$$

we deduce that

$$\frac{\lambda}{\gamma_1\gamma_2\gamma_3}\tau_1 = \frac{(I_1-I_2)(I_2-I_3)(I_3-I_1)}{I_1 I_2 I_3}\frac{\partial U}{\partial\tau_3},$$

$$\frac{\lambda}{\gamma_1\gamma_2\gamma_3}\tau_3 = \frac{(I_1-I_2)(I_2-I_3)(I_3-I_1)}{I_1 I_2 I_3}\frac{\partial W}{\partial\tau_2},$$

$$\frac{\lambda}{\gamma_1\gamma_2\gamma_3}\tau_2 = -\frac{(I_1-I_2)(I_2-I_3)(I_3-I_1)}{I_1 I_2 I_3}$$
$$\times \left(I_1 I_2 I_3\frac{\partial U}{\partial\tau_2} + \frac{\partial W}{\partial\tau_3} - (I_1+I_2+I_3)\frac{\partial U}{\partial\tau_3}\right). \tag{6.39}$$

Denoting $\tilde{\nu} = \dfrac{I_1 I_2 I_3 \lambda}{(I_1-I_2)(I_2-I_3)(I_3-I_1)\gamma_1\gamma_2\gamma_3}$, (6.39) easily yields (6.36). This is the proof of statement (a).

Now we consider the case $\Delta \neq 0$ and $\mu = 0$. The equations of motion for a rigid body without constraints are the Euler–Poisson equations

$$I\dot{\omega} = I\omega\wedge\omega + \gamma\wedge\frac{\partial U}{\partial\gamma}, \qquad \dot{\gamma} = \gamma\wedge\omega.$$

The necessary and sufficient conditions for the existence of a first integral K_4 are obtained from (6.38) with $\lambda = 0$, because the system is free of constraints. Hence, from (6.39) we get the conditions (6.37). This completes the proof of the theorem. □

Corollary 6.4.4. *The following statements hold.*

(a) *The system* (6.10) *with* $\Delta \neq 0$ *and* $\mu \neq 0$ *admits the first integral* Φ_4 *given by*

$$
\begin{aligned}
\Phi_4 = {}& \frac{1}{2}||\gamma \wedge I\omega||^2 - \int_{\tau_2^0}^{\tau_2} \tau_3 \frac{\partial U}{\partial \tau_3}\Big|_{\tau_3 = \tau_3^0} d\tau_2 \\
& + \int_{\tau_3^0}^{\tau_3} \left(|I|\frac{\partial U}{\partial \tau_2} + (\tau_2 - (I_1 + I_2 + I_3)) \frac{\partial U}{\partial \tau_3} \right) d\tau_3
\end{aligned}
\tag{6.40}
$$

if and only if the potential function U *satisfies the second-order linear partial differential equations*

$$
2\frac{\partial U}{\partial \tau_3} + |I|\frac{\partial^2 U}{\partial \tau_2^2} + (\tau_2 - (I_1 + I_2 + I_3)) \frac{\partial^2 U}{\partial \tau_3 \partial \tau_2} + \tau_3 \frac{\partial^2 U}{\partial \tau_3^2} = 0.
\tag{6.41}
$$

(b) *The system* (6.10) *with* $\Delta \neq 0$ *and* $\mu = 0$ *admits the first integral* Φ_4 *given by*

$$
\Phi_4 = \frac{1}{2}||\gamma \wedge I\omega||^2 + \int_{\tau_3^0}^{\tau_3} \left(|I|\frac{\partial U}{\partial \tau_2} \right) d\tau_3
$$

if and only if the potential function U *satisfies the second-order linear partial differential equations*

$$
\frac{\partial U}{\partial \tau_3} = 0, \quad \frac{\partial^2 U}{\partial \tau_2 \partial \tau_2} = 0.
$$

Consequently $U = \alpha\tau_2$, *where* α *is a constant, is the Clebsch–Tisserand potential, and* $W = -|I|\alpha\tau_3$. *Thus the first integral* Φ_4 *is*

$$
\Phi_4 = \frac{1}{2}||\gamma \wedge I\omega||^2 + |I|\alpha\tau_3.
$$

Proof. From (6.36) and considering that $\tau_1 = 1$, we have

$$
\begin{aligned}
\frac{\partial W}{\partial \tau_2} &= \tau_3 \frac{\partial U}{\partial \tau_3}, \\
\frac{\partial W}{\partial \tau_3} &= -|I|\frac{\partial U}{\partial \tau_2} + (-\tau_2 + (I_1 + I_2 + I_3)) \frac{\partial U}{\partial \tau_3}.
\end{aligned}
\tag{6.42}
$$

The compatibility condition of this system is given by equation (6.41). Hence if U satisfies (6.41), then the function W is obtained integrating (6.42). By (6.35), the required first integral takes the form (6.40). In short, statement (a) is proved.

Under the assumptions of statement (b) the potential function U must satisfy the equations (6.37). So $\tilde{\nu} = 0$. Now the proof of (b) easily follows from that of the previous statement. Thus the theorem is proved. $\square$

Corollary 6.4.5. *A particular solution of* (6.41) *is the potential function*

$$U = a_0 + a_1\tau_2 + a_2\left(\tau_2^2 - |I|\tau_3\right)$$
$$+ \frac{\alpha_3}{r_1\left((I_2 + I_3)\tau_1 - \tau_2 - I_2 I_3\tau_3\right)} + \frac{\alpha_4}{r_2\left((I_1 + I_3)\tau_1 - \tau_2 - I_1 I_3\tau_3\right)} \qquad (6.43)$$
$$+ \frac{\alpha_5}{r_3\left((I_2 + I_1)\tau_1 - \tau_2 - I_2 I_1\tau_3\right)},$$

where a_0 *and* α_j *for* $j = 3, 4, 5$ *are constants. Consequently, we have the first integral*

$$\Phi_4 = \frac{1}{2}||\gamma \wedge I\omega||^2 - |I|a_2\left(\tau_2\tau_3 + (I_1 + I_2 + I_3)\tau_3\right) + |I|a_1\tau_3$$
$$+ \frac{\alpha_3\left(\tau_2 - I_2 - I_3\right)}{r_1\left((I_2 + I_3)\tau_1 - \tau_2 - I_2 I_3\tau_3\right)} + \frac{\alpha_4\left(\tau_2 - I_1 - I_3\right)}{r_2\left((I_1 + I_3)\tau_1 - \tau_2 - I_1 I_3\tau_3\right)}$$
$$+ \frac{\alpha_5\left(\tau_2 - I_2 - I_1\right)}{r_3\left((I_2 + I_1)\tau_1 - \tau_2 - I_2 I_1\tau_3\right)}.$$

Proof. The potential function

$$U = a_0 + a_1\tau_2 + a_2(\tau_2^2 - |I|\tau_3) + \frac{a_4}{a_7 + a_5\tau_2 + a_6\tau_3},$$

where

$$a_6 = \frac{a_4(|I|a_4 - a_5(I_1 + I_2 + I_3))}{a_5} \quad \text{for arbitrary } a_0, a_1, a_2, a_3, a_4, a_5 \neq 0,$$

is a particular solution of (6.41). Then the first integral (6.40) becomes

$$\Phi_4 = \frac{1}{2}||\gamma \wedge I\omega||^2 - |I|\alpha\left(\tau_2\tau_3 + (I_1 + I_2 + I_3)\tau_3\right) + |I|\beta\tau_3 + \frac{a_3(a_4\tau_2 + a_6)}{a_6 + a_4\tau_2 + a_5\tau_3}.$$

Consequently, from (6.42) we get that

$$W = \frac{a_4(a_5\tau_2 + a_7)}{a_7 + a_5\tau_2 + a_6\tau_3} + |I|a_2\left(\tau_2\tau_3 + (I_1 + I_2 + I_3\tau_3)\right) - |I|a_1\tau_3.$$

We denote $w = a_7 + a_5\tau_2 + a_6\tau_3$ and consider the following particular cases:

$$a_4 = \frac{\alpha_3}{r_1}, \quad a_5 = r_1, \quad a_6 = r_1 I_1 I_2, \quad a_7 = -r_1(I_3 + I_2) \implies w = \gamma_1^2,$$
$$a_4 = \frac{\alpha_4}{r_2}, \quad a_5 = r_2, \quad a_6 = r_2 I_1 I_3, \quad a_7 = -r_2(I_3 + I_1) \implies w = \gamma_2^2,$$
$$a_4 = \frac{\alpha_5}{r_3}, \quad a_5 = r_3, \quad a_6 = r_3 I_1 I_3, \quad a_7 = -r_3(I_3 + I_1) \implies w = \gamma_3^2,$$

where r_1, r_2, r_3 are constants given in (6.34). Since (6.42) and (6.41) are linear partial differential equations, the linear combinations of the functions

$$U_1 = \frac{1}{\gamma_1^2}, \qquad U_2 = \frac{1}{\gamma_2^2}, \qquad U_3 = \frac{1}{\gamma_3^2},$$

$$W_1 = \frac{I_2(1 - \gamma_2^2) + I_3(1 - \gamma_3^2) + I_1\gamma_1^2}{\gamma_1^2},$$

$$W_2 = \frac{I_1(1 - \gamma_1^2) + I_3(1 - \gamma_3^2) + I_2\gamma_2^2}{\gamma_2^2},$$

$$W_3 = \frac{I_2(1 - \gamma_2^2) + I_1(1 - \gamma_1^2) + I_3\gamma_3^2}{\gamma_3^2},$$

are solutions of these equations. Therefore, for the potential function (6.43), or what is the same

$$U = a_0 + a_1\tau_2 + a_2(\tau_2^2 - |I|\tau_3) + a_3U_1 + a_4U_2 + a_5U_3,$$
$$W = |I|a_2\left(\tau_2\tau_3 + (I_1 + I_2 + I_3\tau_3)\right) + |I|a_1\tau_3 + a_4W_1 + a_5W_2 + a_6W_3,$$

we obtain the first integral

$$\Phi_4 = \frac{1}{2}||\gamma \wedge I\omega||^2 + -|I|a_2\left(\tau_2\tau_3 + (I_1 + I_2 + I_3\tau_3)\right) + |I|a_1\tau_3$$
$$+ \frac{a_3(I_2(1 - \gamma_2^2) + I_3(1 - \gamma_3^2) + I_1\gamma_1^2)}{\gamma_1^2}$$
$$+ \frac{a_4((I_1(1 - \gamma_1^2) + I_3(1 - \gamma_3^2) + I_2\gamma_2^2)}{\gamma_2^2}$$
$$+ \frac{a_5((I_2(1 - \gamma_2^2) + I_1(1 - \gamma_1^2) + I_3\gamma_3^2))}{\gamma_3^2},$$

where α_j for $j = 3, 4, 5$ are constants. $\square$

Remark 6.4.6. Fedorov and Jovanović in [54] claimed that

$$\Phi = \frac{1}{2}||\gamma \wedge I\omega||^2 - W(\tau_2, \tau_3)$$

$$= \frac{1}{2}||\gamma \wedge I\omega||^2 + \alpha_1|I|\langle I\gamma, \gamma\rangle\langle I^{-1}\gamma, \gamma\rangle - \alpha_2|I|\langle I^{-1}\gamma, \gamma\rangle \qquad (6.44)$$

$$+ \alpha_3\left(I_2\frac{\gamma_2^2}{\gamma_1^2} + I_3\frac{\gamma_3^2}{\gamma_1^2}\right) + \alpha_4\left(I_3\frac{\gamma_3^2}{\gamma_2^2} + I_1\frac{\gamma_1^2}{\gamma_2^2}\right) + \alpha_5\left(I_1\frac{\gamma_1^2}{\gamma_3^2} + I_2\frac{\gamma_2^2}{\gamma_3^2}\right)$$

is a first integral of system (6.10) with the potential

$$U = \alpha_1\left(\langle I^2\gamma, \gamma\rangle - \langle I\gamma, \gamma\rangle^2\right) + \alpha_2\langle I\gamma, \gamma\rangle + \frac{\alpha_3}{\gamma_1^2} + \frac{\alpha_4}{\gamma_2^2} + \frac{\alpha_5}{\gamma_3^2}, \qquad (6.45)$$

where α_j for $j = 1, \ldots, 5$ are constants.

In fact, the first integral (6.44) of the system (6.10) for the potential (6.45) in the variables (τ_1, τ_2, τ_3) (see (6.34)) becomes

$$
\begin{aligned}
\Phi &= \frac{1}{2}\|\gamma \wedge I\omega\|^2 - W(\tau_2, \tau_3) \\
&= \frac{1}{2}\|\gamma \wedge I\omega\|^2 + \alpha_1|I|\tau_2\tau_3 - \alpha_2|I|\tau_3 \\
&\quad + \frac{\alpha_3\tau_2}{r_1\left((I_2+I_3)\tau_1 - \tau_2 - I_2I_3\tau_3\right)} + \frac{\alpha_4\tau_2}{r_2\left((I_1+I_3)\tau_1 - \tau_2 - I_1I_3\tau_3\right)} \\
&\quad + \frac{\alpha_5\tau_2}{r_3\left((I_2+I_1)\tau_1 - \tau_2 - I_2I_1\tau_3\right)} - \alpha_3I_1 - \alpha_4I_2 - \alpha_5I_3.
\end{aligned}
$$

In the variables τ_2 and τ_3, the potential (6.45) and the function W defined in (6.44) have the expressions

$$
\begin{aligned}
U &= \alpha_1\left((I_1+I_2+I_3)\tau_2 + |I|\tau_3 - (I_1I_2+I_2I_3+I_3I_1) - \tau_2^2\right) \\
&\quad + \frac{\alpha_3}{r_1\left((I_2+I_3)\tau_1 - \tau_2 - I_2I_3\tau_3\right)} + \frac{\alpha_4}{r_2\left((I_1+I_3)\tau_1 - \tau_2 - I_1I_3\tau_3\right)} \\
&\quad + \frac{\alpha_5}{r_3\left((I_2+I_1)\tau_1 - \tau_2 - I_2I_1\tau_3\right)} + \alpha_2\tau_2, \\
W &= \alpha_1|I|\tau_2\tau_3 - \alpha_2|I|\tau_3 \\
&\quad + \frac{\alpha_3\tau_2}{r_1\left((I_2+I_3)\tau_1 - \tau_2 - I_2I_3\tau_3\right)} + \frac{\alpha_4\tau_2}{r_2\left((I_1+I_3)\tau_1 - \tau_2 - I_1I_3\tau_3\right)} \\
&\quad + \frac{\alpha_5\tau_2}{r_3\left((I_2+I_1)\tau_1 - \tau_2 - I_2I_1\tau_3\right)} - \alpha_3I_1 - \alpha_4I_2 - \alpha_5I_3.
\end{aligned}
\tag{6.46}
$$

Note that U coincides with the potential (6.43). This W does not satisfy (6.36) with U given by (6.46).

6.5 Constrained rigid body with invariant measure

In this section we shall study the equations of motion of the constrained rigid body with the constraint (6.3) with the Lagrange multiplier eliminated. By using these equations we provide a simple proof of the well-known *Veselova theorem* and improve Kozlov's result on the existence of an invariant measure. We give a new approach to solve the Suslov problem in absence of a force field and of an invariant measure.

First we prove the following proposition.

Proposition 6.5.1. *The motion of the rigid body around a fixed point in a force field with potential $U = U(\gamma_1, \gamma_2, \gamma_3)$ and with the constraint (6.3) with $\nu = $ const* *and $\nu = \gamma$ can be written as*

$$\dot{\omega} = \mathcal{X}_\omega$$

$$:= I^{-1} \left(\frac{\langle I\nu, \omega \rangle}{\langle I^{-1}\nu, \nu \rangle} \left(I^{-1}\nu \wedge \omega \right) + \frac{I^{-1}\nu}{\langle I^{-1}\nu, \nu \rangle} \wedge \left(\left(\gamma \wedge \frac{\partial U(\gamma)}{\partial \gamma} \right) \wedge \nu \right) \right), \quad (6.47)$$

$$\dot{\gamma} = \mathcal{X}_\gamma := \gamma \wedge \omega,$$

Proof. Indeed, by applying the method of Lagrange multipliers the equations of motion of a rigid body around a fixed point in a force field with potential $U(\gamma) = U(\gamma_1, \gamma_2, \gamma_3)$ subject to the constraint (6.3) are the equations (6.5).

Using (6.3), the Lagrange multiplier μ can be expressed as a function of ω and γ as follows:

$$\mu = -\frac{\langle \omega, \dot{\nu} \rangle}{\langle \nu, I^{-1}\nu \rangle} - \frac{\left\langle I^{-1}\nu, I\omega \wedge \omega + \gamma \wedge \frac{\partial U}{\partial \gamma} \right\rangle}{\langle \nu, I^{-1}\nu \rangle}. \quad (6.48)$$

Indeed, after differentiating (6.3) we obtain

$$0 = \langle \dot{\omega}, \nu \rangle + \langle \omega, \dot{\nu} \rangle = \left\langle I^{-1} \left(I\omega \wedge \omega + \gamma \wedge \frac{\partial U}{\partial \gamma} + \mu\nu \right), \nu \right\rangle + \langle \omega, \dot{\nu} \rangle.$$

Thus

$$0 = \left\langle I^{-1} \left(I\omega \wedge \omega + \gamma \wedge \frac{\partial U}{\partial \gamma} \right), \nu \right\rangle + \langle \omega, \dot{\nu} \rangle + \mu \left\langle I^{-1}\nu, \nu \right\rangle.$$

Hence, we easily deduce (6.48).

Inserting (6.48) into (6.5) and taking into account that $\langle \omega, \dot{\nu} \rangle = 0$ when ν is a constant vector or $\nu = \gamma$, we obtain that the first group of differential equations (6.5) becomes

$$\dot{\omega} = I^{-1} \left(I\omega \wedge \omega + \gamma \wedge \frac{\partial U}{\partial \gamma} - \frac{1}{\langle \nu, I^{-1}\nu \rangle} \left\langle I^{-1}\nu, I\omega \wedge \omega + \gamma \wedge \frac{\partial U}{\partial \gamma} \right\rangle \nu \right)$$

$$= I^{-1} \left(\left(I\omega \wedge \omega + \gamma \wedge \frac{\partial U}{\partial \gamma} \right) - \frac{1}{\langle \nu, I^{-1}\nu \rangle} \left\langle I^{-1}\nu, I\omega \wedge \omega + \gamma \wedge \frac{\partial U}{\partial \gamma} \right\rangle \nu \right)$$

$$= I^{-1} \left(\frac{I^{-1}\nu}{\langle \nu, I^{-1}\nu \rangle} \wedge \left(\left(I\omega \wedge \omega + \gamma \wedge \frac{\partial U}{\partial \gamma} \right) \wedge \nu \right) \right)$$

$$= \frac{\langle I\nu, \omega \rangle}{\langle \nu, I^{-1}\nu \rangle} I^{-1} \left(I^{-1}\nu \wedge \omega \right) + I^{-1} \left(\frac{I^{-1}\nu}{\langle \nu, I^{-1}\nu \rangle} \wedge \left(\left(\gamma \wedge \frac{\partial U(\gamma)}{\partial \gamma} \right) \wedge \nu \right) \right).$$

In the last equality we have used that $\langle \nu, I\omega \rangle = \langle I\nu, \omega \rangle$. Consequently, we obtain the differential equations (6.47). $\qquad \Box$

Equations (6.47) correspond to system (6.1) with $\mathbf{x} = (\omega_1, \omega_2, \omega_3, \gamma_1, \gamma_2, \gamma_3)$ and

$$\mathcal{X} = (\mathcal{X}_\omega, \mathcal{X}_\gamma) = (\mathcal{X}_{\omega_1}, \mathcal{X}_{\omega_2}, \mathcal{X}_{\omega_3}, \mathcal{X}_{\gamma_1}, \mathcal{X}_{\gamma_2}, \mathcal{X}_{\gamma_3}).$$

Now we calculate the divergence of the vector field $\mathcal{X}$. Since $\sum_{j=1}^{3} \dfrac{\partial \mathcal{X}_{\gamma_j}}{\partial \gamma_j} = 0$, we only need to determine $\sum_{j=1}^{3} \dfrac{\partial \mathcal{X}_{\omega_j}}{\partial \omega_j}$.

We represent the vector field $\mathcal{X}_\omega$ as

$$\mathcal{X}_\omega = \frac{\langle I\nu, \omega \rangle}{\langle I^{-1}\nu, \nu \rangle} \zeta + \frac{I^{-1}\nu}{\langle I^{-1}\nu, \nu \rangle} \wedge \left(\left(\gamma \wedge \frac{\partial U(\gamma)}{\partial \gamma} \right) \wedge \nu \right),$$

where $\zeta = (\zeta_1, \zeta_2, \zeta_3)^T = I^{-1} \left(I^{-1}\nu \wedge \omega \right)$.

Since $\operatorname{div} \zeta = 0$, we have

$$\sum_{j=1}^{3} \frac{\partial \mathcal{X}_{\omega_j}}{\partial \omega_j} = \sum_{j=1}^{3} \frac{\partial}{\partial \omega_j} \left(\frac{\langle I\nu, \omega \rangle}{\langle I^{-1}\nu, \nu \rangle} \right) \zeta_j = \langle I\nu, \xi \rangle$$

$$= -\frac{\langle \nu, (\omega \wedge I^{-1}\nu) \rangle}{\langle I^{-1}\nu, \nu \rangle} = -\frac{\langle \omega, (\nu \wedge I^{-1}\nu) \rangle}{\langle I^{-1}\nu, \nu \rangle}.$$

Thus

$$\operatorname{div}\mathcal{X} = \frac{\langle I\nu, \zeta \rangle}{\langle I^{-1}\nu, \nu \rangle} = -\frac{\langle I^{-1}\nu, (\nu \wedge \omega) \rangle}{\langle I^{-1}\nu, \nu \rangle} = -\frac{\langle \omega, (\nu \wedge I^{-1}\nu) \rangle}{\langle I^{-1}\nu, \nu \rangle}. \tag{6.49}$$

We shall assume that we can write

$$\operatorname{div}\mathcal{X} = -\mathcal{X} (\log M(\gamma)) = -\frac{d}{dt} (\log M(\gamma)).$$

In view of the relations

$$-\left\langle \frac{\partial \log M(\gamma)(\gamma)}{\partial \gamma}, \dot{\gamma} \right\rangle = -\left\langle \frac{\partial \log M(\gamma)(\gamma)}{\partial \gamma}, \gamma \wedge \omega \right\rangle = -\left\langle \omega, \frac{\partial \log M(\gamma)}{\partial \gamma} \wedge \gamma \right\rangle$$

we have

$$\operatorname{div}\mathcal{X} = \frac{\langle \omega, (\nu \wedge I^{-1}\nu) \rangle}{\langle I^{-1}\nu, \nu \rangle} = -\left\langle \omega, \frac{\partial \log M(\gamma)}{\partial \gamma} \wedge \gamma \right\rangle. \tag{6.50}$$

Clearly in view of the Euler–Jacobi Theorem and (6.49) it follows that (6.50) is a necessary and sufficient condition for the existence of an invariant measure in $\{(\omega, \gamma) \in \mathbb{R}^3 \times \mathbb{R}^3\}$ with respect to the action of the vector field (6.47). Now let us examine the condition (6.50) for the Suslov and Veselova problems.

From (6.50) we have

$$\left\langle \omega, \frac{\partial \log M(\gamma)}{\partial \gamma} \wedge \gamma - \frac{I^{-1}\nu \wedge \nu}{\langle I^{-1}\nu, \nu \rangle} \right\rangle = 0.$$

Thus, in view of the constraint $\langle \omega, \nu \rangle = 0$ we deduce the following equations for the function M:

$$\frac{\partial \log M(\gamma)}{\partial \gamma} \wedge \gamma = \frac{I^{-1}\nu \wedge \nu}{\langle I^{-1}\nu, \nu \rangle} + \upsilon\nu, \tag{6.51}$$

where $\upsilon = \upsilon(\gamma)$ is an arbitrary function such that

$$\upsilon = \left\langle \frac{\partial \log M(\gamma)}{\partial \gamma} \wedge \gamma, \frac{\nu}{\langle \nu, \nu \rangle} \right\rangle = \left\langle \frac{\partial \log M(\gamma)}{\partial \gamma} \wedge \gamma, \frac{I^{-1}\nu}{\langle I^{-1}\nu, \nu \rangle} \right\rangle.$$

Thus the function M must satisfy the first-order partial differential equation

$$\left\langle \frac{\partial \log M(\gamma)}{\partial \gamma}, \eta \wedge \gamma \right\rangle = 0, \tag{6.52}$$

with $\eta = (\eta_1, \eta_2, \eta_3) = \nu \wedge \left(I^{-1}\nu \wedge \nu \right)$. The equations for the characteristic curves are

$$\frac{d\gamma_1}{\gamma_3 \eta_2 - \gamma_2 \eta_3} = \frac{d\gamma_2}{\gamma_1 \eta_3 - \gamma_3 \eta_1} = \frac{d\gamma_3}{\gamma_2 \eta_1 - \gamma_1 \eta_2}. \tag{6.53}$$

From these equations we obtain the differential equations

$$\gamma_1 d\gamma_1 + \gamma_2 d\gamma_2 + \gamma_3 d\gamma_3 = 0, \quad \eta_1 d\eta_1 + \eta_2 d\eta_2 + \eta_3 d\eta_3 = 0.$$

Hence, if $\eta_1 d\gamma_1 + \eta_2 d\gamma_2 + \eta_3 d\gamma_3 = d\xi$, then the function $M = M\left(\langle \gamma, \gamma \rangle, \xi \right)$.

Now we study the condition (6.50) for the Suslov and Veselova problems.

Proposition 6.5.2. *The flow of the Suslov system*

$$\dot{\omega} = X_\omega|_{\nu=a}$$
$$= -I^{-1}\left(\frac{\langle Ia, \omega \rangle}{\langle I^{-1}a, a \rangle} \left(I^{-1}a \wedge \omega \right) + \frac{I^{-1}a}{\langle I^{-1}a, a \rangle} \wedge \left(a \wedge \left(\gamma \wedge \frac{\partial U(\gamma)}{\partial \gamma} \right) \right) \right),$$
$$\dot{\gamma} = X_\gamma = \gamma \wedge \omega, \tag{6.54}$$

where $\mathbf{a}$ is a constant vector, preserves the Lebesgue measure on $\{(\omega, \gamma) \in \mathbb{R}^3 \times \mathbb{R}^3\}$ if and only if

$$I\mathbf{a} = \kappa \mathbf{a} \tag{6.55}$$

for some $\kappa \in \mathbb{R}$.

Proof. By the Euler–Jacobi theorem, a necessary and sufficient condition for the existence of a measure invariant under the action of the vector field (6.54) is (see formula (6.50))

$$\operatorname{div} \mathcal{X}\big|_{\nu=a} == \frac{\langle \omega, (a \wedge I^{-1}a) \rangle}{\langle I^{-1}a, a \rangle} = -\left\langle \omega, \frac{\partial \log M(\gamma)}{\partial \gamma} \wedge \gamma \right\rangle.$$

Thus if (6.55) holds, then $\operatorname{div} \mathcal{X}\big|_{\nu=a} = 0$. The convergence is obtained by a "reductio ad absurdum" argument. Let us suppose that (6.55) does not holds. Then from the relation (6.51) for the vector field $\mathcal{X}\big|_{\nu=a}$ it follows that relation

$$\frac{\partial \log M(\gamma)}{\partial \gamma} \wedge \gamma = \frac{I^{-1}a \wedge a}{\langle I^{-1}a, a \rangle} + \nu a, \tag{6.56}$$

Consequently, the function M must satisfy the partial differential equation (6.52) with $\eta = a \wedge (I^{-1}a \wedge a)$. The solutions of (6.53) are the functions $\tau = \langle \gamma, \gamma \rangle$, $\xi = \langle \eta, \gamma \rangle$. Hence the function $M = M(\eta, \xi)$ and inserting in (6.56) we have

$$\frac{\partial \log M(\eta, \xi)}{\partial \xi} (a \wedge (I^{-1}a \wedge a) \wedge \gamma) = \frac{I^{-1}a \wedge a}{\langle I^{-1}a, a \rangle} + \nu a.$$

Thus,

$$\left(\frac{\partial \log M(\eta, \xi)}{\partial \xi} \langle I^{-1}a \wedge a, \gamma \rangle - \nu \right) a$$

$$- \left(\frac{\partial \log M(\eta, \xi)}{\partial \xi} \langle a, \gamma \rangle - \frac{1}{\langle I^{-1}a, a \rangle} \right) I^{-1}a \wedge a = 0.$$

Since the vectors a and $I^{-1}a \wedge a$ are independent, we deduce that the function M must be such that

$$\frac{\partial \log M(\eta, \xi)}{\partial \xi} = \frac{\nu}{\langle I^{-1}a \wedge a, \gamma \rangle} = \frac{1}{\langle I^{-1}a, a \rangle \langle a, \gamma \rangle}.$$

Hence we reached a contradiction because the right hand part of this last equality does not depend on ξ. Consequently, $a \wedge I^{-1}a = 0$, i.e., a is eigenvector of I. The proposition is proved.   □

Proposition 6.5.2 provides an "if and only if" result, which improves the previous result of Kozlov (see Proposition 5 of [91]) which was only "if". Here we have presented a new proof using equations (6.54).

Proposition 6.5.3. *The flow of the Veselova system*

$$\dot{\omega} = \mathcal{X}_\omega\big|_{\nu=\gamma} = I^{-1} \left(\frac{\langle I\gamma, \omega \rangle}{\langle I^{-1}\gamma, \gamma \rangle} (I^{-1}\gamma \wedge \omega) - \frac{I^{-1}\gamma}{\langle I^{-1}\gamma, \gamma \rangle} \wedge \left(\gamma \wedge \left(\gamma \wedge \frac{\partial U(\gamma)}{\partial \gamma} \right) \right) \right),$$

$$\dot{\gamma} = \mathcal{X}_\gamma = \gamma \wedge \omega,$$

preserves the measure $M(\gamma) = \sqrt{\langle I^{-1}\gamma, \gamma \rangle}$ *on* $\{(\omega, \gamma) \in \mathbb{R}^3 \times \mathbb{R}^3\}$.

Proof. Indeed, in the present case condition (6.50) takes the form

$$\operatorname{div} \mathcal{X}|_{\nu=\gamma} = \frac{\langle I^{-1}\nu, \nu \wedge \omega \rangle}{\langle I^{-1}\nu, \nu \rangle} = \frac{\langle I^{-1}\gamma, \gamma \wedge \omega \rangle}{\langle I^{-1}\gamma, \gamma \rangle}$$

$$= -\frac{\langle I^{-1}\gamma, \dot{\gamma} \rangle}{\langle I^{-1}\gamma, \gamma \rangle} = -\frac{d}{dt} \log \left(\sqrt{\langle I^{-1}\gamma, \gamma \rangle} \right).$$

Thus $\operatorname{div} \left(\sqrt{\langle I^{-1}\gamma, \gamma \rangle} \, \mathcal{X}|_{\nu=\gamma} \right) = 0$. Consequently in view of the Euler–Jacobi Theorem we deduce the existence of an invariant measure in $\{(\omega, \gamma) \in \mathbb{R}^3 \times \mathbb{R}^3\}$ with $M = \sqrt{\langle I^{-1}\gamma, \gamma \rangle}$. □

The proof of Proposition 6.5.3 is different from the well-known Veselova result (see [155]) on the existence of an invariant measure on $\{(\omega, \gamma) \in \mathbb{R}^3 \times \mathbb{R}^3\}$.

If (6.8) holds, then in view of the relation $\dfrac{I^{-1}\mathbf{a}}{\langle I^{-1}\mathbf{a}, \mathbf{a} \rangle} = \dfrac{\mathbf{a}}{\langle \mathbf{a}, \mathbf{a} \rangle}$ we obtain that the differential equations (6.54) read

$$\dot{\omega} = -I^{-1} \left(\frac{\mathbf{a}}{\langle \mathbf{a}, \mathbf{a} \rangle} \wedge \left(\mathbf{a} \wedge \left(I\omega \wedge \omega + \gamma \wedge \frac{\partial U}{\partial \gamma} \right) \right) \right)$$

$$= -I^{-1} \left(\langle I\omega, \mathbf{a} \rangle \frac{\mathbf{a} \wedge \omega}{\langle \mathbf{a}, \mathbf{a} \rangle} + \frac{\mathbf{a}}{\langle \mathbf{a}, \mathbf{a} \rangle} \wedge \left(\mathbf{a} \wedge \left(\gamma \wedge \frac{\partial U}{\partial \gamma} \right) \right) \right) \qquad (6.57)$$

$$= -I^{-1} \left(\gamma \wedge \frac{\partial U}{\partial \gamma} - \left\langle \mathbf{a}, \gamma \wedge \frac{\partial U}{\partial \gamma} \right\rangle \frac{\mathbf{a}}{\langle \mathbf{a}, \mathbf{a} \rangle} \right),$$

$$\dot{\gamma} = \gamma \wedge \omega.$$

In the above equalities we have used that $\langle I\omega, \mathbf{a} \rangle = \langle \omega, I\mathbf{a} \rangle = \kappa \langle \omega, \mathbf{a} \rangle = 0$.

We assume that the vector $\mathbf{a}$ coincides with one of the principal axes of inertia and without loss of generality we can choose it as the third axis, i.e., $\mathbf{a} = (0, 0, 1)$ and consequently the constraint becomes $\omega_3 = 0$, and the differential system (6.57) takes the form (6.9).

6.6 Equations of motion of the rigid body without invariant measure

The Suslov system (6.54) in the absence of external forces, i.e., when $U = 0$ becomes

$$\dot{\omega} = -\frac{\langle I\mathbf{a}, \omega \rangle}{\langle I^{-1}\mathbf{a}, \mathbf{a} \rangle} I^{-1} \left(I^{-1}\mathbf{a} \wedge \omega \right), \qquad (6.58)$$

$$\dot{\gamma} = \gamma \wedge \omega.$$

Now assume that $\mathbf{a}$ is not an eigenvector of I. Making the change

$$d\sigma = \frac{\langle I\mathbf{a}, \omega \rangle}{\langle I^{-1}\mathbf{a}, \mathbf{a} \rangle} dt = \rho dt,$$

and denoting by $'$ the differentiation with respect to the new variable σ, we recast that (6.58) as

$$\omega' = I^{-1}\left(I^{-1}\mathbf{a} \wedge \omega\right) = I^{-1}A\omega = K\omega,$$
$$\gamma' = \frac{1}{\rho}\gamma \wedge \omega = B(\omega)\gamma, \tag{6.59}$$

where

$$A = \begin{pmatrix} 0 & -b_3 & b_2 \\ b_3 & 0 & -b_1 \\ -b_2 & b_1 & 0 \end{pmatrix}, \quad \mathbf{b} = I^{-1}, \mathbf{a}$$

$$B(\omega) = \begin{pmatrix} 0 & \dfrac{\omega_3}{\rho} & -\dfrac{\omega_2}{\rho} \\ -\dfrac{\omega_3}{\rho} & 0 & \dfrac{\omega_1}{\rho} \\ \dfrac{\omega_2}{\rho} & -\dfrac{\omega_1}{\rho} & 0 \end{pmatrix}.$$

Consequently, the equations of motion in the Suslov problem when $U = 0$ can be written in the variables ω_1, ω_2, ω_3 as a system of first-order ordinary differential equations with constants coefficients. After the integration of this differential system, we obtain

$$\omega = \omega(\sigma) = e^{K\sigma}\omega_0, \qquad t = t_0 + \langle I^{-1}\mathbf{a}, \mathbf{a}\rangle \int \frac{d\sigma}{\langle I\mathbf{a}, e^{K\sigma}\omega_0\rangle}, \tag{6.60}$$

where ω_0 is an arbitrary nonzero constant vector. In view of the relations

$$K^T\mathbf{a} = A^T I^{-1}\mathbf{a} = -A\mathbf{b} = -\mathbf{b} \wedge \mathbf{b} = \mathbf{0},$$
$$\langle \mathbf{a}, K^n\omega\rangle = \langle K^T\mathbf{a}, K^{n-1}\omega\rangle = 0, \quad \text{for any} \quad n \in \mathbb{N},$$

where we used that $I^T = I$ and $A^T = -A$. From (6.60) we deduce that

$$\langle \mathbf{a}, \omega_0\rangle = \langle \mathbf{a}, e^{-K\sigma}\omega\rangle = \sum_{n=0}^{+\infty} \langle \mathbf{a}, K^n\omega\rangle \frac{(-\sigma)^n}{n!} = 0.$$

Thus the vector ω_0 is an orthogonal vector to $\mathbf{a}$.

After some computations one obtains that

$$K^3 = -\varrho K, \qquad \varrho = \frac{\langle \mathbf{a}, I^{-1}\mathbf{a}\rangle}{|I|}.$$

Hence from (6.60) it follows that

$$\omega = e^{K\sigma}\omega_0 = \sum_{n=0}^{+\infty} \frac{K^n \sigma^n}{n!}\omega_0$$

$$= \omega_0 + \sum_{n=0}^{+\infty} \frac{K^{2n+1}\sigma^{2n+1}}{(2n+1)!}\omega_0 + \sum_{n=0}^{+\infty} \frac{K^{2n+2}\sigma^{2n+2}}{(2n+2)!}\omega_0$$

$$= \omega_0 + \sum_{n=0}^{+\infty} \frac{(-1)^n \varrho^n \sigma^{2n+1}}{(2n+1)!}K\omega_0 + \sum_{n=0}^{+\infty} \frac{(-1)^n \varrho^n \sigma^{2n+2}}{(2n+2)!}K^2\omega_0 \qquad (6.61)$$

$$= \omega_0 + K^2\omega_0 - \frac{\cos\left(\sqrt{\varrho}\sigma\right)}{\varrho}K^2\omega_0 + \frac{\sin\left(\sqrt{\varrho}\sigma\right)}{\sqrt{\varrho}}K\omega_0$$

$$= \mathbf{c}_1 + \mathbf{c}_2 \sin\left(\sqrt{\varrho}\sigma\right) + \mathbf{c}_3 \cos\left(\sqrt{\varrho}\sigma\right),$$

for suitable constant vectors $\mathbf{c}_1, \mathbf{c}_2, \mathbf{c}_3$. Now the dependence $\sigma = \sigma(t)$ is readily derived from the equality

$$\int \frac{d\sigma}{\langle I\mathbf{a}, \omega\rangle} = \int \frac{d\sigma}{\langle \alpha_1 + \alpha_2 \sin\left(\sqrt{\varrho}\sigma\right) + \alpha_3 \cos\left(\sqrt{\varrho}\sigma\right)\rangle} = \frac{t - t_0}{\langle I^{-1}\mathbf{a}, \mathbf{a}\rangle}, \qquad (6.62)$$

where $\alpha_j = \langle I\mathbf{a}, \mathbf{c}_j\rangle$, for $j = 1, 2, 3$.

Now we study the case when the vector $\mathbf{a} = (0, 0, 1)$ (see [15]), and so the vector $\mathbf{b} = I^{-1}\mathbf{a}$ has the components

$$b_1 = \frac{I_{12}I_{23} - I_{13}I_{22}}{|I|}, \quad b_2 = \frac{I_{11}I_{23} - I_{12}I_{13}}{|I|}, \quad b_3 = \frac{I_{11}I_{22} - I_{12}^2}{|I|} > 0,$$

where the last inequality follows from the fact that the matrix I is positive definite. Consequently, $\varrho = \dfrac{I_{11}I_{22} - I_{12}^2}{|I|^2} > 0$. After some computations we obtain the vector (6.61) in the form

$$\omega_1 = \sqrt{I_{22}h}\cos\left(\sqrt{\varrho}\sigma + \alpha\right), \quad \omega_2 = \sqrt{I_{11}h}\sin\left(\sqrt{\varrho}\sigma + \beta\right), \quad \omega_3 = 0,$$

where $h = 1/2\left(I_{11}\omega_{10}^2 + 2I_{12}\omega_{10}\omega_{20} + I_{22}\omega_{20}^2\right)$, and α and β are angles such that

$$\cos\alpha = \frac{\omega_{10}\sqrt{I_{11}I_{22} - I_{12}^2}}{\sqrt{I_{22}h}}, \quad \cos\beta = \frac{I_{11}\omega_{10} + I_{12}\omega_{20}}{\sqrt{I_{11}h}}.$$

Hence, if $I_{12} = 0$, then we can suppose that $\alpha = \beta$ and consequently after integration (6.62) yields

$$\tan\frac{\left(\sqrt{\varrho}\sigma + \vartheta\right)}{2} = \exp\left(\frac{|I|\sqrt{h(I_{11} + I_{22})}}{I_{11}I_{22}}(t - t_0)\right),$$

where $\cos\vartheta = \dfrac{\sqrt{I_{22}}}{\sqrt{I_{11} + I_{22}}}$. Thus, the rigid body motion tends asymptotically towards a uniform *rotation around a straight line* (see [15, 91, 55]).

We observe that the second differential equations (6.59) becomes a *Riccati equation*. Indeed by introducing the two Darboux functions ξ and η (see for instance [50]) we obtain that

$$\gamma_1 = \frac{i(1 + \xi\eta)}{\xi - \eta}, \quad \gamma_2 = \frac{1 - \xi\eta}{\xi - \eta}, \quad \gamma_3 = \frac{\eta + \xi}{\xi - \eta}.$$

After some computations we obtain that γ_j for $j = 1, 2, 3$, are different solutions of the Riccati differential equation

$$\frac{d\zeta}{d\sigma} = \frac{1}{2\varrho\, i}\left((\omega_2 - i\,\omega_1)\zeta^2 - 2\omega_3\zeta - \omega_2 - i\,\omega_1\right).$$

Hence

$$\frac{d\zeta}{d\sigma} = \frac{1}{\rho}\left(\psi_1(\sigma)\zeta^2 + \psi_2(\sigma)\zeta + \overline{\psi}_1(\sigma)\right),$$

$$\psi_j(\sigma) = \alpha_j + \beta_j \sin\left(\sqrt{\rho}\sigma\right) + \kappa_j \cos\left(\sqrt{\varrho}\sigma\right), \quad j = 1, 2,$$

where α_j, β_j and κ_j are suitable complex constants and $\overline{\psi}_1$ is the complex conjugate of function ψ_1. Clearly, in general the integration of this equation is not possible. Thus, whether is possible to find the time dependence $\gamma = \gamma(t)$ by quadratures remains an open problem (see for instance [91, 15]).

We observe that the case when $U = 0$ was studied by Suslov in [147] in a different way.

Chapter 7

Inverse Problem in Vakonomic Mechanics

7.1 Introduction

The mechanical systems free of constraints are called *Lagrangian systems* or *holonomic systems*. The mechanical systems with integrable constraints are called *holonomic constrained mechanical systems*. Finally, the mechanical systems with non-integrable constraints are usually called *nonholonomic mechanical systems*, or *nonholonomic constrained mechanical systems*.

The history of *nonholonomic mechanical systems* is long and complex and goes back to the 19th century, with important contributions by Hertz [72] (1894), Ferrers [56] (1871), Vierkandt [157] (1892), and Chaplygin [22] (1897).

Nonholonomic mechanics is a remarkable generalization of classical Lagrangian and Hamiltonian mechanics. The birth of the theory of dynamics of nonholonomic systems occurred when the Lagrangian–Euler formalism was found to be inapplicable for studying the simple mechanical problem of a rigid body rolling without slipping on a plane.

A long period of time has been needed for finding the correct equations of motion of nonholonomic mechanical systems and studying the deeper questions associated with the geometry and the analysis of these equations. In particular, the integrability theory of the equations of motion for nonholonomic mechanical systems is not so complete as in the case of holonomic systems. This is due to several reasons. First, the equations of motion of a nonholonomic system have more complex structure than the Lagrange ones, which describe the behavior of a holonomic system. Indeed, a holonomic system can be described by a unique function of its state and time, the Lagrangian function. For a nonholonomic system this is not possible. Second, the equations of motion of a nonholonomic system in general have no invariant measures, in contrast to the equations of motion of a holonomic system (see [84, 96, 106, 154]).

One of the most important directions in the development of nonholonomic mechanics is the research aiming at elaborating a general mathematical formalism to describe the behavior of such systems which differ from the Lagrangian and Hamiltonian formalism. A main problem concerning the equations of motion of the nonholonomic mechanics has been centered on whether or not these equations can be derived from the *Hamiltonian principle* in the usual sense, as for the holonomic systems (see for instance [115]). But there is not doubt that the correct equations of motion for nonholonomic systems are given by the d'Alembert–Lagrange principle

The general understanding of the inapplicability of Lagrange equations and variational Hamiltonian principles to the nonholonomic systems is due to *Hertz* who expressed it in his fundamental work *Die Prinzipien der Mechanik in neuen Zusammenhange dargestellt* [72]. Hertz's ideas were developed by *Poincaré* in [129] At the same time, various aspects of nonholonomic systems need to be studied such as

(a) The problem of the realization of nonholonomic constraints (see for instance [85, 86]).

(b) The stability of nonholonomic systems (see for instance [120, 137, 161]).

(c) The role of the so called transpositional relations (see [81, 116, 120, 131, 133]

$$\delta \frac{d\mathbf{x}}{dt} - \frac{d}{dt}\delta \mathbf{x} = \left(\delta \frac{dx_1}{dt} - \frac{d}{dt}\delta x_1, \ldots, \delta \frac{dx_N}{dt} - \frac{d}{dt}\delta x_N \right), \qquad (7.1$$

where $\dfrac{d}{dt}$ denotes the differentiation with respect to time, δ is the virtua variation, and $\mathbf{x} = (x_1, \ldots, x_N)$ is the vector of generalized coordinates.

Indeed, the most general formulation of the Hamiltonian principle is the *Hamilton–Suslov principle*

$$\int_{t_0}^{t_1} \left(\delta \tilde{L} - \sum_{j=1}^{N} \frac{\partial \tilde{L}}{\partial \dot{x}_j} \left(\delta \frac{dx_j}{dt} - \frac{d}{dt}\delta x_j \right) \right) dt = 0, \qquad (7.2$$

suitable for constrained and unconstrained Lagrangian systems, where $\tilde{L}$ is the Lagrangian of the mechanical system under consideration. Clearly, the equations of motion obtained from the Hamilton–Suslov principle depend on the point of view on the transpositional relations (c). This fact shows the importance of these relations.

(d) The relation between nonholonomic mechanical systems and *vakonomic me chanical systems*. More precisely, there was some confusion in the literature between nonholonomic mechanical systems and variational nonholonomic mechanical systems, also called vakonomic mechanical systems. Both kinds of systems involve the same mathematical "ingredients": a Lagrangian function and a set of constraints. But the ways in which the equations of motion are derived differ. As we observed, the equations of motion in nonholonomic me chanics are deduced using the d'Alembert–Lagrange principle. In the case of vakonomic mechanics the equations of motion are obtained through the application of a constrained variational principle. The term vakonomic ("variational axiomatic kind") was coined by Kozlov (see [87, 88, 89]), who proposed this mechanics as an alternative set of equations of motion for a constrained Lagrangian system.

The distinction between the classical differential equations of motion and the equations of motion of variational nonholonomic mechanical systems has a long history going back to the survey article of *Korteweg* (1899) [83] and is discussed in a more modern context in [53, 19, 69, 80, 95, 13, 14, 160]. In these papers the authors have discussed areas of vakonomic and nonholonomic mechanics. In the paper *A Critique of some mathematical models of mechanical systems with differential constraints* [79], Kharlamov studied the Kozlov model and in a concrete example showed that the subset of solutions of the studied nonholonomic systems is not included in the set of solutions of the vakonomic model. In [93] the authors exhibit the main differences between the d'Alembertian and the vakonomic approaches. From the results obtained in several papers it follows that in general the vakonomic model is not applicable to the nonholonomic constrained Lagrangian systems.

The equations of motion for the constrained mechanical systems deduced by Kozlov (see for instance [4]) from the Hamiltonian principle with the Lagrangian $L : \mathbb{R} \times TQ \times \mathbb{R}^M \to \mathbb{R}$ such that $L = L_0 - \sum_{j=1}^{M} \lambda_j L_j$, where $L_j = 0$ for $j = 1, \ldots, M < N$ are the given constraints, and L_0 is the classical Lagrangian, are

$$E_k L = \frac{d}{dt} \frac{\partial L}{\partial \dot{x}_k} - \frac{\partial L}{\partial x_k} = 0 \iff E_k L_0 = \sum_{j=1}^{M} \left(\lambda_j E_k L_j + \frac{d\lambda_j}{dt} \frac{\partial L_j}{\partial \dot{x}_k} \right), \qquad (7.3)$$

for $k = 1, \ldots, N$. Clearly, equations (7.3) differ from the classical equations by the presence of the terms $\lambda_j E_k L_j$. If the constraints are integrable, i.e., $L_j = \frac{d}{dt} g_j(t, \mathbf{x})$, then the vakonomic mechanics reduces to the holonomic one.

In this chapter we solve the *inverse problem of vakonomic mechanics* (see Section 7.7), and we obtain a *modified vakonomic mechanics*. This modification is valid for holonomic and nonholonomic constrained Lagrangian systems. We apply the generalized constrained Hamiltonian principle with non-zero transpositional relations, and we deduce the equations of motion for nonholonomic systems with constraints which in general are nonlinear in the velocity. These equations coincide, except perhaps in a set of Lebesgue measure zero, with the classical differential equations deduced from the d'Alembert–Lagrange principle.

7.2 Hamiltonian principle

We introduce the following results, notations, and definitions which we will us
later on (see [4]).

A *Lagrangian system* is a pair $(Q, \tilde{L})$ consisting of a smooth manifold Q
and a smooth function $\tilde{L} : \mathbb{R} \times TQ \to \mathbb{R}$, where TQ is the tangent bundle of Q
The point $\mathbf{x} = (x_1, \ldots, x_N) \in Q$ represents the *position* (usually its component
are called *generalized coordinates*) of the system and we call each tangent vecto
$\dot{\mathbf{x}} = (\dot{x}_1, \ldots, \dot{x}_N) \in T_{\mathbf{x}}Q$ the *velocity* (usually called *generalized velocity*) of th
system at the point $\mathbf{x}$. A pair $(\mathbf{x}, \dot{\mathbf{x}})$ is called a *state* of the system. In Lagrangia
mechanics it is usual to call Q the *configuration space*, the tangent bundle $T($
phase space, $\tilde{L}$ the *Lagrange function* or *Lagrangian*, and the dimension N of $($
the *number of degrees of freedom*.

Let a_0 and a_1 be two points of Q. The map

$$\gamma : [t_0, t_1] \subset \mathbb{R} \longrightarrow Q, \quad t \longmapsto \gamma(t) = (x_1(t), \ldots, x_N(t)),$$

such that $\gamma(t_0) = a_0$, $\gamma(t_1) = a_1$ is called a *path* from a_0 to a_1. We denote the se
of all these paths by $\Omega(Q, a_0, a_1, t_0, t_1) := \Omega$.

We shall derive one of the simplest and most general variational principles
the *Hamiltonian principle* (see [127]).

The functional $F : \Omega \to \mathbb{R}$ defined by

$$F(\gamma(t)) = \int_{\gamma(t)} \tilde{L} dt = \int_{t_0}^{t_1} \tilde{L}(t, \mathbf{x}(t), \dot{\mathbf{x}}(t)) dt$$

is called the *action*.

Consider a path $\gamma(t) = \mathbf{x}(t) = (x_1(t), \ldots, x_N(t)) \in \Omega$. A *variation* of th
path $\gamma(t)$ is defined as a smooth mapping

$$\gamma^* : [t_0, t_1] \times (-\tau, \tau) \longrightarrow Q, \quad (t, \varepsilon) \longmapsto \gamma^*(t, \varepsilon),$$

where
$$\gamma^*(t, \varepsilon) = \mathbf{x}^*(t, \varepsilon) = (x_1(t) + \varepsilon \delta x_1(t), \ldots, x_N(t) + \varepsilon \delta x_N(t)),$$

satisfying
$$\mathbf{x}^*(t_0, \varepsilon) = a_0, \quad \mathbf{x}^*(t_1, \varepsilon) = a_1, \quad \mathbf{x}^*(t, 0) = \mathbf{x}(t).$$

By definition, we have

$$\delta \mathbf{x}(t) = \left. \frac{\partial \mathbf{x}^*(t, \varepsilon)}{\partial \varepsilon} \right|_{\varepsilon = 0}.$$

This function is called the *virtual displacement* or *virtual variation* correspondin
to the variation of $\gamma(t)$ and it is a function of time; all its components are function
of t of class $C^2(t_0, t_1)$ and vanish at t_0 and t_1 i.e. $\delta \mathbf{x}(t_0) = \delta \mathbf{x}(t_1) = 0$.

A *varied path* is a path which can be obtained as a variation path.

The *first variation* of the functional F at $\gamma(t)$ is

$$\delta F := \left.\frac{\partial F\left(\mathbf{x}^*(t,\varepsilon)\right)}{\partial \varepsilon}\right|_{\varepsilon=0},$$

and it is called the *differential* of the functional F (see [4]). The path $\gamma(t) \in \Omega$ is called the *critical point F* if $\delta F(\gamma(t)) = 0$.

Let $\mathbb{L}$ be the space of all smooth functions $g : \mathbb{R} \times TQ \to \mathbb{R}$. The operator

$$E_\nu : \mathbb{L} \longrightarrow \mathbb{R}, \quad g \longmapsto E_\nu g = \frac{d}{dt}\frac{\partial g}{\partial \dot{x}_\nu} - \frac{\partial g}{\partial x_\nu}, \quad \nu = 1,\ldots,N,$$

is called the *Lagrangian derivative*.

It is easy to show that the Lagrangian derivative satisfies

$$E_\nu \frac{df}{dt} = 0, \tag{7.4}$$

for arbitrary smooth functions $f = f(t,\mathbf{x})$. Note that by (7.4), the Lagrangian derivative is unchanged if we replace the function g by $g + \dfrac{df}{dt}$, for any function $f = f(t,\mathbf{x})$. This reflects the *gauge invariance*. We shall say that the functions $g = g\left(t,\mathbf{x},\dot{\mathbf{x}}\right)$ and $\hat{g} = \hat{g}\left(t,\mathbf{x},\dot{\mathbf{x}}\right)$ are *equivalent* if $g - \hat{g} = \dfrac{df(t,\mathbf{x})}{dt}$, and we shall write $g \simeq \hat{g}$.

Proposition 7.2.1. *The first variation of the action can be calculated as*

$$\delta F = -\int_{t_0}^{t_1} \sum_{k=1}^N \left(E_k \tilde{L} \delta x_k - \frac{\partial \tilde{L}}{\partial \dot{x}_k}\left(\delta \frac{dx_k}{dt} - \frac{d}{dt}\delta x_k\right)\right) dt, \tag{7.5}$$

where $\mathbf{x} = \mathbf{x}(t)$, $\dot{\mathbf{x}} = \dfrac{d\mathbf{x}}{dt}$, *and* $\tilde{L} = \tilde{L}\left(t,\mathbf{x},\dfrac{d\mathbf{x}}{dt}\right)$.

Proof. We have that

$$\delta F = \left.\frac{\partial F\left(\mathbf{x}^*(t,\varepsilon)\right)}{\partial \varepsilon}\right|_{\varepsilon=0}$$

$$= \int_{t_0}^{t_1} \left.\frac{\partial}{\partial \varepsilon}\right|_{\varepsilon=0} L\left(t,\mathbf{x}^*(t,\varepsilon),\frac{d}{dt}\left(\mathbf{x}^*(t,\varepsilon)\right)\right) dt$$

$$= \int_{t_0}^{t_1} \sum_{k=1}^N \left(\frac{\partial L}{\partial x_k}\delta x_k + \frac{\partial L}{\partial \dot{x}_k}\delta \dot{x}_k\right) dt$$

$$= \int_{t_0}^{t_1} \sum_{k=1}^N \left(\frac{\partial L}{\partial x_k}\delta x_k + \frac{\partial L}{\partial \dot{x}_k}\frac{d}{dt}\delta x_k + \frac{\partial L}{\partial \dot{x}_k}\left(\delta\frac{dx_k}{dt} - \frac{d}{dt}\delta x_k\right)\right) dt$$

$$= \sum_{k=1}^{N} \frac{\partial L}{\partial \dot{x}_k} \delta x_k \bigg|_{t=t_0}^{t=t_1} + \int_{t_0}^{t_1} \sum_{k=1}^{N} \left(\left(\frac{\partial L}{\partial x_k} - \frac{d}{dt} \frac{\partial L}{\partial \dot{x}_k} \right) \delta x_k \right.$$
$$\left. + \frac{\partial L}{\partial \dot{x}_k} \left(\delta \frac{dx_k}{dt} - \frac{d}{dt} \delta x_k \right) \right) dt.$$

Hence, since the virtual variation vanishes at the points $t = t_0$ and $t = t_1$, the proposition is established. ⬚

Corollary 7.2.2. *The first variation of the action for a Lagrangian system* $(Q, \tilde{L})$ *can be calculated as*

$$\delta F = - \int_{t_0}^{t_1} \sum_{k=1}^{N} E_k \tilde{L} \left(t, \mathbf{x}, \frac{d\mathbf{x}}{dt} \right) \delta x_k \, dt.$$

Proof. This follows from Proposition 7.2.1 by observing that for the Lagrangian system the transpositional relation is equal to zero (see for instance [113, p. 29]) i.e.,

$$\delta \frac{d\mathbf{x}}{dt} - \frac{d}{dt} \delta \mathbf{x} = 0. \tag{7.6}$$

⬚

The path $\gamma(t) \in \Omega$ is called a *motion* of the Lagrangian systems $(Q, \tilde{L})$ if $\gamma(t)$ is a critical point of the action F, i.e.,

$$\delta F(\gamma(t)) = 0 \Longleftrightarrow \int_{t_0}^{t_1} \delta \tilde{L} \, dt = 0. \tag{7.7}$$

This definition is known as the *Hamiltonian variational principle* or *Hamiltonian least action principle*, or simply *Hamiltonian principle*.

Now we need the *Lagrange lemma* or *fundamental lemma of calculus of variations* (see for instance [2]).

Lemma 7.2.3. *Let* f *be a continuous function on the interval* $[t_0, t_1]$ *such that*

$$\int_{t_0}^{t_1} f(t) \zeta(t) dt = 0$$

for any continuous function $\zeta(t)$ *for which* $\zeta(t_0) = \zeta(t_1) = 0$. *Then* $f(t) \equiv 0$.

Corollary 7.2.4. *The Hamiltonian principle for Lagrangian systems is equivalent to the Lagrangian equations*

$$E_\nu \tilde{L} = \frac{d}{dt} \left(\frac{\partial \tilde{L}}{\partial \dot{x}_\nu} \right) - \frac{\partial \tilde{L}}{\partial x_\nu} = 0, \quad \nu = 1, \ldots, N. \tag{7.8}$$

Proof. Clearly, if (7.8) holds, then by Corollary 7.2.2, $\delta F = 0$. The converse follows from Lemma 7.2.3. $\square$

Formally, the Hamiltonian principle in the form (7.7) is equivalent to the problem of *calculus of variation* [65, 127]. However, despite the superficial similarity, they differ essentially. Namely, in mechanics the symbol δ stands for the *virtual variation*, i.e., it is not an arbitrary variation, but a displacement compatible with the constraints imposed on the system. Thus only in the case of holonomic systems, for which the number of degrees of freedom is equal to the number of generalized coordinates, the virtual variations are arbitrary and the Hamiltonian principle (7.7) is completely equivalent to the corresponding problem of the calculus of variation. An important difference arises for the systems with nonholonomic constraints, when the variations of the generalized coordinates are connected by the additional relations usually called *Chetaev conditions*, which will be given later on.

7.3   d'Alembert–Lagrange principle

Let $L_j : \mathbb{R} \times TQ \to \mathbb{R}$, $j = 1, \ldots, M$, be smooth functions. The equations

$$L_j = L_j\,(t, \mathbf{x}, \dot{\mathbf{x}}) = 0, \quad j = 1, \ldots, M < N,$$

with rank $\left(\dfrac{\partial(L_1, \ldots, L_M)}{\partial(\dot{x}_1, \ldots, \dot{x}_N)} \right) = M$ in all the points of $\mathbb{R} \times TQ$, except perhaps in a set of Lebesgue measure zero, define M *independent constraints* for the Lagrangian systems $(Q, \tilde{L})$.

Let $\mathcal{M}^*$ be the submanifold of $\mathbb{R} \times TQ$ given by

$$\mathcal{M}^* = \{(t, \mathbf{x}, \dot{\mathbf{x}}) \in \mathbb{R} \times TQ : L_j(t, \mathbf{x}, \dot{\mathbf{x}}) = 0, \quad j = 1, \ldots, M\}.$$

A *constrained Lagrangian system* is a triplet $(Q, \tilde{L}, \mathcal{M}^*)$. The number of degree of freedom is $\kappa = \dim Q - M = N - M$.

A constraint is called *integrable* if it can be written in the form

$$L_j = \frac{d}{dt}\,(G_j(t, \mathbf{x})) = 0,$$

for a suitable function G_j. Otherwise the constraint is called *nonintegrable*. According to Hertz [72], the *nonintegrable constraints* are also called *nonholonomic*.

The Lagrangian systems with nonintegrable constraints are usually called (also following to Hertz) *nonholonomic mechanical systems*, or *nonholonomic constrained mechanical systems*, and those with integrable constraints are called *holonomic constrained mechanical systems* or *holonomic constrained Lagrangian systems*. The systems free of constraints are called *Lagrangian systems or holonomic systems*.

For some other authors (see for instance Wikipedia) a classical mechanica system is defined as holonomic if all constraints of the system are holonomic. Fo a constraint to be holonomic it must be expressible as a $f(x_1, \ldots, x_N, t) = 0$, i.e. a holonomic constraint depends only on the coordinates $(x_1, \ldots, x_N)$ and time t. A constraint that cannot be expressed in this form is a nonholomic constraint There is a longer discussion about the relationships between the two definitions.

Sometimes it is also useful to distinguish between constraints that are de pendent on, or independent of time. The former are called *rheonomic*, while the latter are called *scleronomic*. This terminology can also be applied to the me chanical systems themselves. Thus we say that a constrained Lagrangian systems is rheonomic (scleronomic) if the constraints and Lagrangian are time dependen (independent).

Constraints of the form

$$L_k = \sum_{j=1}^{N} a_{kj} \dot{x}_j + a_k = 0, \quad \text{for} \quad k = 1, \ldots, M, \tag{7.9}$$

where $a_{kj} = a_{kj}(t, \mathbf{x})$, $a_k = a_k(t, \mathbf{x})$, are called *linear constraints with respect t the velocity*. For simplicity we shall call these *linear constraints*.

We observe that (7.9) admits an equivalent representation as a system o *Pfaffian equations* (for more details see [126])

$$\omega_k := \sum_{j=1}^{N} a_{kj} dx_j + a_k \, dt = 0.$$

We shall consider only two classes of systems of equations, the equations of con straints linear in the velocity $(\dot{x}_1, \ldots, \dot{x}_N)$, or linear in the differential $(dx_1, \ldots dx_N, dt)$. In order to study the integrability or nonintegrability of the constraints the last representation, a Pfaffian system, is the more useful. This is related to the fact that for the given 1-forms we have the *Frobenius Theorem*, which provides necessary and sufficient conditions for the 1-forms to be closed, and consequently for the given set of constraints to be integrable.

The constraints $L_j(t, \mathbf{x}, \dot{\mathbf{x}}) = 0$ are called *perfect* or *ideal constraints* if they satisfy the Chetaev conditions (see [27])

$$\sum_{k=1}^{N} \frac{\partial L_\alpha}{\partial \dot{x}_k} \delta x_k = 0, \quad \alpha = 1, \ldots, M.$$

In what follows, we shall consider only perfect constraints.

If the constraints admit the representation

$$\dot{x}_\alpha = \Phi_\alpha \left(\mathbf{x}, \dot{x}_{M+1}, \ldots, \dot{x}_N \right) \quad \alpha = 1, \ldots, M, \tag{7.10}$$

hen the Chetaev conditions take the form

$$\delta x_\alpha = \sum_{k=M+1}^{N} \frac{\partial \Phi_\alpha}{\partial \dot{x}_k} \delta x_k.$$

The virtual variations of the variables x_α for $\alpha = 1, \ldots, M$ are called *dependent variations*, while those of the variable x_β for $\beta = M + 1, \ldots, N$ are called *independent variations*.

We say that the path $\gamma(t) = \mathbf{x}(t)$ is *compatible* with the given perfect constraints if $L_j(t, \mathbf{x}(t), \dot{\mathbf{x}}(t)) = 0$.

The *compatible path* $\gamma(t)$ is called a *motion* of the constrained Lagrangian systems $(Q, \tilde{L}, \mathcal{M}^*)$ if for all $t \in [t_0, t_1]$

$$\sum_{\nu=1}^{N} E_\nu \tilde{L}(t, \mathbf{x}(t), \dot{\mathbf{x}}(t)) \, \delta x_\nu(t) = 0,$$

for all virtual displacements $\delta \mathbf{x}(t)$ of the path $\gamma(t)$. This definition is known as the *l'Alembert–Lagrange principle*.

The following result is well known (see for instance [4, 12, 70, 120]).

Proposition 7.3.1. *The d'Alembert–Lagrange principle for constrained Lagrangian systems is equivalent to the Lagrange differential equations with multipliers*

$$E_j \tilde{L} = \frac{d}{dt} \frac{\partial \tilde{L}}{\partial \dot{x}_j} - \frac{\partial \tilde{L}}{\partial x_j} = \sum_{\alpha=1}^{M} \mu_\alpha \frac{\partial L_\alpha}{\partial \dot{x}_j}, \, j = 1, \ldots, N, \tag{7.11}$$

$$L_j(t, \mathbf{x}, \dot{\mathbf{x}}) = 0, \qquad\qquad j = 1, \ldots, M,$$

where μ_α for $\alpha = 1, \ldots, M$ are the Lagrange multipliers.

7.4 The varied path and transpositional relations

n general, the varied path produced in the Hamiltonian principle is not an admissible path if the perfect constraints are nonholonomic, i.e., the mechanical system cannot travel along the varied path without violating the constraints. We prove he following result, which shall play an important role in all the assertions below.

Proposition 7.4.1. *If the varied path is an admissible path, then*

$$\sum_{k=1}^{N} \frac{\partial L_\alpha}{\partial \dot{x}_k} \left(\delta \frac{dx_k}{dt} - \frac{d}{dt} \delta x_k \right) = \sum_{k=1}^{N} E_k L_\alpha \delta x_k, \qquad \alpha = 1, \ldots, M. \tag{7.12}$$

Proof. Indeed, the original path $\gamma(t) = \mathbf{x}(t)$ does by definition satisfy the Chetaev conditions and the constraints, i.e., $L_j\left(t, \mathbf{x}(t), \dot{\mathbf{x}}(t)\right) = 0$. Now suppose that the varied path $\gamma^*(t) = \mathbf{x}(t) + \varepsilon\delta\mathbf{x}(t)$ also satisfies the constraints, i.e.,

$$L_j\left(t, \mathbf{x} + \varepsilon\delta\mathbf{x}, \dot{\mathbf{x}} + \varepsilon\delta\dot{\mathbf{x}}\right) = L_j\left(t, \mathbf{x}(t), \dot{\mathbf{x}}(t)\right) + \varepsilon\delta\, L_\alpha\left(t, \mathbf{x}(t), \dot{\mathbf{x}}(t)\right) + \cdots = 0.$$

Then restricting only to the terms of first order in ε and using the Chetaev conditions we have (for simplicity, we omit the argument)

$$0 = \delta\, L_\alpha = \sum_{k=1}^{N}\left(\frac{\partial L_\alpha}{\partial x_k}\delta x_k + \frac{\partial L_\alpha}{\partial \dot{x}_k}\delta\dot{x}_k\right),$$

$$0 = \sum_{k=1}^{N}\frac{\partial L_\alpha}{\partial \dot{x}_k}\,\delta x_k, \tag{7.13}$$

for $\alpha = 1,\ldots, M$. The Chetaev conditions are satisfied at each instant, so

$$\frac{d}{dt}\left(\sum_{k=1}^{N}\frac{\partial L_\alpha}{\partial \dot{x}_k}\delta x_k\right) = \sum_{k=1}^{N}\frac{d}{dt}\left(\frac{\partial L_\alpha}{\partial \dot{x}_k}\right)\delta x_k + \sum_{k=1}^{N}\frac{\partial L_\alpha}{\partial \dot{x}_k}\frac{d}{dt}\delta x_k = 0.$$

Subtracting these relations from (7.13) we obtain (7.12). Consequently if the varied path is admissible, then relations (7.12) must hold. $\quad\square$

Now we shall suppose that the following relation holds

$$\delta\frac{d\mathbf{x}}{dt} - \frac{d}{dt}\delta\mathbf{x} = A\left(t, \mathbf{x}, \dot{\mathbf{x}}, \ddot{\mathbf{x}}\right)\delta\mathbf{x}, \tag{7.14}$$

where $A = A\left(t, \mathbf{x}, \dot{\mathbf{x}}, \ddot{\mathbf{x}}\right) = \left(A_{\nu j}\left(t, \mathbf{x}, \dot{\mathbf{x}}, \ddot{\mathbf{x}}\right)\right)$ is an $N \times N$ matrix,

From (7.12) and (7.14) it follows that the elements of the matrix A satisfy

$$\sum_{m=1}^{N}\delta x_m\left(E_m L_\alpha - \sum_{k=1}^{N}A_{km}\frac{\partial L_\alpha}{\partial \dot{x}_k}\right) = \sum_{m=1}^{N}\delta x_m D_m L_\alpha = 0, \quad \alpha = 1,\ldots, M. \tag{7.15}$$

This property will be used below.

Corollary 7.4.2. *For the holonomic constrained Lagrangian systems the relations* (7.12) *hold if and only if*

$$\sum_{k=1}^{N}\frac{\partial L_\alpha}{\partial \dot{x}_k}\left(\delta\frac{dx_k}{dt} - \frac{d}{dt}\delta x_k\right) = 0, \quad \alpha = 1,\ldots, M. \tag{7.16}$$

Proof. Indeed, for such systems the constraints are integrable, consequently in view of (7.4) we have $E_k L_\alpha = 0$ for $k = 1,\ldots, N$ and $\alpha = 1,\ldots, M$. Thus, from (7.12), we obtain (7.16). $\quad\square$

Clearly the equalities (7.16) are satisfied if (7.6) holds. We observe that in general for holonomic constrained Lagrangian systems relation (7.6) cannot hold (see Example 7.10.1).

Now we introduce the following important concept: the *transpositional relations*.

As we observed in the previous section, for nonholonomic constrained Lagrangian systems the curves, obtained doing a virtual variation in the motion of the system, in general are not *kinematically possible trajectories* when (7.6) is not fulfilled. This leads to the conclusion that the Hamiltonian principle cannot be applied to nonholonomic systems, as it is usually employed for holonomic systems. The essence of the problem of the applicability of this principle for nonholonomic systems remains unclarified (see [120]). In order to clarify this situation, it is sufficient to note that the question of the applicability of the principle of stationary action to nonholonomic systems is intimately related to the question of transpositional relations.

The key point is that the Hamiltonian principle assumes that the operation of differentiation with respect to time $\dfrac{d}{dt}$ and the virtual variation δ commute in all the generalized coordinate systems.

For the holonomic constrained Lagrangian systems relations (7.6) cannot hold (see Corollary 7.4.2). For a nonholonomic systems the form of the Hamiltonian principle will depend on the point of view adopted on the transpositional relations.

What are then the correct transpositional relations? Until now, there is no generally accepted point of view concerning to the commutativity of the operation of differentiation with respect to time and the virtual variation when nonintegrable constraints are present. Two points of view have been maintained. According to one (supported, for example, by Volterra, Hamel, Hölder, Lurie, Pars, ...), the operations $\dfrac{d}{dt}$ and δ commute for all the generalized coordinates, regardless of whether the systems are holonomic or nonholonomic, i.e.,

$$\delta \frac{dx_k}{dt} - \frac{d}{dt}\delta x_k = 0, \quad \text{for} \quad k = 1, \ldots, N.$$

According to the other point of view (supported by Suslov, Voronets, Levi-Civita, Amaldi, ...), the operations $\dfrac{d}{dt}$ and δ commute always for holonomic systems, and for nonholonomic systems with constraints of the form

$$\dot{x}_\alpha = \sum_{j=M+1}^{N} a_{\alpha j}(t, \mathbf{x})\dot{x}_j + a_\alpha(t, \mathbf{x}), \quad \text{for} \quad \alpha = 1, \ldots, M. \tag{7.17}$$

The transpositional relations are equal to zero only for the generalized coordinates $x_{M+1}, \ldots, x_N$, (for which their virtual variations are independent). For the remaining coordinates $x_1, \ldots, x_M$ (for which their virtual variations are dependent), the transpositional relations must be derived on the basis of the equations of the nonholonomic constraints, and cannot be identically zero, i.e.,

$$\delta \frac{dx_k}{dt} - \frac{d}{dt}\delta x_k = 0, \quad \text{for} \quad k = M+1, \ldots, N,$$

$$\delta \frac{dx_k}{dt} - \frac{d}{dt}\delta x_k \neq 0, \quad \text{for} \quad k = 1, \ldots, M.$$

The second point of view gained general acceptance and the first point of view was considered erroneous (for more details, see [120]). The meaning of the transpositional relations (7.1) can be found in [81, 113, 116, 120].

In the results given in the following section a key role is played by the equalities (7.12). These equalities and the examples below will demonstrate that the second point of view is correct only for the so called Voronets–Chaplygin systems and in general for locally nonholonomic systems. There exist many examples for which the independent virtual variations generate non–zero transpositional relations. Thus we propose a third point of view on the transpositional relations: the virtual variations can generate the transpositional relations given by the formula (7.14) where the elements of the matrix A satisfies the conditions (see formula (7.15))

$$D_\nu L_\alpha = E_\nu L_\alpha - \sum_{k=1}^{N} A_{k\nu} \frac{\partial L_\alpha}{\partial \dot{x}_k} = 0, \quad \text{for} \quad \nu = 1, \ldots, M, \quad \alpha = 1, \ldots, M. \quad (7.18)$$

We observe that here the $L_\alpha = 0$ are constraints which in general are nonlinear in the velocity.

7.5   Hamilton–Suslov principle

After the introduction of nonholonomic mechanics by Hertz, there arose the problem of extending to nonholonomic mechanics the results of holonomic mechanics. Hertz [72] was the first to study the problem of applying the Hamiltonian principle to systems with nonintegrable constraints. In [72] Hertz wrote: "Application of Hamilton's principle to any material systems does not exclude that between selected coordinates of the systems rigid constraints exist, but it still requires that these relations could be expressed by integrable constraints. The appearance of nonintegrable constraints is unacceptable. In this case Hamilton's principle is not valid." Appell [3] in agreement with Hertz's ideas affirmed that it is not possible to apply the Hamiltonian principle for systems with nonintegrable constraints.

Suslov [147] claimed that "Hamilton's principle is not applicable to systems with nonintegrable constraints, as derived based on this equation are different from the corresponding equations of Newtonian mechanics".

The applications of the most general differential principle, i.e., the d'Alembert–Lagrange and their equivalent *Gauss and Appel principles*, is complicated due to the presence of the terms containing the second-order derivative. On the other hand, the most general integral variational principle of Hamilton is not valid for nonholonomic constrained Lagrangian systems. The generalization of the Hamiltonian principle for nonholonomic mechanical systems was carried out by Voronets and Suslov (see for instance [147, 159]). As we will see later on this principle, shows the importance of the transpositional relations for determining the correct equations of motion for nonholonomic constrained Lagrangian systems.

Proposition 7.5.1. *The d'Alembert–Lagrange principle for the contrained Lagrangian systems $\sum_{k=1}^{N} \delta x_k E_k \tilde{L} = 0$ is equivalent to the Hamilton–Suslov principle (7.2), where we assume that $\delta x_\nu(t)$, $\nu = 1, \ldots, N$, are arbitrary smooth functions defined in the interior of the interval $[t_0, t_1]$ and vanishing at its endpoints, i.e., $\delta x_\nu(t_0) = \delta x_\nu(t_1) = 0$.*

Proof. From the d'Alembert–Lagrange principle we obtain the identity

$$
\begin{aligned}
0 = -\sum_{k=1}^{N} \delta x_k E_k \tilde{L} &= \sum_{k=1}^{N} \delta x_k \frac{\partial \tilde{L}}{\partial x_k} - \sum_{k=1}^{N} \delta x_k \frac{d}{dt} \frac{\partial \tilde{L}}{\partial \dot{x}_k} \\
&= \sum_{k=1}^{N} \left(\delta x_k \frac{\partial \tilde{L}}{\partial x_k} + \delta \dot{x}_k \frac{\partial \tilde{L}}{\partial \dot{x}_k} \right) - \sum_{k=1}^{N} \left(\left(\delta \frac{dx_k}{dt} - \frac{d}{dt} \delta x_k \right) \frac{\partial \tilde{L}}{\partial \dot{x}_k} - \frac{d}{dt} \left(\frac{\partial \tilde{L}}{\partial \dot{x}_k} \delta x_k \right) \right) \\
&= \delta \tilde{L} - \sum_{k=1}^{N} \left(\left(\delta \frac{dx_k}{dt} - \frac{d}{dt} \delta x_k \right) \frac{\partial \tilde{L}}{\partial \dot{x}_k} - \frac{d}{dt} \left(\frac{\partial \tilde{L}}{\partial \dot{x}_k} \delta x_k \right) \right),
\end{aligned}
$$

where $\delta \tilde{L}$ is a variation of the Lagrangian $\tilde{L}$. After integration and assuming that $\delta x_k(t_0) = 0$, $\delta x_k(t_1) = 0$ we easily obtain (7.2), which represents the most general formulation of the Hamiltonian principle (*Hamilton–Suslov principle*) suitable for constrained and unconstrained Lagrangian systems. $\qquad \square$

Suslov determined the transpositional relations only for the case when the constraints are of Voronets type, i.e., given by the formula (7.17). Assume that

$$
\delta \frac{dy_k}{dt} - \frac{d}{dt} \delta y_k = 0, \quad \text{for} \quad k = M+1, \ldots, N.
$$

Voronets and Suslov deduced that

$$
\delta \frac{dx_k}{dt} - \frac{d}{dt} \delta x_k = \sum_{k=1}^{N} B_{kr} \delta y_r - \delta a_k
$$

for certain suitable functions $B_{kr} = B_{kr}(t, \mathbf{x}, \mathbf{y}, \dot{\mathbf{x}}, \dot{\mathbf{y}})$, for $r = M+1, \ldots, N$ and $k = 1, \ldots, M$.

Thus we obtain

$$\int_{t_0}^{t_1} \left(\delta \tilde{L} - \sum_{k=1}^{N} \frac{\partial \tilde{L}}{\partial \dot{x}_j} \left(\sum_{k=1}^{N} B_{kr} \delta y_r - \delta a_k \right) \right) dt = 0,$$

This is the Hamiltonian principle for nonholonomic systems in the Suslov form (see for instance [147]). We observe that the same result was deduced by Voronets in [159].

It is important to observe that Suslov and Voronets required a priori that the independent virtual variations produce the zero transpositional relations. At the sometimes these authors consider only linear constraints with respect to the velocity of the type (7.17).

7.6   Modificated vakonomic mechanics (MVM)

As we observed in the Introduction to this chapter, the main objective here is to construct the variational equations of motion describing the behavior of the constrained Lagrangian systems for which the equalities (7.12) hold in the most general possible way. We shall show that the d'Alembert–Lagrange principle is not the only way to deduce the equations of motion for the constrained Lagrangian systems. Instead of it we can apply the generalization of the Hamiltonian principle, whereby the motions of such systems are extremals of the *variational Lagrange problem* (see for instance [65]), i.e., the problem of determining the critical points of the action in the class of curves with fixed endpoints and satisfying the constraints. As we will see, the solution of this problem will give second-order differential equations which coincide with the well-known classical equations of mechanics except perhaps in a set of Lebesgue measure zero.

From the previous section we deduce that in order to generalize the Hamiltonian principle to nonholonomic systems we must take into account the following relations:

(A) $\delta L_\alpha = \sum_{j=1}^{N} \left(\frac{\partial L_\alpha}{\partial x_j} \delta x_j + \frac{\partial L_\alpha}{\partial \dot{x}_j} \delta \dot{x}_j \right) = 0 \quad$ for $\quad \alpha = 1, \ldots, M,$

(B) $\sum_{j=1}^{N} \frac{\partial L_\alpha}{\partial \dot{x}_j} \delta x_j = 0 \quad$ for $\quad \alpha = 1, \ldots, M,$

(C) $\delta \dfrac{dx_j}{dt} - \dfrac{d}{dt} \delta x_j = 0 \quad$ for $\quad j = 1, \ldots, N,$

where $L_\alpha = 0$ for $\alpha = 1, \ldots, M$ are the constraints.

Many authors consider that (C) is always fulfilled (see for instance [113, 126]), together with the conditions (A) and (B). However, these conditions are ncompatible in the case of the nonintegrable constrains. We observe that these uthors deduced that the Hamiltonian principle is not applicable to nonholonomic ystems.

To obtain a generalization of the Hamiltonian principle to nonholonomic mechanical systems, some of the three conditions above must be excluded.

In particular, for the *Hölder principle* condition (A) is excluded and one etains (B) and (C) (see [74]). For the Hamilton–Suslov principle condition (A) nd (B) hold, but (C) only holds for the independent variations.

Here we extend the Hamiltonian principle by supposing that conditions (A) nd (B) hold, but (C) does not hold. Instead of (C) we assume that transpositional elation are given by (7.14) where the elements of the matrix A satisfy the relations 7.18).

.7 Inverse problem in vakonomic mechanics. Main results

n this section we state and solve the following *inverse problem in vakonomic mechanics* (see [107])

We consider constrained Lagrangian systems with configuration space Q and phase space TQ.

Let $L : \mathbb{R} \times TQ \times \mathbb{R}^M \to \mathbb{R}$ be a smooth function such that

$$L\left(t, \mathbf{x}, \dot{\mathbf{x}}, \Lambda\right) = L_0\left(t, \mathbf{x}, \dot{\mathbf{x}}\right) - \sum_{j=1}^{M} \lambda_j L_j\left(t, \mathbf{x}, \dot{\mathbf{x}}\right) - \sum_{j=M+1}^{N} \lambda_j^0 L_j\left(t, \mathbf{x}, \dot{\mathbf{x}}\right), \qquad (7.19)$$

where $\Lambda = (\lambda_1, \ldots, \lambda_M)$ are the additional coordinates (Lagrange multipliers), $L_j : \mathbb{R} \times TQ \to \mathbb{R}$, $(t, \mathbf{x}, \dot{\mathbf{x}}) \mapsto L_j\left(t, \mathbf{x}, \dot{\mathbf{x}}\right)$ are smooth functions for $j = 0, \ldots, N$, with L_0 non-singular, i.e., $\det\left(\dfrac{\partial^2 L_0}{\partial \dot{x}_k \partial \dot{x}_j}\right) \neq 0$, and $L_j = 0$, for $j = 1, \ldots, M$, are the constraints, satisfying

$$\mathrm{rank}\left(\frac{\partial(L_1, \ldots, L_M)}{\partial(\dot{x}_1, \ldots, \dot{x}_N)}\right) = M \qquad (7.20)$$

n all the points of $\mathbb{R} \times TQ$, except perhaps in a set of Lebesgue measure zero, L_j nd λ_j^0 are arbitrary functions and constants, respectively, for $j = M + 1, \ldots, N$.

We must determine the smooth functions L_j, the constants λ_j^0 for $j = M + 1, \ldots, N$, and the matrix A in such a way that the differential equations escribing the behavior of the constrained Lagrangian system are obtained from the Hamiltonian principle

$$
\begin{aligned}
0 &= \int_{t_0}^{t_1} \delta L \\
&= \int_{t_0}^{t_1} \left(\frac{\partial L}{\partial x_j} \delta x_j + \frac{\partial L}{\partial \dot{x}_j} \frac{d}{dt} \delta x_j + \sum_{j=1}^{N} \frac{\partial L}{\partial \dot{x}_j} \left(\delta \frac{dx_j}{dt} - \frac{d}{dt} \delta x_j \right) \right) dt,
\end{aligned}
\tag{7.21}
$$

with transpositional relations given by (7.14).

We give the solution of this problem in two steps. First, we obtain the di: ferential equations along the solutions satisfying (7.21). Second, we shall contras the obtained equations and the classical differential equations which describe th behavior of constrained mechanical systems. The solution of this inverse probler is presented in Section 4.

Note that the function L is singular, due to the absence of $\dot{\lambda}$.

We observe that the arbitrariness of the functions L_j, of the constants λ for $j = M+1, \ldots, N$, and of the matrix A will play a fundamental role in th construction of the mathematical model proposed in this section.

The solution of this inverse problem is given by the following theorem.

Theorem 7.7.1. *We assume that $\delta x_\nu(t)$, $\nu = 1, \ldots, N$, are arbitrary function defined in the interval $[t_0, t_1]$, smooth in the interior of $[t_0, t_1]$, and vanishing its endpoints, i.e., $\delta x_\nu(t_0) = \delta x_\nu(t_1) = 0$. If (7.14) holds, then the path $\gamma(t)$ = $(x_1(t), \ldots, x_N(t))$ compatible with the constraints $L_j(t, \mathbf{x}, \dot{\mathbf{x}}) = 0$, for $j = 1, \ldots, M$ satisfies (7.21) with L given by the formula (7.19) if and only if it is a solution of the differential equations*

$$
D_\nu L := E_\nu L - \sum_{j=1}^{N} A_{\nu j} \frac{\partial L}{\partial \dot{x}_j} = 0, \qquad \frac{\partial L}{\partial \lambda_k} = -L_k = 0,
\tag{7.22}
$$

for $\nu = 1, \ldots, N$, and $k = 1, \ldots, M$, where $E_\nu = \dfrac{d}{dt} \dfrac{\partial}{\partial \dot{x}_\nu} - \dfrac{\partial}{\partial x_\nu}$. System (7.22) equivalent to the following two differential systems:

$$
D_\nu L_0 = \sum_{j=1}^{M} \left(\lambda_j D_\nu L_j + \frac{d\lambda_j}{dt} \frac{\partial L_j}{\partial \dot{x}_\nu} \right) + \sum_{j=M+1}^{N} \lambda_j^0 D_\nu L_j, \quad L_k = 0 \Longleftrightarrow
$$

$$
E_\nu L_0 = \sum_{k=1}^{N} A_{jk} \frac{\partial L_0}{\partial \dot{x}_k} + \sum_{j=1}^{M} \left(\lambda_j D_\nu L_j + \frac{d\lambda_j}{dt} \frac{\partial L_j}{\partial \dot{x}_\nu} \right) + \sum_{j=M+1}^{N} \lambda_j^0 D_\nu L_j
\tag{7.23}
$$

$$
L_k = 0,
$$

for $\nu = 1, \ldots, N$ and $k = 1, \ldots, M$.

The following results are consequences of Theorem 7.7.1.

Theorem 7.7.2. *Using the notation of Theorem 7.7.1, let*

$$L = L(t, \mathbf{x}, \dot{\mathbf{x}}, \Lambda) = L_0(t, \mathbf{x}, \dot{\mathbf{x}}) - \sum_{j=1}^{M} \lambda_j L_j(t, \mathbf{x}, \dot{\mathbf{x}}) - \sum_{j=M+1}^{N} \lambda_j^0 L_j(t, \mathbf{x}, \dot{\mathbf{x}}) \quad (7.24)$$

be the Lagrangian, $L_j(t, \mathbf{x}, \dot{\mathbf{x}}) = 0$ be the independent constraints for $j = 1, \ldots,$ $M < N$, λ_k^0 be arbitrary constants for $k = M+1, \ldots, N$, and $L_k : \mathbb{R} \times TQ \to \mathbb{R}$, $k = M+1, \ldots, N$, be arbitrary functions such that

$$|W_1| = \det W_1 = \det\left(\frac{\partial(L_1, \ldots, L_N)}{\partial(\dot{x}_1, \ldots, \dot{x}_N)}\right) \neq 0,$$

except perhaps in the set $|W_1| = 0$ of Lebesgue measure zero. We determine the matrix A satisfying

$$W_1 A = \Omega_1 := \begin{pmatrix} E_1 L_1 & \cdots & E_N L_1 \\ \vdots & \cdots & \vdots \\ \vdots & \cdots & \vdots \\ E_1 L_N & \cdots & E_N L_N \end{pmatrix}. \quad (7.25)$$

Then the differential equations (7.23) become

$$D_\nu L_0 = \sum_{\alpha=1}^{M} \dot{\lambda}_\alpha \frac{\partial L_\alpha}{\partial \dot{x}_\nu} \quad for \quad \nu = 1, \ldots, N$$

$$\iff \frac{d}{dt}\frac{\partial L_0}{\partial \dot{\mathbf{x}}} - \frac{\partial L_0}{\partial \mathbf{x}} = \left(W_1^{-1}\Omega_1\right)^T \frac{\partial L_0}{\partial \dot{\mathbf{x}}} + W_1^T \frac{d\lambda}{dt}, \quad (7.26)$$

where

$$\frac{\partial}{\partial \dot{\mathbf{x}}} = \left(\frac{\partial}{\partial \dot{x}_1}, \ldots, \frac{\partial}{\partial \dot{x}_N}\right)^T, \frac{\partial}{\partial \mathbf{x}} = \left(\frac{\partial}{\partial x_1}, \ldots, \frac{\partial}{\partial x_N}\right)^T, \lambda = (\lambda_1, \ldots, \lambda_M, 0, \ldots, 0)^T,$$

and the transpositional relation (7.14) reads

$$\delta\frac{d\mathbf{x}}{dt} - \frac{d}{dt}\delta\mathbf{x} = \left(W_1^{-1}\Omega_1\right)\delta\mathbf{x}. \quad (7.27)$$

Theorem 7.7.3. *Using the notation of Theorem 7.7.1, let*

$$L(t, \mathbf{x}, \dot{\mathbf{x}}, \Lambda) = L_0(t, \mathbf{x}, \dot{\mathbf{x}}) - \sum_{j=1}^{M} \lambda_j L_j(t, \mathbf{x}, \dot{\mathbf{x}}) - \sum_{j=M+1}^{N-1} \lambda_j^0 L_j(t, \mathbf{x}, \dot{\mathbf{x}}) \quad (7.28)$$

be the Lagrangian, $L_j(t, \mathbf{x}, \dot{\mathbf{x}}) = 0$ be the independent constraints for $j = 1, \ldots,$ $M < N$, λ_j^0 be arbitrary constants, for $j = M+1, \ldots, N-1$ and $\lambda_N^0 = 0$,

$L_j : \mathbb{R} \times TQ \to \mathbb{R}$ *for* $j = M+1, \ldots, N-1$ *arbitrary functions, and* $L_N = L_0$
such that

$$|W_2| = \det W_2 = \det\left(\frac{\partial(L_1, \ldots, L_{N-1}, L_0)}{\partial(\dot{x}_1, \ldots, \dot{x}_N)}\right) \neq 0,$$

except perhaps in the set $|W_2| = 0$ *of Lebesgue measure zero. We determine th*
matrix A satisfying

$$W_2 A = \Omega_2 := \begin{pmatrix} E_1 L_1 & \cdots & E_N L_1 \\ \vdots & \cdots & \vdots \\ E_1 L_{N-1} & \cdots & E_N L_{N-1} \\ 0 & \cdots & 0 \end{pmatrix}. \tag{7.29}$$

Then the differential equations (7.23) *become*

$$\frac{d}{dt}\frac{\partial L_0}{\partial \dot{\mathbf{x}}} - \frac{\partial L_0}{\partial \mathbf{x}} = W_2^T \frac{d}{dt}\tilde{\lambda}, \tag{7.30}$$

where $\lambda := \tilde{\lambda} = (\tilde{\lambda}_1, \ldots, \tilde{\lambda}_M, 0, \ldots, 0)^T$, *and the transpositional relation* (7.14
reads

$$\delta\frac{d\mathbf{x}}{dt} - \frac{d}{dt}\delta\mathbf{x} = \left(W_2^{-1}\Omega_2\right)\delta\mathbf{x}, \tag{7.31}$$

The results will be illustrated below with precise examples.

Now we solve the inverse problem of vakonomic mechanics and prove the
main results.

First we shall determine the equations of motion of constrained Lagrangian
systems using the Hamiltonian principle with non-zero transpositional relations
whereby the motions of the systems are *extremals* of the *variational Lagrang*
problem (see for instance [65]), i.e., are *critical points of the action functional*

$$\int_{t_0}^{t_1} L_0\left(t, \mathbf{x}, \dot{\mathbf{x}}\right) dt,$$

in the class of paths with fixed endpoints satisfying the independent constraints

$$L_j\left(t, \mathbf{x}, \dot{\mathbf{x}}\right) = 0, \quad \text{for} \quad j = 1, \ldots, M.$$

In the classical solution of the Lagrange problem we usually apply the *Lagrange*
multipliers which consists in the following. We introduce the additional coordinate
$\Lambda = (\lambda_1, \ldots, \lambda_M)$ and the Lagrangian $\widehat{L} : \mathbb{R} \times TQ \times \mathbb{R}^M \to \mathbb{R}$ given by

$$\widehat{L}\left(t, \mathbf{x}, \dot{\mathbf{x}}, \Lambda\right) = L_0\left(t, \mathbf{x}, \dot{\mathbf{x}}\right) - \sum_{j=1}^{M}\lambda_j L_j\left(t, \mathbf{x}, \dot{\mathbf{x}}\right).$$

Under this choice we reduce the Lagrange problem to a variational problem with
out constraints, i.e., we must determine the extremals of the action functional

$\int_{t_0}^{t_1} \widehat{L} \, dt$. We shall study a slight *modification of the method of Lagrange multipli-ers*. Namely, we introduce the additional coordinates $\Lambda = (\lambda_1, \ldots, \lambda_M)$, and the Lagrangian on $\mathbb{R} \times TQ \times \mathbb{R}^M$ given by the formula (7.19), where we assume that λ_j^0 are arbitrary constants, and L_j are arbitrary functions for $j = M + 1, \ldots, N$.

We need to determine the critical points of the action functional

$$\int_{t_0}^{t_1} L\left(t, \mathbf{x}, \dot{\mathbf{x}}, \Lambda\right) dt,$$

i.e., the paths $\gamma(t)$ such that $\int_{t_0}^{t_1} \delta\left(L\left(t, \mathbf{x}, \dot{\mathbf{x}}, \Lambda\right)\right) dt = 0$ under the additional condition that the transpositional relations are given by (7.14).

The solution of the inverse problem stated in Section 2 runs as follows. The differential equations obtained from (7.21) are given by (7.22) (see Theorem 7.7.1). We choose the arbitrary functions L_j in such a away that the matrices W_1 and V_2 given in Theorems 7.7.2 and 7.7.3 are non-singular, except perhaps in a set of Lebesgue measure zero. The constants λ_j^0 for $j = M + 1, \ldots, N$ are arbitrary in Theorem 7.7.2, and λ_j^0 for $j = 1, \ldots, N - 1$ are arbitrary and $\lambda_N^0 = 0$ in Theorem 7.7.3. The matrix A is determined from the equalities (7.25) and (7.29) of Theorems 7.7.2 and 7.7.3, respectively.

Remark 7.7.4. It is interesting to observe that in the solutions of the inverse problem, the constants λ_j^0 for $j = M + 1, \ldots, N$ are arbitrary, except in Theorem 7.7.3, in which $\lambda_N^0 = 0$. Clearly, if $L_j\left(t, \mathbf{x}, \dot{\mathbf{x}}\right) = \dfrac{d}{dt} f_j(t, \mathbf{x})$ for $j = M + 1, \ldots, N$, then the $L \simeq \widehat{L}$. Using the arbitrariness of the constants λ_j^0 we can always take $\lambda_k^0 = 0$ if $L_k\left(t, \mathbf{x}, \dot{\mathbf{x}}\right) \neq \dfrac{d}{dt} f_k(t, \mathbf{x})$. Consequently, we can always suppose that $L \simeq \widehat{L}$. Thus the only difference between the classical and the modified method of Lagrange multipliers concerns the transpositional relations: for the classical method, the virtual variations produce null transpositional relations (i.e., the matrix A is the zero matrix), while for the modified method in general A is determined by the formulae (7.14) and (7.15).

A very important subscase is obtained when the constraints are given in the form (*constraints of Voronets–Chapliguin type*) $\dot{x}_\alpha - \Phi_\alpha\left(t, \mathbf{x}, \dot{x}_{M+1}, \ldots, \dot{x}_N\right) = 0$, for $\alpha = 1, \ldots, M$. As we shall show, under these assumptions the arbitrary functions are determined as follows: $L_j = \dot{x}_j$ for $j = M + 1, \ldots, N$. Consequently, the action of the modified method of Lagrange multipliers and the action of the classical method of Lagrange multipliers are equivalent. In view of (7.10), this equivalence always holds locally for any constrained Lagrangian systems.

Proof of Theorem 7.7.1. We have

$$
\int_{t_0}^{t_1} \delta L \, dt = \int_{t_0}^{t_1} \sum_{k=1}^{M} \left(\frac{\partial L}{\partial \lambda_k} \delta \lambda_k \right) dt + \int_{t_0}^{t_1} \sum_{j=1}^{N} \left(\frac{\partial L}{\partial x_j} \delta x_j + \frac{\partial L}{\partial \dot{x}_j} \delta \frac{dx_j}{dt} \right) dt
$$

$$
= \int_{t_0}^{t_1} \sum_{k=1}^{M} (-L_k \delta \lambda_k) \, dt
$$

$$
+ \int_{t_0}^{t_1} \sum_{j=1}^{N} \left(\frac{\partial L}{\partial x_j} \delta x_j + \frac{\partial L}{\partial \dot{x}_j} \frac{d}{dt} \delta x_j + \frac{\partial L}{\partial \dot{x}_j} \left(\delta \frac{dx_j}{dt} - \frac{d}{dt} \delta x_j \right) \right) dt
$$

$$
= \int_{t_0}^{t_1} \sum_{k=1}^{M} (-L_k \delta \lambda_k) \, dt + \int_{t_0}^{t_1} \sum_{j=1}^{N} \frac{d}{dt} \left(\frac{\partial T}{\partial \dot{x}_j} \delta x_j \right) dt
$$

$$
- \int_{t_0}^{t_1} \sum_{j=1}^{N} \left(\left(-\frac{\partial L}{\partial x_j} + \frac{d}{dt} \left(\frac{\partial L}{\partial \dot{x}_j} \right) \right) \delta x_j + \frac{\partial L}{\partial \dot{x}_j} \left(\delta \frac{dx_j}{dt} - \frac{d}{dt} \delta x_j \right) \right) dt.
$$

Consequently,

$$
\int_{t_0}^{t_1} \delta L \, dt \bigg|_{L_\nu = 0} = \int_{t_0}^{t_1} \sum_{j=1}^{N} \left(\frac{d}{dt} \left(\frac{\partial T}{\partial \dot{x}_j} \delta x_j \right) - \left(E_j L - \sum_{k=1}^{N} A_{jk} \frac{\partial L}{\partial \dot{x}_k} \right) \delta x_j \right) dt
$$

$$
= \sum_{j=1}^{N} \frac{\partial T}{\partial \dot{x}_j} \delta x_j \bigg|_{t=t_0}^{t=t_1} - \int_{t_0}^{t_1} \sum_{j=1}^{N} \left(E_j L - \sum_{k=1}^{N} A_{jk} \frac{\partial L}{\partial \dot{x}_k} \right) \delta x_j dt
$$

$$
= - \int_{t_0}^{t_1} \sum_{j=1}^{N} \left(E_j L - \sum_{k=1}^{N} A_{jk} \frac{\partial L}{\partial \dot{x}_k} \right) \delta x_j dt = 0,
$$

where $\nu = 1, \ldots, M$. Here we use the fact that $\delta \mathbf{x}(t_0) = \delta \mathbf{x}(t_1) = 0$. Hence, if (7.22) holds, then (7.21) is satisfied. The converse is proved by choosing

$$
\delta x_k(t) = \begin{cases} \zeta(t), & \text{if } k = 1, \\ 0, & \text{otherwise,} \end{cases}
$$

where $\zeta(t)$ is a positive function in the interval (t_0^*, t_1^*), and equal to zero in the intervals $[t_0, t_0^*]$ and $[t_1^*, t_1]$, and by applying Corollary 7.2.3.

From the definition (7.22) we have that

$$
D_\nu(fg) = D_\nu f \, g + f \, D_\nu g + \frac{\partial f}{\partial \dot{x}_\nu} \frac{dg}{dt} + \frac{df}{dt} \frac{\partial g}{\partial \dot{x}_\nu}, \quad D_\nu a = 0,
$$

where a is a constant.

Now let us write (7.22) in the more convenient way

$$0 = D_\nu L = D_\nu \left(L_0 - \sum_{j=1}^{M} \lambda_j L_j - \sum_{j=M+1}^{N} \lambda_j^0 L_j \right)$$

$$= D_\nu L_0 - \sum_{j=1}^{M} D_\nu (\lambda_j L_j) - \sum_{j=M+1}^{N} \lambda_j^0 D_\nu L_j$$

$$= D_\nu L_0 - \sum_{j=M+1}^{N} \lambda_j^0 D_\nu L_j$$

$$- \sum_{j=1}^{M} \left(D_\nu \lambda_j \, L_j + \lambda_j D_\nu \, L_j + \frac{d\lambda_j}{dt} \frac{\partial L_j}{\partial \dot{x}_\nu} + \frac{dL_j}{dt} \frac{\partial \lambda_j}{\partial \dot{x}_\nu} \right).$$

From these relations and since the constraints $L_j = 0$ for $j = 1, \ldots, M$, we easily obtain equations (7.23), or equivalently

$$E_\nu L_0 = \sum_{k=1}^{N} A_{jk} \frac{\partial L_0}{\partial \dot{x}_k} + \sum_{j=1}^{M} \left(\lambda_j D_\nu L_j + \frac{d\lambda_j}{dt} \frac{\partial L_j}{\partial \dot{x}_\nu} \right) + \sum_{j=M+1}^{N} \lambda_j^0 D_\nu L_j. \qquad (7.32)$$

Thus the theorem is proved.   □

Now we show that the differential equations (7.32) for suitable functions L_j, constants λ_j^0 for $j = M+1, \ldots, N$, and for a suitable matrix A describe the motion of the constrained Lagrangian system.

Proof of Theorem 7.7.2. The matrix equation (7.25) can be rewritten in components as

$$\sum_{j=1}^{N} A_{kj} \frac{\partial L_\alpha}{\partial \dot{x}_j} = E_k L_\alpha \iff D_k L_\alpha = 0, \quad \alpha, k = 1, \ldots, N. \qquad (7.33)$$

Consequently, the differential equations (7.32) become

$$E_\nu L_0 = \sum_{k=1}^{N} \left(A_{\nu k} \frac{\partial L_0}{\partial \dot{x}_k} + \frac{d\lambda_k}{dt} \frac{\partial L_k}{\partial \dot{x}_\nu} \right) \iff D_\nu L_0 = \sum_{j=1}^{M} \frac{d\lambda_j}{dt} \frac{\partial L_j}{\partial \dot{x}_\nu}, \qquad (7.34)$$

which coincide with the first system in (7.26).

Thanks to the condition $|W_1| \neq 0$, we can solve equation (7.25) with respect to A and obtain $A = W_1^{-1} \Omega_1$. Hence, by considering (7.33) we obtain the second system in (7.26) and the transpositional relation (7.27).   □

Proof of Theorem 7.7.3. The matrix equation (7.29) is equivalent to the systems

$$\sum_{j=1}^{N} A_{kj} \frac{\partial L_\alpha}{\partial \dot{x}_j} = E_k L_\alpha \iff D_k L_\alpha = 0,$$

$$\sum_{j=1}^{N} A_{kj} \frac{\partial L_0}{\partial \dot{x}_j} = 0,$$

for $k = 1, \ldots, N$, and $\alpha = 1, \ldots, N - 1$. Thus, by considering that $\lambda_N^0 = 0$, the system (7.32) takes the form

$$E_\nu L_0 = \sum_{j=1}^{M} \frac{d\tilde{\lambda}_j}{dt} \frac{\partial L_j}{\partial \dot{x}_\nu}.$$

Hence we obtain the system (7.30). On the other hand from (7.29) we have that $A = W_2^{-1} \Omega_2$. Hence we deduce that the transpositional relation (7.14) can be rewritten in the form (7.31). $\quad\square$

The mechanics based on the Hamiltonian principle with non-zero transpositional relations given by formula (7.14), Lagrangian (7.19), and equations of motion (7.22) is called here the *modified of the vakonomic mechanics*, denoted MVM.

From the proofs of Theorems 7.7.2 and 7.7.3 it follows that the relations (7.15) holds identically in MVM.

Corollary 7.7.5. *The differential equations (7.26) are invariant under the change*

$$L_0 \longrightarrow L_0 - \sum_{j=1}^{N} a_j L_j,$$

where the a_j's are constants for $j = 1, \ldots, N$.

Proof. Indeed, from (7.34) and (7.33) it follows that

$$D_\nu \left(L_0 - \sum_{j=1}^{N} a_j L_j \right) = D_\nu L_0 - \sum_{j=1}^{N} a_j D_\nu L_j = D_\nu L_0 = \sum_{j=1}^{M} \frac{d\lambda_j}{dt} \frac{\partial L_j}{\partial \dot{x}_\nu}. \quad\square$$

Remark 7.7.6. The following interesting facts follow from Theorems 7.7.2 and 7.7.3.

(1) The equations of motion obtained from Theorem 7.7.2 are more general than the equations obtained from Theorem 7.7.3. Indeed, in (7.26) there are $N - M$ arbitrary functions, whereas in (7.30) there are $N - M - 1$ arbitrary functions

(2) If the constraints are linear in the velocity, then between the Lagrange multipliers μ, $\dfrac{d\lambda}{dt}$, and $\dfrac{d\tilde{\lambda}}{dt}$ are connected by the relation

$$\mu = \frac{d\tilde{\lambda}}{dt} = \left(W_2^{-1}\right)^T \left(W_1^T \frac{d\lambda}{dt} + W_2^{-1}\Omega_1^T W_1^{-T}\frac{\partial L_0}{\partial \dot{\mathbf{x}}}\right),$$

where W_1 and W_2 are the matrices defined in Theorems 7.7.2 and 7.7.3.

(3) If the constraints are linear in the velocity, then one of the important questions, which arise in MVM is related with the arbitrariness of the functions L_j for $j = M+1, \ldots, N$. Specifically, is it possible to determine these functions in such a way that $|W_1|$ or $|W_2|$ is non-zero everywhere in $\mathcal{M}^*$? If we have a positive answer to this question, then the equations of motion of MVM describe the *global behavior* of the constrained Lagrangian systems, i.e., the obtained motions completely coincide with the motions obtained from the classical mathematical models. Thus if $|W_1| \neq 0$ and $|W_2| \neq 0$ everywhere in $\mathcal{M}^*$, we have the equivalence

$$D_\nu L_0 = \sum_{j=1}^{M} \frac{d\lambda_j}{dt}\frac{\partial L_j}{\partial \dot{x}_\nu} \Longleftrightarrow E_\nu L_0 = \sum_{j=1}^{M} \frac{d\tilde{\lambda}_j}{dt}\frac{\partial L_j}{\partial \dot{x}_\nu}$$
$$\Longleftrightarrow E_\nu L_0 = \sum_{j=1}^{M} \mu_j \frac{\partial L_j}{\partial \dot{x}_\nu}. \tag{7.35}$$

f the constraints are nonlinear in the velocity and $|W_2| \neq 0$ everywhere in $\mathcal{M}^*$, we have the equivalence

$$E_\nu L_0 = \sum_{j=1}^{M} \frac{d\tilde{\lambda}_j}{dt}\frac{\partial L_j}{\partial \dot{x}_\nu} \Longleftrightarrow E_\nu L_0 = \sum_{j=1}^{M} \mu_j \frac{\partial L_j}{\partial \dot{x}_\nu}. \tag{7.36}$$

The equivalence concerning the equations $D_\nu L_0 = \sum\limits_{j=1}^{M} \dfrac{d\lambda_j}{dt}\dfrac{\partial L_j}{\partial \dot{x}_\nu}$ fails in general in his case because the term $\Omega_1^T W_1^{-T}\dfrac{\partial L_0}{\partial \dot{\mathbf{x}}}$ depends on $\ddot{\mathbf{x}}$.

7.8 Study of the Appell–Hamel mechanical systems by applying MVM

As a general rule, the constraints studied in classical mechanics are linear in the velocities, i.e., L_j can be written in the form (7.9). However, Appell and Hamel (see [3, 71]) in 1911, considered an artificial example of nonlinear nonholonomic constrains. Numerous of investigations have been devoted to the derivation of

the equations of motion of mechanical systems with nonlinear nonholonomic con
straints, see for instance [27, 71, 120, 123]. The works of these authors do no
contain examples of systems with nonlinear nonholonomic constraints differing
essentially from the example given by Appell and Hamel.

Corollary 7.8.1. *The equivalence (7.35) also holds for the Appell–Hamel system
i.e., for the constrained Lagrangian systems*

$$\left(\mathbb{R}^3, \quad \tilde{L} = \frac{1}{2}(\dot{x}^2 + \dot{y}^2 + \dot{z}^2) - gz, \quad \{\dot{z} - a\sqrt{\dot{x}^2 + \dot{y}^2} = 0\} \right),$$

where a and g are positive constants.

Proof. The classical equations (7.11) for the Appell–Hamel system are

$$\ddot{x} = -\frac{a\dot{x}}{\sqrt{\dot{x}^2 + \dot{y}^2}}\mu, \qquad \ddot{y} = -\frac{a\dot{y}}{\sqrt{\dot{x}^2 + \dot{y}^2}}\mu, \qquad \ddot{z} = -g + \mu, \tag{7.37}$$

where μ is the Lagrange multiplier.

Now we apply Theorem 7.7.3. In order to ensure that $|W_2| \neq 0$ everywhere
we choose the functions L_j for $j = 1, 2, 3$ as

$$L_1 = \dot{z} - a\sqrt{\dot{x}^2 + \dot{y}^2} = 0, \quad L_2 = \arctan \frac{\dot{x}}{\dot{y}}, \quad L_3 = L_0 = \tilde{L}.$$

In this case the matrices W_2, and Ω_2 are

$$W_2 = \begin{pmatrix} -\dfrac{a\dot{x}}{\sqrt{\dot{x}^2 + \dot{y}^2}} & -\dfrac{a\dot{y}}{\sqrt{\dot{x}^2 + \dot{y}^2}} & 1 \\ \dfrac{\dot{y}}{\dot{x}^2 + \dot{y}^2} & -\dfrac{\dot{x}}{\dot{x}^2 + \dot{y}^2} & 0 \\ \dot{x} & \dot{y} & \dot{z} \end{pmatrix}, \quad |W_2|_{L_1=0} = 1 + a^2,$$

$$\Omega_2 = \begin{pmatrix} -\dot{y}q & \dot{x}q & 0 \\ \dfrac{\ddot{y}\left(\dot{x}^2 - \dot{y}^2\right) - 2\dot{x}\dot{y}\ddot{x}}{(\dot{x}^2 + \dot{y}^2)^2} & \dfrac{\ddot{x}\left(\dot{x}^2 - \dot{y}^2\right) + 2\dot{x}\dot{y}\ddot{y}}{(\dot{x}^2 + \dot{y}^2)^2} & 0 \\ 0 & 0 & 0 \end{pmatrix},$$

and the matrix $A|_{L_1=0}$ is

$$\begin{pmatrix} -\dfrac{\dot{y}\left(a^2\dot{y}\dot{x}\ddot{x} + \left((a^2 + 1)\dot{y}^2 + \dot{x}^2\right)\ddot{y}\right)}{(1 + a^2)\left(\dot{x}^2 + \dot{y}^2\right)} & \dfrac{\left(a^2\dot{x}^2 + (a^2 + 1)\left(\dot{y}^2 + \dot{x}^2\right)^2\right)\dot{y}\ddot{x} - a^2\dot{x}^3\ddot{y}}{(1 + a^2)\left(\dot{x}^2 + \dot{y}^2\right)^2} & 0 \\ \dfrac{\left(a^2\dot{y}^2 + (a^2 + 1)\left(\dot{y}^2 + \dot{x}^2\right)\right)\dot{x}\ddot{y} - a^2\dot{y}^3\ddot{x}}{(1 + a^2)\left(\dot{x}^2 + \dot{y}^2\right)} & -\dfrac{\dot{x}\left(a^2\dot{x}\dot{y}\ddot{y} + \left((a^2 + 1)\dot{x}^2 + \dot{y}^2\right)\ddot{x}\right)}{(1 + a^2)\left(\dot{x}^2 + \dot{y}^2\right)^2} & 0 \\ \dfrac{\dot{y}a\left(\dot{y}\ddot{x} - \dot{x}\ddot{y}\right)}{(1 + a^2)\left(\dot{x}^2 + \dot{y}^2\right)^{3/2}} & -\dfrac{\dot{x}a\left(\dot{y}\ddot{x} - \dot{x}\ddot{y}\right)}{(1 + a^2)\left(\dot{x}^2 + \dot{y}^2\right)^{3/2}} & 0 \end{pmatrix}$$

By considering that $|W_2|_{L_1=0} = 1 + a^2$, we obtain that the equations (7.30) in this case describe the global behavior of the Appell–Hamel systems and take the form

$$\ddot{x} = -\frac{a\dot{x}}{\sqrt{\dot{x}^2 + \dot{y}^2}}\tilde{\lambda}, \qquad \ddot{y} = -\frac{a\dot{y}}{\sqrt{\dot{x}^2 + \dot{y}^2}}\tilde{\lambda}, \qquad \ddot{z} = -g + \tilde{\lambda}. \tag{7.38}$$

Clearly, this system coincides with the classical differential equations (7.37) with $\tilde{\lambda} = \mu$.

After we differentiate the constraint $\dot{z} - a\sqrt{\dot{x}^2 + \dot{y}^2} = 0$ along the solutions of (7.38), we obtain

$$0 = \ddot{z} - a\frac{\ddot{x}}{\sqrt{\dot{x}^2 + \dot{y}^2}} + a\frac{\ddot{y}}{\sqrt{\dot{x}^2 + \dot{y}^2}} = -g + (1 + a^2)\tilde{\lambda}.$$

Therefore, $\tilde{\lambda} = \dfrac{g}{1 + a^2}$, and the equations of motion (7.38) become

$$\ddot{x} = -\frac{ag}{1+a^2}\frac{\dot{x}}{\sqrt{\dot{x}^2 + \dot{y}^2}}, \qquad \ddot{y} = -\frac{ag}{1+a^2}\frac{\dot{y}}{\sqrt{\dot{x}^2 + \dot{y}^2}}, \qquad \ddot{z} = -\frac{a^2 g}{1+a^2}.$$

In this case the Lagrangian (7.28) is given by

$$L = \frac{1}{2}(\dot{x}^2 + \dot{y}^2 + \dot{z}^2) - gz - \frac{g(t+C)}{1+a^2}(\dot{z} - a\sqrt{\dot{x}^2 + \dot{y}^2}) - \lambda_2^0 \arctan\frac{\dot{x}}{\dot{y}},$$

where C and λ_2^0 are an arbitrary constants.

Under the condition $L_1 = 0$ we obtain that the transpositional relations are

$$\delta\frac{dx}{dt} - \frac{d}{dt}\delta x = \frac{\dot{y}\left((1+a^2)(\dot{x}^2 + \dot{y}^2)(\ddot{x}\delta y - \ddot{y}\delta x) + a^2\dot{x}(\dot{y}\ddot{x} - \dot{x}\ddot{y})(\dot{x}\delta y - \dot{y}\delta x)\right)}{(1+a^2)(\dot{x}^2 + \dot{y}^2)^2},$$

$$\delta\frac{dy}{dt} - \frac{d}{dt}\delta y = \frac{\dot{x}\left((1+a^2)(\dot{x}^2 + \dot{y}^2)(\ddot{y}\delta x - \ddot{x}\delta y) + a^2\dot{y}(\dot{y}\ddot{x} - \dot{x}\ddot{y})(\dot{x}\delta y - \dot{y}\delta x)\right)}{(1+a^2)(\dot{x}^2 + \dot{y}^2)^2},$$

$$\delta\frac{dz}{dt} - \frac{d}{dt}\delta z = \frac{a(\dot{y}\ddot{x} - \dot{x}\ddot{y})(\dot{x}\delta y - \dot{x}\delta y)}{(1+a^2)(\dot{x}^2 + \dot{y}^2)^{3/2}}. \tag{7.39}$$

In this example the independent virtual variations δx and δy produce non-zero transpositional relations. This result is not in accordance with with the Suslov point on view on the transpositional relations.

Now we apply Theorem 7.7.2. The functions L_0, L_1, L_2, and L_3 are determined as follows:

$$L_0 = \tilde{L}, \qquad L_1 = \dot{z} - a\sqrt{\dot{x}^2 + \dot{y}^2}, \qquad L_2 = \dot{y}, \qquad L_3 = \dot{x}.$$

Thus the matrices W_1 and Ω_1 are

$$W_1 = \begin{pmatrix} -\dfrac{a\dot{x}}{\sqrt{\dot{x}^2 + \dot{y}^2}} & -\dfrac{a\dot{y}}{\sqrt{\dot{x}^2 + \dot{y}^2}} & 1 \\ 0 & 1 & 0 \\ 1 & 0 & 0 \end{pmatrix}, \quad \Omega_1 = \begin{pmatrix} \dot{y}q & -\dot{x}q & 0 \\ 0 & 0 & 0 \\ 0 & 0 & 0 \end{pmatrix},$$

where $q = \dfrac{a(\ddot{x}\dot{y} - \ddot{x}\dot{y})}{\sqrt{\dot{x}^2 + \dot{y}^2}^3}$. Therefore $|W_1| = -1$.

Hence, after some computations we obtain from (7.25) that

$$A = \begin{pmatrix} 0 & 0 & 0 \\ 0 & 0 & 0 \\ \dot{y}q & -\dot{x}q & 0 \end{pmatrix}.$$

The equations of motion (7.26) become

$$\ddot{x} = -\frac{a^2\dot{y}}{\dot{x}^2 + \dot{y}^2}(\dot{y}\ddot{x} - \dot{x}\ddot{y}) - \frac{a\dot\lambda}{\sqrt{\dot{x}^2 + \dot{y}^2}}\dot{x},$$

$$\ddot{y} = -\frac{a^2\dot{x}}{\dot{x}^2 + \dot{y}^2}(\dot{x}\ddot{y} - \dot{y}\ddot{x}) - \frac{a\dot\lambda}{\sqrt{\dot{x}^2 + \dot{y}^2}}\dot{y}, \qquad (7.40)$$

$$\ddot{z} = -g + \dot\lambda.$$

By solving these equations with respect to $\ddot{x}$, $\ddot{y}$, and $\ddot{z}$ we obtain

$$\ddot{x} = -\frac{a\dot{x}}{\sqrt{\dot{x}^2 + \dot{y}^2}}\dot\lambda, \qquad \ddot{y} = -\frac{a\dot{y}}{\sqrt{\dot{x}^2 + \dot{y}^2}}\dot\lambda, \qquad \ddot{z} = -g + \dot\lambda,$$

We observe in this case that $|W_1| = -1$, consequently these equations, obtained from Theorem 7.7.2, describe the global behavior of the Appell–Hamel system, i.e., coincide with the classical equations (7.37) with $\dot\lambda = \dot\lambda = \mu = \dfrac{g}{1+a^2}$.

The transpositional relations (7.27) can be written as

$$\delta\frac{dx}{dt} - \frac{d}{dt}\delta x = 0, \quad \delta\frac{dy}{dt} - \frac{d}{dt}\delta y = 0, \quad \delta\frac{dz}{dt} - \frac{d}{dt}\delta z = q\left(\dot{y}\delta x - \dot{x}\delta y\right). \qquad (7.41)$$

It follows that the independent virtual variations δx and δy produce non–zero transpositional relations (7.39) and zero transpositional relations (7.41).

The Lagrangian (7.24) in this case takes the form

$$L = \frac{1}{2}(\dot{x}^2 + \dot{y}^2 + \dot{z}^2) - gz - \frac{g(t+C)}{1+a^2}(\dot{z} - a\sqrt{\dot{x}^2 + \dot{y}^2}) - \lambda_2^0\dot{y} - \lambda_3^0\dot{x}$$

$$\simeq \frac{1}{2}(\dot{x}^2 + \dot{y}^2 + \dot{z}^2) - gz - \frac{g(t+C)}{1+a^2}(\dot{z} - a\sqrt{\dot{x}^2 + \dot{y}^2}).$$

Now (7.12) yields

$$\delta\frac{dz}{dt} - \frac{d}{dt}\delta z = q\left(\dot{y}\delta x - \dot{x}\delta y\right) + \frac{a\dot{x}}{\sqrt{\dot{x}^2 + \dot{y}^2}}\left(\delta\frac{dx}{dt} - \frac{d}{dt}\delta x\right)$$
$$+ \frac{a\dot{y}}{\sqrt{\dot{x}^2 + \dot{y}^2}}\left(\delta\frac{dy}{dt} - \frac{d}{dt}\delta y\right).$$

Therefore this relation holds identically for (7.39) and (7.41).

In the next sections we show the importance of the equations of motion (7.26) and (7.30), contrasting them with the classical differential equations of nonholonomic mechanics.

7.9 MVM versus vakonomic mechanics

Now we show that the equations of the *vakonomic mechanics* (7.3) can be obtained from the equations (7.23). More precisely, if in (7.14) we require that all the virtual variations of the coordinates produce null transpositional relations, i.e., A is the zero matrix, and we require that $\lambda_j^0 = 0$ for $j = M + 1, \ldots, N$, then from (7.23), since $D_k L = E_k L$, we obtain the *vakonomic equations* (7.3), i.e.,

$$D_\nu L_0 = \sum_{j=1}^{M}\left(\lambda_j D_\nu L_j + \frac{d\lambda_j}{dt}\frac{\partial L_j}{\partial\dot{x}_\nu}\right) + \sum_{j=M+1}^{N}\lambda_j^0 D_\nu L_j \implies$$

$$E_\nu L_0 = \sum_{j=1}^{M}\left(\lambda_j E_\nu L_j + \frac{d\lambda_j}{dt}\frac{\partial L_j}{\partial\dot{x}_\nu}\right), \quad \nu = 1, \ldots, N$$

In the following example, in order to contrast Theorems 7.7.2 with the vakonomic model, we study the *skate or knife edge on an inclined plane*.

Example 7.9.1. To set up the problem, consider a plane Ξ with cartesian coordinates x and y, slanted at an angle α. We assume that the y-axis is horizontal, while the x-axis is directed downward from the horizontal; let (x, y) be the coordinates of the point of contact of the skate with the plane. The angle φ represents the orientation of the skate measured from the x-axis. The skate is moving under the influence of the gravity. The acceleration due to gravity is denoted by g. The skate has mass m, and its moment inertia about a vertical axis through its contact point is denoted by J (see page 108 of [120] for a picture). The equation of the nonintegrable constraint is

$$L_1 = \dot{x}\sin\varphi - \dot{y}\cos\varphi = 0. \tag{7.42}$$

With these notations, the Lagrangian function of the skate is

$$\hat{L} = \frac{m}{2}\left(\dot{x}^2 + \dot{y}^2\right) + \frac{J}{2}\dot{\varphi}^2 + mg\,x\sin\alpha.$$

Thus we have the constrained mechanical system

$$\left(\mathbb{R}^2 \times \mathbb{S}^1, \quad \hat{L} = \frac{m}{2}\left(\dot{x}^2 + \dot{y}^2\right) + \frac{J}{2}\dot{\varphi}^2 + mg\,x\,\sin\alpha, \quad \{\dot{x}\sin\varphi - \dot{y}\cos\varphi = 0\}\right).$$

By an appropriate choice of mass, length, and time units, we reduce the Lagrangian $\hat{L}$ to

$$L_0 = \frac{1}{2}\left(\dot{x}^2 + \dot{y}^2 + \dot{\varphi}^2\right) + x\,g\sin\alpha,$$

where for simplicity we keep the same notations for the all variables. The question is, *what is the motion of the point of contact?* To answer this we shall use the vakonomic equations (7.3) and the equations (7.26) proposed in Theorems 7.7.2 and 7.7.3.

The study of the skate applying the MVM *model.*

We determine the motion of the point of contact of the skate using Theorem 7.7.2. We choose the arbitrary functions L_2 and L_3 as

$$L_2 = \dot{x}\cos\varphi + \dot{y}\sin\varphi, \quad L_3 = \dot{\varphi},$$

in order that the determinant $|W_1| \neq 0$ everywhere in the configuration space.

The Lagrangian (7.24) becomes

$$L(x, y, \varphi, \dot{x}, \dot{y}, \dot{\varphi}, \Lambda) = \frac{1}{2}\left(\dot{x}^2 + \dot{y}^2 + \dot{\varphi}^2\right) + g\sin\alpha x - \lambda(\dot{x}\sin\varphi - \dot{y}\cos\varphi) - \lambda_3^0\dot{\varphi}$$

$$\simeq \frac{1}{2}\left(\dot{x}^2 + \dot{y}^2 + \dot{\varphi}^2\right) + g\sin\alpha x - \lambda(\dot{x}\sin\varphi - \dot{y}\cos\varphi),$$

where $\lambda := \lambda_1$.

The matrices W_1 and Ω_1 are

$$W_1 = \begin{pmatrix} \sin\varphi & -\cos\varphi & 0 \\ \cos\varphi & \sin\varphi & 0 \\ 0 & 0 & 1 \end{pmatrix}, \quad |W_1| = 1,$$

$$\Omega_1 = \begin{pmatrix} \dot{\varphi}\cos\varphi & \dot{\varphi}\sin\varphi & -L_2 \\ -\dot{\varphi}\sin\varphi & \dot{\varphi}\cos\varphi & -L_1 \\ 0 & 0 & 0 \end{pmatrix},$$

and so $A = W_1^{-1}\Omega_1$ becomes

$$A = \begin{pmatrix} 0 & \dot{\varphi} & -\sin\varphi L_2 - \cos\varphi L_1 \\ -\dot{\varphi} & 0 & \cos\varphi L_2 - \sin\varphi L_1 \\ 0 & 0 & 0 \end{pmatrix}\Bigg|_{L_1=0} = \begin{pmatrix} 0 & \dot{\varphi} & -\dot{y} \\ -\dot{\varphi} & 0 & \dot{x} \\ 0 & 0 & 0 \end{pmatrix}.$$

Hence the equation (7.26) and transpositional relations (7.27) take the form

$$\ddot{x} + \dot{\varphi}\dot{y} = g\sin\alpha + \dot{\lambda}\sin\varphi, \quad \ddot{y} - \dot{\varphi}\dot{x} = -\dot{\lambda}\cos\varphi, \quad \ddot{\varphi} = 0, \tag{7.43}$$

and

$$\delta \frac{dx}{dt} - \frac{d\delta x}{dt} = \dot{y}\delta\varphi - \dot{\varphi}\delta y,$$

$$\delta \frac{dy}{dt} - \frac{d\delta y}{dt} = \dot{\varphi}\delta x - \dot{x}\delta\varphi, \tag{7.44}$$

$$\delta \frac{d\varphi}{dt} - \frac{d\delta\varphi}{dt} = -L_2 (\delta x \sin\varphi - \delta y \cos\varphi) = 0,$$

respectively; here we have applied Chetaev's condition $\sin\varphi\,\delta x - \cos\varphi\,\delta y = 0$.

The initial conditions

$$x_0 = x|_{t=0}\,,\ y_0 = y|_{t=0}\,,\ \varphi_0 = \varphi|_{t=0}\,,\ \dot{x}_0 = \dot{x}|_{t=0}\,,\ \dot{y}_0 = \dot{y}|_{t=0}\,,\ \dot{\varphi}_0 = \dot{\varphi}|_{t=0}\,,$$

satisfy the constraint, i.e.,

$$\sin\varphi_0\,\dot{x}_0 - \cos\varphi_0\,\dot{y}_0 = 0.$$

Differentiating the constraint along the solutions of the equations of motion (7.43), and using (7.42) we obtain

$$0 = \sin\varphi\,\ddot{x} - \cos\varphi\,\ddot{y} + \dot{\varphi}(\cos\varphi\,\dot{x} + \sin\varphi\,\dot{y})$$
$$= \sin\varphi(g\sin\alpha + \dot{\lambda}\sin\varphi - \dot{\varphi}\dot{y}) - \cos\varphi(-\dot{\lambda}\cos\varphi + \dot{\varphi}\dot{x}) + \dot{\varphi}(\cos\varphi\,\dot{x} + \sin\varphi\,\dot{y}).$$

Hence $\dot{\lambda} = -g\sin\alpha\sin\varphi$. Now the differential equations (7.43) read

$$\ddot{x} + \dot{\varphi}\dot{y} = g\sin\alpha\cos^2\varphi, \quad \ddot{y} - \dot{\varphi}\dot{x} = g\sin\alpha\sin\varphi\cos\varphi, \quad \ddot{\varphi} = 0. \tag{7.45}$$

We study the motion of the skate in the following three cases:

(i) $\dot{\varphi}|_{t=0} = \omega = 0.$

(ii) $\dot{\varphi}|_{t=0} = \omega \neq 0.$

(iii) $\alpha = 0.$

For the first case ($\omega = 0$), after the change of variables

$$X = \cos\varphi_0\,x - \sin\varphi_0\,y, \quad Y = \cos\varphi_0\,x + \sin\varphi_0\,y,$$

the differential equations (7.45) and the constraint become

$$\ddot{X} = 0, \quad \ddot{Y} = g\sin\alpha\cos\varphi_0, \quad \varphi = \varphi_0, \quad \dot{X} = 0,$$

respectively. Consequently

$$X = X_0, \quad Y = g\sin\alpha\cos\varphi_0\frac{t^2}{2} + \dot{Y}_0 t + Y_0, \quad \varphi = \varphi_0,$$

thus the trajectories are straight lines.

For the second case ($\omega \neq 0$), we take

$$\varphi_0 = \dot{y}_0 = \dot{x}_0 = x_0 = y_0 = 0$$

in order to simplify the computations. Since $\dot{\varphi} = \dot{\varphi}|_{t=0} = \omega$, if we denote by $'$ the differentiation with respect to φ, we recast (7.45) as

$$x'' + y' = \frac{g\sin\alpha}{\omega^2}\cos^2\varphi, \quad x'' - x' = \frac{g\sin\alpha}{\omega^2}\sin\varphi\cos\varphi, \quad \varphi' = 1.$$

These equations are easy to integrate and we obtain

$$x = -\frac{g\sin\alpha}{4\omega^2}\cos(2\varphi), \quad y = -\frac{g\sin\alpha}{4\omega^2}\sin(2\varphi) + \frac{g}{2\omega^2}\varphi, \quad \varphi = \omega t,$$

which are recognized as the equations of a *cycloid*. Hence the point of contact of the skate follows a cycloid along the plane, but does not slide down the plane.

For the third case ($\alpha = 0$), if $\varphi_0 = 0$, $\omega \neq 0$ we obtain that the solutions of the differential system (7.45) are

$$x = \dot{y}_0 \cos\varphi + \dot{x}_0 \sin\varphi + a, \quad y = \dot{y}_0 \sin\varphi + \dot{y}_0 \cos\varphi + b, \quad \varphi = \varphi_0 + \omega t,$$

where $a = x_0 - \dfrac{\dot{y}_0}{\omega}$, $b = y_0 + \dfrac{\dot{x}_0}{\omega}$, which are recognized as the equation of the circle with center at (a, b) and radius $\dfrac{\dot{x}_0^2 + \dot{y}_0^2}{\omega^2}$.

If $\alpha = 0$ and $\varphi_0 = 0$, $\omega = 0$, then the solutions are

$$x = \dot{x}_0 t + x_0, \quad y = \dot{y}_0 t + y_0.$$

All these solutions coincide with the solutions obtained from the Lagrangian equations (7.11) with multipliers (see [4])

$$\ddot{x} = g\sin\alpha + \mu\sin\varphi, \quad \ddot{y} - \dot{\varphi}\dot{x} = -\mu\cos\varphi, \quad \ddot{\varphi} = 0,$$

with $\mu = \dot{\lambda} = -g\sin\alpha\sin\varphi$.

Now we study the skate motions using the equations of motion (7.31) of Theorem 7.7.3.

We choose the functions L_2 and L_3 as

$$L_2 = \dot{\varphi}\cos\varphi = \frac{d}{dt}\sin\varphi, \quad L_3 = L_0 = \tilde{L}.$$

Consequently, the Lagrangian L (see (7.28)) and the matrices W_2 and Ω_2

$$L(x, y, \varphi, \dot{x}, \dot{y}, \dot{\varphi}, \Lambda) = \frac{(\dot{x}^2 + \dot{y}^2 + \dot{\varphi}^2)}{2} + g \sin \alpha x - \lambda(\dot{x} \sin \varphi - \dot{y} \cos \varphi)$$

$$-\lambda_2^0 \dot{\varphi} \cos \varphi \simeq \frac{1}{2}(\dot{x}^2 + \dot{y}^2 + \dot{\varphi}^2) + g \sin \alpha x - \lambda(\dot{x} \sin \varphi - \dot{y} \cos \varphi), \quad \lambda = \lambda_1,$$

$$W_2 = \begin{pmatrix} \sin \varphi & -\cos \varphi & 0 \\ \dot{x} & \dot{y} & \dot{\varphi} \\ 0 & 0 & \cos \varphi \end{pmatrix}, \quad |W_2|_{L_1=0} = \dot{x} \neq 0,$$

$$\Omega_2 = \begin{pmatrix} \dot{\varphi} \cos \varphi & \dot{\varphi} \sin \varphi & -l \\ 0 & 0 & 0 \\ 0 & 0 & 0 \end{pmatrix};$$

note that here the determinant $|W_2|$ can vanish along $\dot{x} = 0$.

Therefore,

$$A = \begin{pmatrix} \dfrac{\dot{y}\dot{\varphi} \cos \varphi}{l} & -\dfrac{\dot{y}\dot{\varphi} \sin \varphi}{l} & -\dfrac{\dot{y}\dot{\varphi} \cos \varphi}{l} \\ -\dfrac{\dot{x}\dot{\varphi} \cos \varphi}{l} & \dfrac{\dot{x}\dot{\varphi} \sin \varphi}{l} & \dfrac{\dot{x}\dot{\varphi} \cos \varphi}{l} \\ 0 & 0 & 0 \end{pmatrix} \Bigg|_{L_1=0}$$

$$= \begin{pmatrix} \dot{\varphi} \cos \varphi \sin \varphi & -\dot{\varphi} \sin^2 \varphi & -\dot{\varphi} \cos \varphi \sin \varphi \\ -\dot{\varphi} \cos^2 \varphi & \dot{\varphi} \sin \varphi \cos \varphi & \dot{\varphi} \cos^2 \varphi \\ 0 & 0 & 0 \end{pmatrix},$$

where $l = \dot{x} \cos \varphi + \dot{y} \sin \varphi$. The differential equations (7.30) and the transpositional relations (7.31) take the form

$$\ddot{x} = g \sin \alpha + \dot{\lambda} \sin \varphi, \quad \ddot{y} = -\dot{\lambda} \cos \varphi, \quad \ddot{\varphi} = 0, \tag{7.46}$$

and

$$\delta \frac{dx}{dt} - \frac{d\delta x}{dt} = \dot{y}\Phi, \quad \delta \frac{dy}{dt} - \frac{d\delta y}{dt} = -\dot{x}\Phi, \quad \delta \frac{d\varphi}{dt} - \frac{d\delta \varphi}{dt} = 0, \tag{7.47}$$

respectively, where

$$\Phi = \cos \varphi \left(\dot{\varphi} \delta x - \dot{x} \delta \varphi \right) + \sin \varphi \left(\dot{\varphi} \delta y - \dot{y} \delta \varphi \right).$$

Differentiating the constraint (7.42) along the solutions of the equations (7.46) we obtain

$$\dot{\lambda} = -g \sin \alpha \sin \varphi - (\dot{x} \cos \varphi + \dot{y} \sin \varphi) \dot{\varphi}.$$

Thus in view of the constraint we have

$$\sin \varphi \dot{\lambda} = -g \sin \alpha \sin^2 \varphi - \dot{\varphi}\dot{y}, \quad \cos \varphi \dot{\lambda} = -g \sin \alpha \sin \varphi \cos \varphi + \dot{\varphi}\dot{x}.$$

Inserting these expressions into the (7.46), we deduce the differential system
(7.45), under the condition $|W| = \dot{x} \neq 0$. Now we establish what motions o
(7.45) are excluded by the condition $\dot{x} \neq 0$.

From the constraint we have that if $\dot{x} = 0$, then $\dot{y} \cos \varphi = 0$. Thus we have
three possibilities: (i) $\dot{y} = 0$, and (ii) $\varphi = \pm\dfrac{\pi}{2}$.

For case (i) from (7.46) we obtain the equations

$$g \sin \alpha + \dot{\lambda} \sin \varphi = 0, \quad -\dot{\lambda} \cos \varphi = 0, \quad \ddot{\varphi} = 0.$$

Hence, if $\varphi = \pm\dfrac{\pi}{2}$, then $\lambda = \mp g \sin \alpha$. Thus we exclude the motions $\left(x_0, \ y_0, \pm\dfrac{\pi}{2}\right)$
which are stationary points of the systems (7.46). If $\varphi \neq \pm\dfrac{\pi}{2}$, then $\lambda = g \sin \alpha = 0$
Thus we exclude the motion

$$x = x_0, \quad y = y_0, \quad \varphi = \omega t + \varphi_0,$$

which correspond to the uniform rotation of the skate around the point (x_0, y_0).

For case (ii) we have

$$\dot{\lambda} = \mp g \sin \alpha, \quad \ddot{y} = 0,$$

thus we exclude the motions

$$x = x_0, \quad y = \dot{y}_0 t + y_0, \quad \varphi = \pm\dfrac{\pi}{2},$$

which correspond to straight lines.

Clearly, it is possible to choose the functions L_2 and L_3 in a different way
Now we analyze the case when

$$L_2 = \dot{x} \cos \varphi + \dot{y} \sin \varphi, \quad L_3 = L_0 = \tilde{L},$$

and $\lambda_2^0 = 0$. Thus the Lagrangian L given in (7.28) and the matrices W_2, Ω_2 and
A become

$$L(x, y, \varphi, \dot{x}, \dot{y}, \dot{\varphi}, \Lambda) = \frac{1}{2}\left(\dot{x}^2 + \dot{y}^2 + \dot{\varphi}^2\right) + g \sin \alpha x - \lambda(\dot{x} \sin \varphi - \dot{y} \cos \varphi), \quad \lambda = \lambda_1,$$

$$W_2 = \begin{pmatrix} \sin \varphi & -\cos \varphi & 0 \\ \dot{x} & \dot{y} & \dot{\varphi} \\ \cos \varphi & \sin \varphi & 0 \end{pmatrix}, \quad |W_2| = \dot{\varphi} = \omega \neq 0,$$

$$\Omega_2 = \begin{pmatrix} \dot{\varphi} \cos \varphi & \dot{\varphi} \sin \varphi & -L_3 \\ 0 & 0 & 0 \\ -\sin \varphi \dot{\varphi} & \cos \varphi \dot{\varphi} & -L_1 \end{pmatrix},$$

$$A = \begin{pmatrix} 0 & \dot{\varphi} & -L_2 \sin \varphi - \cos \varphi L_1 \\ -\dot{\varphi} & 0 & -L_1 \sin \varphi - \cos \varphi L_3 \\ \dot{y} & -\dot{x} & 0 \end{pmatrix}.$$

The equations of motion coincide with (7.46) under the condition that $\dot\varphi = \omega \neq 0$. Consequently, we exclude all the solutions for which $\omega = 0$, which we obtained after integrating the equations (7.46).

The transpositional relations are

$$\delta\frac{dx}{dt} - \frac{d\delta x}{dt} = \dot\varphi\delta y - \dot y\delta\varphi, \quad \delta\frac{dy}{dt} - \frac{d\delta y}{dt} = \dot x\delta\varphi - \dot\varphi\delta x, \quad \delta\frac{d\varphi}{dt} - \frac{d\delta\varphi}{dt} = \dot y\delta x - \dot x\delta y. \quad (7.48)$$

As we can observe from this example, all the virtual variations produce non-zero transpositional relations.

From (7.12) we deduce the following equation that the transpositional relationsmust satisfy:

$$\sin\varphi\left(\delta\frac{dx}{dt} - \frac{d\delta x}{dt}\right) - \cos\varphi\left(\delta\frac{dy}{dt} - \frac{d\delta y}{dt}\right)$$
$$= \cos\varphi\,(\dot x\delta\varphi - \dot\varphi\delta x) + \sin\varphi\,(\dot y\delta\varphi - \dot\varphi\delta y).$$

After some computations it is easy to show that the transpositional relations determined by the equations (7.44), (7.47), and (7.48) satisfies identically the previous equation.

The study of the skate applying the vakonomic model.

Now we examine instead of Theorem 7.7.2 the vakomic model for the motion of the skate.

We consider the Lagrangian

$$L(x, y, \varphi, \dot x, \dot y, \dot\varphi, \Lambda) = \frac{1}{2}\left(\dot x^2 + \dot y^2 + \dot\varphi^2\right) + g\,x\,\sin\alpha - \lambda(\dot x\sin\varphi - \dot y\cos\varphi).$$

The equations of motion (7.3) for the skate are

$$\frac{d}{dt}\left(\dot x - \lambda\sin\varphi\right) = 0, \quad \frac{d}{dt}\left(\dot y + \lambda\cos\varphi\right) = 0, \quad \ddot\varphi = -\lambda\left(\dot x\cos\varphi + \dot y\sin\varphi\right).$$

We shall study only the case when $\alpha = 0$. After integration we obtain the differential systems

$$\dot x = \lambda\sin\varphi + a = \cos\varphi\,(a\cos\varphi + b\sin\varphi),$$
$$\dot y = -\lambda\cos\varphi + b = \sin\varphi\,(a\cos\varphi + b\sin\varphi),$$
$$\ddot\varphi = (b\cos\varphi - a\sin\varphi)(a\cos\varphi + b\sin\varphi) \quad (7.49)$$
$$= (b_1^2 + a_2^2)\sin(\varphi + \alpha)\cos(\varphi + \alpha),$$
$$\lambda = b\cos\varphi - a\sin\varphi,$$

where $a = \dot x_0 - \lambda_0\sin\varphi_0$, $b = \dot y_0 + \lambda_0\cos\varphi_0$, and $\lambda_0 = \lambda|_{t=0}$ is an arbitrary parameter. Integration of the third equation yields

$$\int_0^\varphi \frac{d\varphi}{\sqrt{1 - \kappa^2\sin^2\varphi}} = t\sqrt{\frac{h + a^2 + b^2}{2}}, \quad (7.50)$$

where h is an arbitrary constant which we choose in such a way that $\kappa^2 =$
$$\frac{2(a^2 + b^2)}{h + a^2 + b^2} < 1.$$

From (7.50) we get

$$\sin\varphi = \text{sn}\left(t\sqrt{\frac{h + a^2 + b^2}{2}}\right), \quad \cos\varphi = \text{cn}\left(t\sqrt{\frac{h + a^2 + b^2}{2}}\right),$$

where sn and cn are the *Jacobi elliptic functions*. Hence, if we take $\dot{x}_0 = 1$, $\dot{y}_0 =$ $\varphi_0 = 0$, then the solutions of the differential equations (7.49) are

$$
\begin{aligned}
x &= x_0 + \int_{t_0}^{t} \left(\text{cn}\left(t\sqrt{\frac{h + 1 + \lambda_0^2}{2}}\right) \text{sn}\left(t\sqrt{\frac{h + 1 + \lambda_0^2}{2}}\right) \right. \\
&\qquad \left. + \lambda_0 \, \text{sn}\left(t\sqrt{\frac{h + 1 + \lambda_0^2}{2}}\right) \right) \, dt, \\
y &= y_0 + \int_{t_0}^{t} \text{sn}\left(t\sqrt{\frac{h + 1 + \lambda_0^2}{2}}\right) \lambda_0 \, \text{sn}\left(t\sqrt{\frac{h + 1 + \lambda_0^2}{2}}\right) \, dt, \\
\varphi &= \text{am}\left(t\sqrt{\frac{h + 1 + \lambda_0^2}{2}}\right).
\end{aligned}
\tag{7.51}
$$

It is interesting to compare these remarkable motions with the motions that w obtained above. For the same initial conditions the skate moves sideways alon circles. By considering that the solutions (7.51) depend on the arbitrary param eter λ_0, we obtain that for the given initial conditions we have nonuniqueness o solutions for the differential equations in the vakonomic model. Consequently, the *principle of determinacy* is not valid for vakonomic mechanics with nonintegrabl constraints (see the Corollary of page 36 in [4]).

7.10 MVM versus Lagrangian and constrained Lagrangian systems

The Lagrangian equations that describe the motion of Lagrangian systems ca be obtained from Theorem 7.7.2 by supposing that $M = 0$, i.e., there are n constraints. We choose the arbitrary functions L_α for $\alpha = 1, \dots, N$ as

$$L_\alpha = dx_\alpha / dt, \quad \alpha = 1, \dots, N.$$

Hence the Lagrangian (7.24) takes the form

$$L = L_0 - \sum_{j=1}^{N} \lambda_j^0 \frac{dx_j}{dt} \simeq L_0.$$

In this case we have that $|W_1| = 1$.

By using the property (7.4) of the Lagrangian derivative, we obtain that Ω_1 is the zero matrix. Hence A_1 is the zero matrix. As a consequence, the equations (7.26) become

$$D_\nu L = E_\nu L = E_\nu \left(L_0 - \sum_{j=1}^{N} \lambda_j^0 \dot{x}_j \right) = E\nu L_0 = 0,$$

because $L \simeq L_0$. The transpositional relations (7.27) are then $\delta\dfrac{d\mathbf{x}}{dt} - \dfrac{d\delta\mathbf{x}}{dt} = 0$, which are the well-known relations in Lagrangian mechanics (see formula (7.6)).

Now we show that the MVM contains as particular case the constrained Lagrangian systems.

From the equivalences (7.35) we have that in the case when the constraints are linear in the velocity, the MVM equations of motions coincide with the Lagrangian equations with multipliers (7.11), except perhaps in a set of Lebesgue measure zero $|W_2| = 0$ or $|W_1| = 0$. When the constraints are nonlinear in the velocity, we have the equivalence (7.36). Consequently, the MVM equations of motions coincide with the Lagrangian equations with multipliers (7.11), except perhaps in the set of Lebesgue measure zero $|W_2| = 0$.

We illustrate this result by the following example.

Example 7.10.1. Consider the constrained Lagrangian system

$$\left(\mathbb{R}^2, \; L_0 = \frac{1}{2}(\dot{x}^2 + \dot{y}^2) - U(x,y), \; \{2(x\dot{x} + y\dot{y}) = 0\} \right).$$

In order to apply Theorem 7.7.2 we choose the arbitrary function L_1 and L_2 as follows:

(a) $$L_1 = 2(x\dot{x} + y\dot{y}), \quad L_2 = -y\dot{x} + x\dot{y}.$$

Thus the matrices W_1 and Ω_1 are

$$W_1 = \begin{pmatrix} 2x & 2y \\ -y & x \end{pmatrix}, \quad |W_1| = 2x^2 + 2y^2 = 2, \quad \Omega_1 = \begin{pmatrix} 0 & 0 \\ -2\dot{y} & 2\dot{x} \end{pmatrix}.$$

Consequently, equations (7.26) describe everywhere the motion for the given constrained Lagrangian system.

Equations (7.26) now read

$$\ddot{x} = \left. -\frac{\partial U}{\partial x} + 2\dot{y}(y\dot{x} - x\dot{y}) + 2x\dot{\lambda} \right|_{L_1=0} = -\frac{\partial U}{\partial x} + x\left(\dot{\lambda} - 2(\dot{x}^2 + \dot{y}^2)\right),$$

$$\ddot{y} = \left. -\frac{\partial U}{\partial y} - 2\dot{x}(y\dot{x} - x\dot{y}) + 2y\dot{\lambda} \right|_{L_1=0} = -\frac{\partial U}{\partial y} + y\left(\dot{\lambda} - 2(\dot{x}^2 + \dot{y}^2)\right),$$

The corresponding transpositional relations take the form

$$\delta\frac{dx}{dt} - \frac{d\delta x}{dt} = 2y\left(\dot{y}\delta x - \dot{x}\delta y\right), \quad \delta\frac{dy}{dt} - \frac{d\delta y}{dt} = -2x\left(\dot{y}\delta x - \dot{x}\delta y\right). \quad (7.52$$

(b) If we choose $L_2 = \dfrac{y\dot{x}}{x^2 + y^2} - \dfrac{x\dot{y}}{x^2 + y^2} = \dfrac{d}{dt}\arctan\dfrac{x}{y}$, then

$$W_1 = \left(\begin{array}{cc}\dfrac{2x}{y} & -\dfrac{2y}{x} \\ \overline{x^2 + y^2} & \overline{x^2 + y^2}\end{array}\right), \quad |W_1| = -2, \quad \Omega_1 = \left(\begin{array}{cc}0 & 0 \\ 0 & 0\end{array}\right).$$

Equations (7.26) and the transpositional relations become

$$\ddot{x} = -\frac{\partial U}{\partial x} + 2x\dot{\lambda}, \quad \ddot{y} = -\frac{\partial U}{\partial y} + 2y\dot{\lambda},$$

and

$$\delta\frac{dx}{dt} - \frac{d\delta x}{dt} = 0, \quad \delta\frac{dy}{dt} - \frac{d\delta y}{dt} = 0, \quad (7.53$$

respectively.

This example shows that for a holonomic constrained Lagrangian system the transpositional relations can be non-zero (see (7.52)), or can be zero (see (7.53)). We observe that conditions (7.12) imply the relation

$$x\left(\delta\frac{dx}{dt} - \frac{d\delta x}{dt}\right) + y\left(\delta\frac{dy}{dt} - \frac{d\delta y}{dt}\right) = 0.$$

Here equality holds identically if (7.53) and (7.52) takes place.

The equations of motions (7.11) in this case are

$$\ddot{x} = -\frac{\partial U}{\partial x} + 2x\mu, \quad \ddot{y} = -\frac{\partial U}{\partial y} + 2y\mu,$$

with $\mu = \dot{\lambda} - 2(\dot{x}^2 + \dot{y}^2)$.

Example 7.10.2. To contrast the MVM with the classical model, we apply Theorems 7.7.2 to the *Gantmacher systems* (for more details see [62, 139]).

Two particles m_1 and m_2 with equal masses are linked by a metal rod of fixed length l and small mass. The system can move only in the vertical plane and so the velocity of the midpoint of the rod is directed along the rod. The task is to determine the trajectories of the particles m_1 and m_2.

Let (q_1, r_1) and (q_2, r_2) be the coordinates of the points m_1 and m_2, respectively. Clearly $(q_1 - q_2)^2 + (r_1 - r_2)^2 = l^2$. Thus we have a constrained

Lagrangian system in the configuration space $\mathbb{R}^4$, with the Lagrangian function $L = \frac{1}{2}\left(\dot{q}_1^2 + \dot{q}_2^2 + \dot{r}_1^2 + \dot{r}_2^2\right) - g/2(r_1 + r_2)$, and with the linear constraints

$$(q_2 - q_1)(\dot{q}_2 - \dot{q}_1) + (r_2 - r_1)(\dot{r}_2 - \dot{r}_1) = 0, \quad (q_2 - q_1)(\dot{r}_2 + \dot{r}_1) - (r_2 - r_1)(\dot{q}_2 + \dot{q}_1) = 0.$$

Performing the change of coordinates

$$x_1 = \frac{q_2 - q_1}{2}, \quad x_2 = \frac{r_1 - r_2}{2}, \quad x_3 = \frac{r_2 + r_1}{2}, \quad x_4 = \frac{q_1 + q_2}{2},$$

we obtain $x_1^2 + x_2^2 = \frac{1}{4}\left((q_1 - q_2)^2 + (r_1 - r_2)^2\right) = \frac{l^2}{4}$. Hence we have the constrained Lagrangian mechanical system

$$\left(\mathbb{R}^4, \quad \tilde{L} = \frac{1}{2}\sum_{j=1}^{4}\dot{x}_j^2 - gx_3, \quad \{x_1\dot{x}_1 + x_2\dot{x}_2 = 0, \quad x_1\dot{x}_3 - x_2\dot{x}_4 = 0\}\right).$$

The equations of motion (7.11) obtained from the d'Alembert–Lagrange principle are

$$\ddot{x}_1 = \mu_1 x_1, \quad \ddot{x}_2 = \mu_1 x_2, \quad \ddot{x}_3 = -g + \mu_2 x_1, \quad \ddot{x}_4 = -\mu_2 x_2, \qquad (7.54)$$

where μ_1, μ_2 are the Lagrange multipliers such that

$$\mu_1 = -\frac{\dot{x}_1^2 + \dot{x}_2^2}{x_1^2 + x_2^2}, \quad \mu_2 = \frac{\dot{x}_2\dot{x}_4 - \dot{x}_1\dot{x}_3 + gx_1}{x_1^2 + x_2^2}. \qquad (7.55)$$

To apply Theorem 7.7.2, we have the constraints

$$L_1 = x_1\dot{x}_1 + x_2\dot{x}_2 = 0, \quad L_2 = x_1\dot{x}_3 - x_2\dot{x}_4 = 0,$$

and we choose the arbitrary functions L_3 and L_4 as

$$L_3 = -x_1\dot{x}_2 + x_2\dot{x}_1, \quad L_4 = x_2\dot{x}_3 + x_1\dot{x}_4.$$

For this choice we have that

$$W_1 = \begin{pmatrix} x_1 & x_2 & 0 & 0 \\ 0 & 0 & x_1 & -x_2 \\ x_2 & -x_1 & 0 & 0 \\ 0 & 0 & x_2 & x_1 \end{pmatrix}, \quad \Omega_1 = \begin{pmatrix} 0 & 0 & 0 & 0 \\ -\dot{x}_3 & \dot{x}_4 & \dot{x}_1 & -\dot{x}_2 \\ -2\dot{x}_2 & 2\dot{x}_1 & 0 & 0 \\ -\dot{x}_4 & -\dot{x}_3 & \dot{x}_2 & \dot{x}_1 \end{pmatrix}.$$

Therefore, $|W_1| = (x_1^2 + x_2^2)^2 = \frac{l^2}{4} \neq 0$. The matrix A in this case is

$$\begin{pmatrix} \dfrac{2x_2\dot{x}_2}{x_1^2 + x_2^2} & -\dfrac{2x_2\dot{x}_1}{x_1^2 + x_2^2} & 0 & 0 \\[2mm] -\dfrac{2x_1\dot{x}_2}{x_1^2 + x_2^2} & \dfrac{2x_1\dot{x}_1}{x_1^2 + x_2^2} & 0 & 0 \\[2mm] \dfrac{x_1\dot{x}_3 + x_2\dot{x}_4}{x_1^2 + x_2^2} & \dfrac{x_1\dot{x}_4 - x_2\dot{x}_3}{x_1^2 + x_2^2} & \dfrac{x_1\dot{x}_1 + x_2\dot{x}_2}{x_1^2 + x_2^2} & \dfrac{x_2\dot{x}_1 - x_1\dot{x}_2}{x_1^2 + x_2^2} \\[2mm] \dfrac{x_1\dot{x}_4 - x_2\dot{x}_3}{x_1^2 + x_2^2} & \dfrac{x_1\dot{x}_3 - x_2\dot{x}_4}{x_1^2 + x_2^2} & \dfrac{x_2\dot{x}_1 - x_1\dot{x}_2}{x_1^2 + x_2^2} & \dfrac{x_1\dot{x}_1 + x_2\dot{x}_2}{x_1^2 + x_2^2} \end{pmatrix}.$$

Consequently, the differential equations (7.26) take the form

$$
\ddot{x}_1 = \left. \left(\frac{2x_2\dot{x}_1\dot{x}_2 - 2x_1\dot{x}_2^2 - x_1\dot{x}_3^2 - x_1\dot{x}_4^2}{x_1^2 + x_2^2} + x_1\dot{\lambda}_1 \right) \right|_{L_1 = L_2 = 0}
$$

$$
= x_1 \left(\dot{\lambda}_1 - \frac{2\dot{x}_1^2 + 2\dot{x}_2^2 + \dot{x}_3^2 + \dot{x}_4^2}{x_1^2 + x_2^2} \right),
$$

$$
\ddot{x}_2 = \left. -\left(\frac{-2x_1\dot{x}_1\dot{x}_2 + 2x_2\dot{x}_2^2 + x_2\dot{x}_3^2 + x_2\dot{x}_4^2}{x_1^2 + x_2^2} + x_2\dot{\lambda}_1 \right) \right|_{L_1 = L_2 = 0}
$$

$$
= x_2 \left(\dot{\lambda}_1 - \frac{2\dot{x}_1^2 + 2\dot{x}_2^2 + \dot{x}_3^2 + \dot{x}_4^2}{x_1^2 + x_2^2} \right),
$$

$$
\ddot{x}_3 = \left. \left(\frac{\dot{x}_3 \left(x_1\dot{x}_1 + x_2\dot{x}_2 \right) - \dot{x}_4 \left(x_2\dot{x}_1 - x_1\dot{x}_2 \right)}{x_1^2 + x_2^2} + x_1\dot{\lambda}_2 - g \right) \right|_{L_1 = L_2 = 0}
$$

$$
= \frac{\dot{x}_4 \left(x_2\dot{x}_1 - x_1\dot{x}_2 \right)}{x_1^2 + x_2^2} + x_1\dot{\lambda}_2 - g,
$$

$$
\ddot{x}_4 = \left. \left(\frac{\dot{x}_4 \left(x_1\dot{x}_1 + x_2\dot{x}_2 \right) - \dot{x}_3 \left(x_2\dot{x}_1 - x_1\dot{x}_2 \right)}{x_1^2 + x_2^2} - x_2\dot{\lambda}_2 \right) \right|_{L_1 = L_2 = 0}
$$

$$
= -\frac{\dot{x}_3 \left(x_2\dot{x}_1 - x_1\dot{x}_2 \right)}{x_1^2 + x_2^2} - x_2\dot{\lambda}_2. \tag{7.56}
$$

Differentiating the constraints, we obtain that the multipliers $\dot{\lambda}_1$ and $\dot{\lambda}_2$ are

$$
\dot{\lambda}_1 = \frac{\dot{x}_1^2 + \dot{x}_2^2 + \dot{x}_3^2 + \dot{x}_4^2}{x_1^2 + x_2^2} = \mu_1 + \frac{\dot{x}_3^2 + \dot{x}_4^2}{x_1^2 + x_2^2}, \quad \dot{\lambda}_2 = \frac{gx_1}{x_1^2 + x_2^2} = \mu_2 + \frac{\dot{x}_1\dot{x}_3 - \dot{x}_2\dot{x}_4}{x_1^2 + x_2^2}.
$$

Inserting these values into (7.56) yields

$$
\ddot{x}_1 = -\frac{x_1 \left(\dot{x}_1^2 + \dot{x}_2^2 \right)}{x_1^2 + x_2^2}, \qquad \ddot{x}_2 = \frac{x_2 \left(\dot{x}_1^2 + \dot{x}_2^2 \right)}{x_1^2 + x_2^2},
$$

$$
\ddot{x}_3 = -g + \frac{x_1 \left(\dot{x}_2\dot{x}_4 - \dot{x}_1\dot{x}_3 + gx_1 \right)}{x_1^2 + x_2^2}, \quad \ddot{x}_4 = -\frac{x_2 \left(\dot{x}_2\dot{x}_4 - \dot{x}_1\dot{x}_3 + gx_1 \right)}{x_1^2 + x_2^2}.
$$

These equations coincide with equations (7.54) everywhere because $|W_1| = \dfrac{l^2}{4}$ where l is the length of the rod.

The transpositional relations in this case are

$$
\delta\frac{dx_1}{dt} - \frac{d\delta x_1}{dt} = -\frac{2x_2}{x_1^2 + x_2^2} \left(\dot{x}_1\delta x_2 - \dot{x}_2\delta x_1 \right),
$$

$$
\delta\frac{dx_2}{dt} - \frac{d\delta x_2}{dt} = \frac{2x_1}{x_1^2 + x_2^2} \left(\dot{x}_1\delta x_2 - \dot{x}_2\delta x_1 \right),
$$

$$\delta \frac{dx_3}{dt} - \frac{d\delta x_3}{dt} = \frac{x_1}{x_1^2 + x_2^2} \left(\dot{x}_1 \delta x_3 - \dot{x}_3 \delta x_1 + \dot{x}_4 \delta x_2 - \dot{x}_2 \delta x_4 \right)$$

$$+ \frac{x_2}{x_1^2 + x_2^2} \left(\dot{x}_1 \delta x_4 - \dot{x}_4 \delta x_1 + \dot{x}_2 \delta x_3 - \dot{x}_3 \delta x_2 \right),$$

$$\delta \frac{dx_4}{dt} - \frac{d\delta x_4}{dt} = - \frac{x_2}{x_1^2 + x_2^2} \left(\dot{x}_1 \delta x_3 - \dot{x}_3 \delta x_1 + \dot{x}_4 \delta x_2 - \dot{x}_2 \delta x_4 \right)$$

$$+ \frac{x_1}{x_1^2 + x_2^2} \left(\dot{x}_1 \delta x_4 - \dot{x}_4 \delta x_1 + \dot{x}_2 \delta x_3 - \dot{x}_3 \delta x_2 \right).$$

This example shows again that the virtual variations produce non-zero transpositional relations.

Remark 7.10.3. From the previous example we observe that the virtual variations produce zero or non-zero transpositional relations, depending on the arbitrary functions which appear in the construction of the proposed mathematical model. Thus, the following question arises: Can the arbitrary functions L_j for $j = M + 1, \ldots, N$ be choosen in such a way that for the nonholonomic systems only the independent virtual variations will generate null transpositional relations?

We provide a positive answer to this question locally for any constrained Lagrangian systems, and globally for the *Chaplygin–Voronets mechanical systems*, and for the generalization of these systems studied in the next section.

7.11 MVM versus Voronets–Chaplygin systems

It was pointed out by Chaplygin [22] that in many conservative nonholonomic systems the generalized coordinates

$$(\mathbf{x}, \mathbf{y}) := (x_1, \ldots, x_{s_1}, y_1, \ldots, y_{s_2}), \quad s_1 + s_2 = N,$$

can be chosen in such a way that the Lagrangian function and the constraints take the simplest form. In particular, Voronets in [159] studied the constrained Lagrangian systems with Lagrangian $\tilde{L} = \tilde{L}(\mathbf{x}, \mathbf{y}, \dot{\mathbf{x}}, \dot{\mathbf{y}})$ and constraints

$$\dot{x}_j = \sum_{k=1}^{s_2} a_{jk}(t, \mathbf{x}, \mathbf{y}) \dot{y}_k + a_j(t, \mathbf{x}), \quad \text{for} \quad j = 1, \ldots, s_1. \tag{7.57}$$

Such systems are called *Voronets mechanical systems*.

We shall apply equations (7.26) to study the following generalization of the Voronets systems:

The constrained Lagrangian mechanical systems of the form

$$\left(Q, \quad \tilde{L}(t, \mathbf{x}, \mathbf{y}, \dot{\mathbf{x}}, \dot{\mathbf{y}}), \quad \{\dot{x}_\alpha - \Phi_\alpha(t, \mathbf{x}, \mathbf{y}, \dot{\mathbf{y}}) = 0, \quad \alpha = 1, \ldots, s_1\} \right), \tag{7.58}$$

are called *generalized Voronets mechanical systems*.

An example of generalized Voronets systems is the Appell–Hamel system analyzed in the previous section.

Corollary 7.11.1. *Every nonholonomic constrained Lagrangian mechanical system is locally a generalized Voronets mechanical system.*

Proof. Indeed, the independent constraints can be locally represented in the form (7.10). Thus upon introducing the coordinates

$$x_j = x_j, \quad \text{for} \quad j = 1, \ldots, M, \quad x_{M+k} = y_k, \quad \text{for} \quad k = 1, \ldots, N - M,$$

we see that any constrained Lagrangian mechanical system is locally a generalized Voronets mechanical system.   ⊏

Now we prove the following result

Theorem 7.11.2. *Under the assumptions of Theorem 7.7.2, assume that*

$$x_\alpha = x_\alpha, \quad x_\beta = y_\beta \quad \mathbf{x} = (x_1, \ldots, x_{s_1}) \quad \mathbf{y} = (y_1, \ldots, y_{s_2}),$$
$$L_\alpha = \dot{x}_\alpha - \Phi_\alpha(\mathbf{x}, \mathbf{y}, \dot{\mathbf{x}}, \dot{\mathbf{y}}) = 0, \quad L_\beta = \dot{y}_\beta,$$

for $\alpha = 1, \ldots, s_1 = M$ and $\beta = s_1 + 1, \ldots, s_1 + s_2 = N$.

Then $|W_1| = 1$ and the differential equations (7.26) take the form

$$E_j L_0 = \sum_{\alpha=1}^{s_1} \left(E_j L_\alpha \frac{\partial L_0}{\partial \dot{x}_\alpha} \right) + \dot{\lambda}_j, \qquad j = 1, \ldots, s_1,$$

$$E_k L_0 = \sum_{\alpha=1}^{s_1} \left(E_k L_\alpha \frac{\partial L_0}{\partial \dot{x}_\alpha} + \dot{\lambda}_\alpha \frac{\partial L_\alpha}{\partial \dot{y}_k} \right), \quad k = 1, \ldots, s_2. \tag{7.59}$$

or, equivalently (excluding the Lagrange multipliers)

$$E_k L_0 = \sum_{\alpha=1}^{s_1} \left(E_k L_\alpha \frac{\partial L_0}{\partial \dot{x}_\alpha} + \left(E_\alpha L_0 - \sum_{\beta=1}^{s_1} \left(E_\alpha L_\beta \frac{\partial L_0}{\partial \dot{x}_\beta} \right) \right) \frac{\partial L_\alpha}{\partial \dot{y}_k} \right), \tag{7.60}$$

for $k = 1, \ldots, s_2$. In particular, if we choose $L_0 = \tilde{L}(\mathbf{x}, \mathbf{y}, \dot{\mathbf{x}}, \dot{\mathbf{y}}) - \tilde{L}(\mathbf{x}, \mathbf{y}, \Phi, \dot{\mathbf{y}}) = \tilde{L} - L^*$, where $\Phi = (\Phi_1, \ldots, \Phi_{s_1})$, then (7.60) holds if

$$E_k \tilde{L} = \sum_{\alpha=1}^{s_1} E_\alpha \tilde{L} \frac{\partial L_\alpha}{\partial \dot{y}_k}, \quad k = 1, \ldots, s_2,$$

and

$$E_k(L^*) = \sum_{\alpha=1}^{s_1} \left(\frac{d}{dt} \left(\frac{\partial \Phi_\alpha}{\partial \dot{y}_k} \right) - \left(\frac{\partial \Phi_\alpha}{\partial y_k} + \sum_{\nu=1}^{s_1} \frac{\partial \Phi_\alpha}{\partial x_\nu} \frac{\partial \Phi_\nu}{\partial \dot{y}_k} \right) \right) \Psi_\alpha + \sum_{\nu=1}^{s_1} \frac{\partial L^*}{\partial x_\nu} \frac{\partial \Phi_\nu}{\partial \dot{y}_k}, \tag{7.61}$$

where $\Psi_\alpha = \left. \dfrac{\partial \tilde{L}}{\partial \dot{x}_\alpha} \right|_{\dot{x}_1 = \Phi_1, \ldots, \dot{x}_{s_1} = \Phi_{s_1}}$.

The transpositional relations (7.27) *in this case are*

$$\delta \frac{dx_\alpha}{dt} - \frac{d}{dt}\delta x_\alpha = \sum_{k=1}^{s_2} \left(\sum_{j=1}^{s_1} E_j(L_\alpha)\frac{\partial L_j}{\partial \dot{y}_k} + E_k(L_\alpha) \right) \delta y_k,$$

$$\delta \frac{dy_m}{dt} - \frac{d}{dt}\delta y_m = 0,$$

(7.62)

where $\alpha = 1,\ldots,s_1$ *and* $m = 1,\ldots,s_2$.

Proof. For simplicity we shall study only scleronomic generalized Voronets systems.

To determine equations (7.26), we suppose that

$$L_\alpha = \dot{x}_\alpha - \Phi_\alpha(\mathbf{x},\mathbf{y},\dot{\mathbf{y}}) = 0, \quad \alpha = 1,\ldots,s_1. \tag{7.63}$$

It is evident from the form of the constraint equations that the virtual variations $\delta \mathbf{y}$ are independent by definition. The remaining variations $\delta \mathbf{x}$ can be expressed in terms of them by the relations (Chetaev's conditions)

$$\delta x_\alpha - \sum_{j=1}^{s_2} \frac{\partial L_\alpha}{\partial \dot{y}_j}\delta y_j = 0, \quad \alpha = 1,\ldots,s_1. \tag{7.64}$$

Now let us apply Theorem 7.7.2. To construct the matrix W_1, we first determine $L_{s_1+1},\ldots,L_{s_1+s_2} = L_N$ as follows:

$$L_{s_1+j} = \dot{y}_j, \quad j = 1,\ldots,s_2.$$

Hence, the Lagrangian (7.19) becomes

$$L = L_0 - \sum_{j=1}^{s_1} \lambda_j \left(\dot{x}_\alpha - \Phi_\alpha(x,y,\dot{y}) \right) - \sum_{j=s_1+1}^{N} \lambda_j^0 \dot{y}_j$$

$$\simeq L_0 - \sum_{j=1}^{s_1} \lambda_j \left(\dot{x}_\alpha - \Phi_\alpha(x,y,\dot{y}) \right).$$

The matrices W_1 and W_1^{-1} are

$$\begin{pmatrix}
1 & \ldots & 0 & 0 & a_{11} & \ldots & a_{s_21} \\
0 & \ldots & 0 & 0 & a_{12} & \ldots & a_{s_22} \\
\vdots & \ldots & \vdots & \vdots & \vdots & \ldots & \vdots \\
0 & \ldots & \vdots & 1 & a_{1s_1} & \ldots & a_{s_2s_1} \\
0 & \ldots & 0 & 0 & 1 & \ldots & 0 \\
\vdots & \ldots & \vdots & \vdots & \vdots & \ldots & \vdots \\
0 & \ldots & 0 & 0 & 0 & \ldots & 1
\end{pmatrix},$$

and

$$\begin{pmatrix}
1 & \dots & 0 & 0 & -a_{11} & \dots & -a_{s_2 1} \\
0 & \dots & 0 & 0 & -a_{12} & \dots & -a_{s_2 2} \\
\vdots & \dots & \vdots & \vdots & \vdots & \dots & \vdots \\
0 & \dots & \vdots & 1 & a_{1 s_1} & \dots & -a_{s_2 s_1} \\
0 & \dots & 0 & 0 & 1 & \dots & 0 \\
\vdots & \dots & \vdots & \vdots & \vdots & \dots & \vdots \\
0 & \dots & 0 & 0 & 0 & \dots & 1
\end{pmatrix}, \tag{7.65}$$

respectively, where $a_{\alpha j} = \dfrac{\partial L_\alpha}{\partial \dot{y}_j}$, and the matrices Ω_1 and A are

$$A = \Omega_1 := \begin{pmatrix}
E_1(L_1) & \dots & E_{s_1}(L_1) & E_{s_1+1}(L_1) & \dots & E_N(L_1) \\
\vdots & \dots & \vdots & \dots & \dots & \vdots \\
E_1(L_{s_1}) & \dots & E_{s_1}(L_{s_1}) & E_{s_1+1}(L_{s_1}) & \dots & E_N(L_{s_1}) \\
0 & \dots & 0 & \dots & 0 & 0 \\
\vdots & \dots & \vdots & \dots & \dots & \vdots \\
0 & \dots & 0 & \dots & 0 & 0
\end{pmatrix}.$$

Consequently, the differential equations (7.26) take the form (7.59).

The transpositional relations (7.27) in view of (7.64) take the form (7.62). As we can observe from (7.62), the independent virtual variations $\delta \mathbf{y}$ for the system with the constraints (7.63) produce null transpositional relations. The fact that the transpositional relations are null follows automatically and it is not necessary to assume it a priori, and it is valid in general for the constraints which are nonlinear in the velocity variables.

We observe that the relations (7.12) in this case read

$$\delta \frac{dx_\alpha}{dt} - \frac{d}{dt}\delta x_\alpha + \sum_{m=1}^{s_2} \frac{\partial L_\alpha}{\partial \dot{y}_m}\left(\delta \frac{dy_m}{dt} - \frac{d}{dt}\delta y_m\right) = \sum_{k=1}^{s_1} E_k(L_\alpha)\delta x_k + \sum_{k=1}^{s_2} E_k(L_\alpha)\delta y_k$$

for $\alpha = 1, \dots, s_1$. Clearly, by (7.62) these relations hold identically.

From differential equations (7.59), eliminating the Lagrange multipliers we obtain the equations (7.60). After some computations we obtain

$$\frac{d}{dt}\left(\frac{\partial L_0}{\partial \dot{y}_k} - \sum_{\alpha=1}^{s_1} \frac{\partial L_\alpha}{\partial \dot{y}_k}\frac{\partial L_0}{\partial \dot{x}_\alpha}\right) - \left(\frac{\partial L_0}{\partial y_k} - \sum_{\alpha=1}^{s_1} \frac{\partial L_\alpha}{\partial y_k}\frac{\partial L_0}{\partial \dot{x}_\alpha}\right)$$
$$+ \sum_{\alpha=1}^{s_1}\left(\frac{\partial L_0}{\partial x_\alpha} - \sum_{\beta=1}^{s_1} \frac{\partial L_\beta}{\partial x_\alpha}\frac{\partial L_0}{\partial \dot{x}_\beta}\right)\frac{\partial L_\alpha}{\partial \dot{y}_k} = 0, \tag{7.66}$$

for $k = 1, \dots, s_2$.

By introducing the function $\Theta = L_0|_{L_1=\cdots=L_{s_1}=0}$, equations (7.66) can be written as

$$\frac{d}{dt}\left(\frac{\partial \Theta}{\partial \dot{y}_k}\right) - \left(\frac{\partial \Theta}{\partial y_k}\right) + \sum_{\alpha=1}^{s_1}\left(\frac{\partial \Theta}{\partial x_\alpha}\right)\frac{\partial L_\alpha}{\partial \dot{y}_k} = 0, \tag{7.67}$$

for $k = 1, \ldots, s_2$. Here we consider that $\dfrac{d}{dt}\left(\dfrac{\partial L_\beta}{\partial \dot{x}_\alpha}\right) = 0$, for α, $\beta = 1, \ldots, s_1$.

We shall study the case when equations (7.67) hold identically, i.e., $\Theta = 0$. We choose

$$L_0 = \tilde{L}(\mathbf{x}, \mathbf{y}, \dot{\mathbf{x}}, \dot{\mathbf{y}}) - \tilde{L}(\mathbf{x}, \mathbf{y}, \Phi, \dot{\mathbf{y}}) = \tilde{L} - L^*, \tag{7.68}$$

$\tilde{L}$ being the Lagrangian of (7.58). Now we establish the relations between equations (7.59) and the classical Voronets differential equations with the Lagrangian function $L^* = \tilde{L}|_{L_1=\cdots=L_{s_1}=0}$. The functions $\tilde{L}$ and L^* are determined in such a way that equations (7.60) hold due to the equalities

$$E_k\tilde{L} = \sum_{\alpha=1}^{s_1} E_\alpha \tilde{L}\frac{\partial L_\alpha}{\partial \dot{y}_k},$$

and

$$E_k L^* = -\sum_{\alpha=1}^{s_1}\left(-E_k(L_\alpha) + \sum_{\nu=1}^{s_1} E_\nu(L_\alpha)\frac{\partial L_\nu}{\partial \dot{y}_k}\right)\frac{\partial \tilde{L}}{\partial \dot{x}_\alpha} - \sum_{\nu=1}^{s_1} E_\nu(L^*)\frac{\partial L_\nu}{\partial \dot{y}_k},$$

for $k = 1, \ldots, s_2$, which in view of the equalities $(d/dt)(\partial L^*/\partial \dot{x}_\nu) = 0$ for $\nu = 1, \ldots, s_1$, take the form (7.61). $\square$

Proposition 7.11.3. *The differential equations* (7.61) *describe the motion of the nonholonomic system with the constraints* $L_\alpha = \dot{x}_\alpha - \Phi_\alpha(\mathbf{x}, \mathbf{y}, \dot{\mathbf{y}}) = 0$ *for* $\alpha = 1, \ldots, s_1$. *In particular, if the constraints are given by the formula* (7.57), *then systems* (7.61) *becomes*

$$E_k(L^*) = \sum_{\alpha=1}^{s_1}\left(\frac{da_{\alpha k}}{dt} - \left(\frac{\partial a_{\alpha m}}{\partial y_k} + \sum_{\nu=1}^{s_1}\frac{\partial a_{\alpha m}}{\partial x_\nu}a_{\nu k}\right)\dot{y}_m\right)\Psi_\alpha + \sum_{\nu=1}^{s_1}\frac{\partial L^*}{\partial x_\nu}a_{\nu k},$$

which are the classical Voronets differential equations. Consequently, the equations (7.61) *are an extension of the Voronets differential equations for the case when the constraints are nonlinear in the velocities.*

Proof. Equations (7.61) describe the motion of the constrained generalized Voronets system with Lagrangian L^* and constraints (7.63). The classical Voronets equations for scleronomic systems are easy to obtain from (7.61) if we take

$$\Phi_\alpha = \sum_{k=1}^{s_2} a_{\alpha k}(\mathbf{x}, \mathbf{y})\dot{y}_k. \qquad \square$$

Finally, Corollary 7.11.1 shows that the differential equations (7.61) describe locally the motions of any constrained Lagragian systems.

7.12   MVM versus Chaplygin systems

The constrained Lagrangian mechanical systems with Lagrangian $\tilde{L} = \tilde{L}(\mathbf{y}, \dot{\mathbf{x}}, \dot{\mathbf{y}})$ and constraints (4.50) is called the *Chaplygin mechanical systems*.

The constrained Lagrangian systems of the form

$$\left(Q, \quad \tilde{L}(\mathbf{y}, \dot{\mathbf{x}}, \dot{\mathbf{y}}), \qquad \{\dot{x}_\alpha - \Phi_\alpha(\mathbf{y}, \dot{\mathbf{y}}) = 0, \quad \alpha = 1, \ldots, s_1\} \right)$$

are called the *generalized Chaplygin systems*. Note that now the Lagrangian doe not depend on $\mathbf{x}$ and the constraints do not depend on $\mathbf{x}$ and $\dot{\mathbf{x}}$. So, the *generalize Chaplygin systems* are a particular case of the *generalized Voronets system*.

Proposition 7.12.1. *The differential equations* (7.61) *describe the motion of th constrained Lagrangian system with the constraints* $L_\alpha = \dot{x}_\alpha - \Phi_\alpha(\mathbf{y}, \dot{\mathbf{y}}) = 0$ *an Lagrangian* $L^* = L^*(\mathbf{y}, \dot{\mathbf{y}})$. *Under these assumptions equations* (7.61) *take th form*

$$E_k(L^*) = \sum_{\alpha=1}^{s_1} \left(\frac{d}{dt} \left(\frac{\partial \Phi_\alpha}{\partial \dot{y}_k} \right) - \frac{\partial \Phi_\alpha}{\partial y_k} \right) \Psi_\alpha. \tag{7.69}$$

In particular, if the constraints are given by the formulas

$$\dot{x}_\alpha = \sum_{k=1}^{s_2} a_{\alpha k}(\mathbf{y}) \dot{y}_k, \quad \alpha = 1, \ldots, s_1, \tag{7.70}$$

then the system (7.69) *becomes*

$$E_k L^* = \sum_{j=1}^{s_1} \sum_{r=1}^{s_2} \left(\frac{\partial \alpha_{jk}}{\partial y_r} - \frac{\partial \alpha_{jr}}{\partial y_k} \right) \dot{y}_r \Psi_j, \tag{7.71}$$

for $k = 1, \ldots, s_2$, *which is the system of equations which Chaplygin published i the Proceeding of the Society of the Friends of Natural Science in* 1897.

Consequently equations (7.69) *are an extension of the classical Chaplygi equations to the case when the constraints are nonlinear.*

Proof. To determine the differential equations which describe the behavior of th generalized Chaplygin systems we apply Theorem 7.7.2, with

$$L_0 = L_0(\mathbf{y}, \dot{\mathbf{x}}, \dot{\mathbf{y}}), \quad L_\alpha = \dot{x}_\alpha - \Phi_\alpha(\mathbf{y}\dot{\mathbf{y}}), \quad L_\beta = \dot{y}_\beta,$$

for $\alpha = 1, \ldots, s_1$ and $\beta = s_1 + 1, \ldots, s_2$.

Then the matrix W_1 is given by the formula (7.65) and

$$
A = \Omega_1 := \begin{pmatrix}
E_1(L_1) & \cdots & E_{s_1}(L_1) & E_{s_1+1}(L_1) & \cdots & E_N(L_1) \\
\vdots & \cdots & \vdots & \cdots & \cdots & \vdots \\
E_1(L_{s_1}) & \cdots & E_{s_1}(L_{s_1}) & E_{s_1+1}(L_{s_1}) & \cdots & E_N(L_{s_1}) \\
0 & \cdots & 0 & \cdots & 0 & 0 \\
\vdots & \cdots & \vdots & \cdots & \cdots & \vdots \\
0 & \cdots & 0 & \cdots & 0 & 0
\end{pmatrix}
$$

$$
= \begin{pmatrix}
0 & \cdots & 0 & E_{s_1+1}(L_1) & \cdots & E_N(L_1) \\
\vdots & \cdots & \vdots & \cdots & \cdots & \vdots \\
0 & \cdots & 0 & E_{s_1+1}(L_{s_1}) & \cdots & E_N(L_{s_1}) \\
0 & \cdots & 0 & \cdots & 0 & 0 \\
\vdots & \cdots & \vdots & \cdots & \cdots & \vdots \\
0 & \cdots & 0 & \cdots & 0 & 0
\end{pmatrix}. \tag{7.72}
$$

Therefore, the differential equations (7.26) take the form

$$
E_j L_0 = \frac{d}{dt}\left(\frac{\partial L_0}{\partial \dot{x}_\alpha}\right) = \dot{\lambda}_j \qquad j = 1, \ldots, s_1,
$$
$$
E_k L_0 = \sum_{\alpha=1}^{s_1}\left(E_k L_\alpha \frac{\partial L_0}{\partial \dot{x}_\alpha} + \dot{\lambda}_\alpha \frac{\partial L_\alpha}{\partial \dot{y}_k}\right) \qquad k = 1, \ldots, s_2. \tag{7.73}
$$

The transpositional relations are

$$
\delta\frac{dx_\alpha}{dt} - \frac{d}{dt}\delta x_\alpha = \sum_{k=1}^{s_2} E_k(L_\alpha)\delta y_k, \qquad \alpha = 1, \ldots, s_1,
$$
$$
\delta\frac{dy_m}{dt} - \frac{d}{dt}\delta y_m = 0, \qquad m = 1, \ldots, s_2. \tag{7.74}
$$

By excluding the Lagrange multipliers from (7.73) we obtain the equations

$$
E_k L_0 = \sum_{\alpha=1}^{s_1}\left(E_k(L_\alpha)\frac{\partial L_0}{\partial \dot{x}_\alpha} + \frac{d}{dt}\left(\frac{\partial L_0}{\partial \dot{x}_\alpha}\right)\frac{\partial L_\alpha}{\partial \dot{y}_k}\right),
$$

for $k = 1, \ldots, s_2$.

In this case equations (7.68) take the form

$$
\frac{d}{dt}\left(\frac{\partial \Theta}{\partial \dot{y}_k}\right) - \left(\frac{\partial \Theta}{\partial y_k}\right) = 0,
$$

Analogously to the Voronets case, we study the subcase when $\Theta = 0$. We choose $L_0 = \tilde{L}(\mathbf{y}, \dot{\mathbf{x}}, \dot{\mathbf{y}}) - \tilde{L}(\mathbf{y}, \Phi, \dot{\mathbf{y}}) := \tilde{L} - L^*$. We assume that the functions $\tilde{L}$ and L^*

are such that

$$E_k L^* = -\sum_{\alpha=1}^{s_1} E_k(L_\alpha)\frac{\partial \tilde{L}}{\partial \dot{x}_\alpha}\Psi_\alpha, \tag{7.75}$$

where $\Psi_\alpha = \left.\dfrac{\partial \tilde{L}}{\partial \dot{x}_\alpha}\right|_{L_1=\cdots=L_{s_1}=0}$ and

$$E_k(\tilde{L}) = \sum_{\alpha=1}^{s_1}\frac{d}{dt}\left(\frac{\partial \tilde{L}}{\partial \dot{x}_\alpha}\right)\frac{\partial L_\alpha}{\partial \dot{y}_k},$$

for $k = 1, \ldots, s_2$.

By inserting $\dot{x}_j = \displaystyle\sum_{k=1}^{s_2} a_{jk}(\mathbf{y})\dot{y}_k$, $j = 1,\ldots,s_1$, into equations (7.75) we obtain the system (7.71). Hence, the system (7.75) is an extension of the classical Chaplygin equations when the constraints are nonlinear.   ⬜

Using (7.20) and the Implicit Function Theorem, we can locally express the constraints (reordering coordinates if is necessary) as (7.10). We note that Propositions 7.11.3 and 7.12.1 are also valid for every constrained mechanical system with constraints locally given by (7.10). This follows from Theorem 7.11.2 changing the notations, see Corollary 7.11.1.

For the generalized Chaplygin systems the Lagrangian L takes the form

$$L = \tilde{L}(\mathbf{y},\dot{\mathbf{x}},\dot{\mathbf{y}}) - \tilde{L}(\mathbf{y},\Phi,\dot{\mathbf{y}}) - \sum_{j=1}^{s_1}\left(\frac{\partial L^*}{\partial \dot{x}_j} + C_j\right)(\dot{x}_j - \Phi_j(\mathbf{y},\dot{\mathbf{y}})) - \sum_{j=}^{s_2}\lambda_j^0\dot{y}_j, \tag{7.76}$$

for $j = 1,\ldots,s_1$, where the constants C_j for $j = 1,\ldots,s_1$ are arbitrary. Indeed, from (7.73) it follows that

$$\lambda_j = \frac{\partial L_0}{\partial \dot{x}_j} + C_j = \frac{\partial L^*}{\partial \dot{x}_j} + C_j.$$

By inserting in (7.19) $L_0 = \tilde{L} - L^*$ and these λ_j for $j = 1,\ldots,s_1$ we obtain the function L of (7.76).

We note that Voronets and Chaplygin equations with nonlinear constraints in the velocity were also obtained by Rumiantsev and Sumbatov (see [138, 146]).

Example 7.12.2. We shall illustrate the above results by the following example.

In Appel's and Hamel's investigations the following mechanical system was analyzed. A weight of mass m hangs on a thread which passes around pulleys and is wound round a drum of radius a. The drum is fixed to a wheel of radius b which rolls without sliding on a horizontal plane, touching it at the point B with the coordinates (x_B, y_B). The legs of the frame that support the pulleys and keep the plane of the wheel vertical slide on the horizontal plane without friction. Let θ be the angle between the plane of the wheel and the Ox axis, φ be the angle of rotation of the wheel in its own plane, and (x, y, z) the coordinates of the mass m. Clearly,

$$\dot{z} = b\dot{\varphi}, \quad b > 0.$$

The coordinates of the point B and the coordinates of the mass are related as (see page 223 of [120] for a picture)

$$x = x_B + \rho\cos\theta, \quad y = y_B + \rho\sin\theta.$$

The condition of rolling without sliding leads to the equations of nonholonomic constraints

$$\dot{x}_B = a\cos\theta\dot{\varphi}, \quad \dot{y}_B = a\sin\theta\dot{\varphi}, \quad b > 0.$$

We observe that the constraint $\dot{z} = b\dot{\varphi}$ admits the representation

$$\dot{z} = \frac{b}{a}\sqrt{\dot{x}^2 + \dot{y}^2 - \rho^2\dot{\theta}^2}.$$

Denoting by m_1, A, and C the mass and the moments of inertia of the wheel and neglecting the mass of the frame, we obtain the following expression for the Lagrangian function:

$$L = \frac{m + m_1}{2}\left(\dot{x}^2 + \dot{y}^2\right) + \frac{m}{2}\dot{z}^2 + m_1\rho\dot{\theta}\left(\sin\theta\dot{x} - \cos\theta\dot{y}\right) + \frac{A + m_1\rho^2}{2}\dot{\theta}^2 + \frac{C}{2}\dot{\varphi}^2 - mgz.$$

The equations of the constraints are

$$\dot{x} - a\cos\theta\dot{\varphi} + \rho\sin\theta\dot{\theta} = 0, \quad \dot{y} - a\sin\theta\dot{\varphi} - \rho\cos\theta\dot{\theta} = 0, \quad \dot{z} - b\dot{\varphi} = 0.$$

Let us study the motion of this constrained Lagrangian in the coordinates

$$x_1 = x, \quad x_2 = y, \quad x_3 = \dot{\varphi}, \quad y_1 = \theta, \quad y_2 = z,$$

i.e., study the nonholonomic system with the Lagrangian

$$\tilde{L} = \tilde{L}\left(y_1, y_2, \dot{x}_1, \dot{x}_2, \dot{x}_3, \dot{y}_1, \dot{y}_2\right)$$
$$= \frac{m + m_1}{2}\left(\dot{x}_1^2 + \dot{x}_2^2\right) + \frac{C}{2}\dot{x}_3^2 + \frac{J}{2}\dot{y}_1^2 + \frac{m}{2}\dot{y}_2^2$$
$$+ m_1\rho\dot{y}_1\left(\sin y_1\dot{x}_1 - \cos y_1\dot{x}_2\right) - \frac{mg}{b}y_2,$$

and with the constraints

$$l_1 = \dot{x}_1 - \frac{a}{b}\dot{y}_2\cos y_1 - \rho\dot{y}_1\sin y_1 = 0,$$
$$l_2 = \dot{x}_2 - \frac{a}{b}\dot{y}_2\sin y_1 + \rho\dot{y}_1\cos y_1 = 0,$$
$$l_3 = \dot{x}_3 - \frac{1}{b}\dot{y}_2 = 0.$$

Thus we are dealing with a classical Chaplygin system. To determine the dif
ferential equations (7.75) and the transpositional relations (7.74) we define the
functions

$$L^* = -\tilde{L}|_{l_1=l_2=l_3=0} = \frac{m(a^2+b^2)m + a^2 m_1 + C}{2b^2}\dot{y}_2^2 + \frac{m\rho^2+J}{2}\dot{y}_1^2 - \frac{mg}{b}y_2,$$

$$L_1 = l_1, \quad L_2 = l_2, \quad L_3 = l_3, \quad L_4 = \dot{y}_1, \quad L_5 = \dot{y}_2.$$

After some computations we obtain that the matrix A (see formulae (7.72)) in
this case becomes

$$A = \begin{pmatrix} 0 & 0 & 0 & -\frac{a}{b}\dot{y}_2 \sin y_1 & \frac{a}{b}\dot{y}_1 \sin y_1 \\ 0 & 0 & 0 & \frac{a}{b}\dot{y}_2 \cos y_1 & -\frac{a}{b}\dot{y}_1 \cos y_1 \\ 0 & 0 & 0 & 0 & 0 \\ 0 & 0 & 0 & 0 & 0 \\ 0 & 0 & 0 & 0 & 0 \end{pmatrix},$$

and so the differential equations (7.75) take the form

$$\left(m\rho^2 + J\right)\ddot{y}_1 + \frac{a\rho m}{b}\dot{y}_1\dot{y}_2 = 0,$$

$$\left((m+m_1)a^2 + mb^2\right)\ddot{y}_2 - mab\rho\dot{y}_1^2 = -mgb.$$

Assuming that $(m+2m_1)\rho^2 + J \neq 0$ and taking into account the existence of the
first integrals

$$C_2 = \dot{y}_1 \exp\left(-\frac{a\rho m y_2}{b(m\rho^2+J)}\right),$$

$$h = \frac{((m+m_1)a^2 + mb^2)}{2}\dot{y}_2^2 + \frac{b^2(m\rho^2+J)}{2}\dot{y}_1^2 + mgby_2,$$

we obtain, after the integration of these first integrals, that

$$\int \frac{\sqrt{(m+m_1)a^2 + mb^2}\,dy_2}{\sqrt{2h - 2mgby_2 - b^2(m\rho^2+J)C_3 \exp\left(\frac{a\rho m y_2}{bm\rho^2+J}\right)}} = t + C_1,$$

$$y_1(t) = C_3 + C_2 \int \exp\left(2\frac{a\rho m y_2(t)}{bm\rho^2+J}\right)dt.$$

Consequently, if $\rho = 0$, then

$$y_1 = C_3 + C_2 t, \qquad \int \frac{\sqrt{(m+m_1)a^2 + mb^2}\,dy_2}{\sqrt{2h - 2mgby_2 - JC_3}} = t + C_1.$$

Hamel in [71] neglected the mass of the wheel ($m_1 = J = C = 0$). Under this condition the previous equations become

$$\rho^2 \ddot{y}_1 + \frac{a\rho}{b} \dot{y}_1 \dot{y}_2 = 0,$$
$$(a^2 + b^2)\ddot{y}_2 - ab\rho \dot{y}_1^2 = -gb.$$

Appell and Hamel obtained their example of nonholonomic system with nonlinear constraints by means of the passage to the limit $\rho \to 0$. However, as a result of this limiting process, the order of the system of differential equations is reduced, i.e., the system becomes degenerate. In [120] the authors study the motion of the nondegenerate system for $\rho > 0$ and $\rho < 0$. From these studies it follows that the motion of the nondegenerate system ($\rho \neq 0$) and those of the degenerate system ($\rho \to 0$) differ essentially. Thus the Appell–Hamel example with nonlinear constraints is incorrect.

The transpositional relations (7.74) become

$$\delta \frac{dx_1}{dt} - \frac{d\delta x_1}{dt} = \frac{a}{b} \sin y_1 \left(\frac{dy_1}{dt} \delta y_2 - \frac{dy_2}{dt} \delta y_1 \right),$$
$$\delta \frac{dx_2}{dt} - \frac{d\delta x_2}{dt} = \frac{a}{b} \cos y_1 \left(\frac{dy_1}{dt} \delta y_2 - \frac{dy_2}{dt} \delta y_1 \right),$$
$$\delta \frac{dx_3}{dt} - \frac{d\delta x_3}{dt} = 0, \quad \delta \frac{dy_1}{dt} - \frac{d\delta y_1}{dt} = 0, \quad \delta \frac{dy_2}{dt} - \frac{d\delta y_2}{dt} = 0.$$

Clearly, these relations are independent of ϱ, A, C, and m_1.

7.13 General aspects of the MVM

Here we list a number of important aspects of the modificated vakonomic mechanics obtained from Theorems 7.7.2 and 7.7.3.

I) *Conjecture on the existence of nonlinear constraints*

We have

Conjecture 7.13.1. *The existence of mechanical systems with constraints nonlinear in the velocity must be investigated outside of Newtonian mechanics.*

Conjecture 7.13.1 is supported by the following facts.

(a) As a general rule, the constraints studied in classical mechanics are linear in the velocities. However Appell and Hamel in 1911, considered an artificial example with a constraint nonlinear in the velocity. As it follows from [120] (see Example 7.12.2) this constraint does not make sense in Newtonian mechanics.

(b) The idea developed for some authors (see for instance [11]) to construct a theory in Newtonian mechanics, by allowing that the field of force depends on

the acceleration, i.e., function of $\ddot{\mathbf{x}}$ as well as of the position $\mathbf{x}$, velocity $\dot{\mathbf{x}}$, and
the time t is inconsistent with one of the *fundamental postulates of the Newtonian
mechanics*: when two forces act simultaneously on a particle the effect is the same
as that of a single force equal to the resultant of both forces (for more detail
see [126] pages 11–12). Consequently *forces depending on acceleration* are not
admissible in Newtonian dynamics. This does not preclude their appearance in
electrodynamics, where this postulate does not hold.

(c) Let T be the kinetic energy of a constrained Lagrangian system. We
consider the generalization of the Newton law: *the acceleration* (see [151, 125])

$$\frac{d}{dt}\frac{\partial T}{\partial \dot{\mathbf{x}}} - \frac{\partial T}{\partial \mathbf{x}}$$

is equal to the force $\mathbf{F}$. Then in the differential equations (7.26) with $L_0 = T$ we
obtain that the force field $\mathbf{F}$ generated by the constraints is

$$\mathbf{F} = \left(W_1^{-1}\Omega_1\right)^T \frac{\partial T}{\partial \mathbf{x}} + W_1^T \frac{d}{dt}\lambda := \mathbf{F}_1 + \mathbf{F}_2.$$

The force field $\mathbf{F}_2 = W_1^T \dfrac{d}{dt}\lambda = (F_{21}, \dots, F_{2N})$ is called the *reaction force of the
constraints*. What is the meaning of the force

$$\mathbf{F}_1 = \left(W_1^{-1}\Omega_1\right)^T \frac{\partial T}{\partial \mathbf{x}} ?$$

If the constraints are nonlinear in the velocity, then $\mathbf{F}_1$ depends on $\ddot{\mathbf{x}}$. Con
sequently, in Newtonian mechanics such a force field does not exist. Therefore
the existence of nonlinear constraints in the velocity and the meaning of force $\mathbf{F}$
must be sought outside of the Newtonian model.

For example, for the Appel–Hamel constrained Lagrangian systems studied
in the previous section we have that

$$\mathbf{F}_1 = \left(-\frac{a^2\dot{x}}{\dot{x}^2 + \dot{y}^2}(\dot{x}\ddot{y} - \dot{y}\ddot{x}), \frac{a^2\dot{y}}{\dot{x}^2 + \dot{y}^2}(\dot{x}\ddot{y} - \dot{y}\ddot{x}), 0\right).$$

For the generalized Voronets systems and locally for any nonholonomic constrained
Lagrangian systems, from the equations (7.59) we obtain that the force field $\mathbf{F}$
has the components

$$\begin{aligned}
F_{k1} &= \sum_{\alpha=1}^{s_1} E_k L_\alpha \frac{\partial L_0}{\partial \dot{x}_\alpha}\\
&= \sum_{j=1}^{N}\sum_{\alpha=1}^{s_1}\left(\frac{\partial^2 L_\alpha}{\partial \dot{x}_k \dot{x}_j}\frac{\partial L_0}{\partial \dot{x}_\alpha}\ddot{x}_j + \frac{\partial^2 L_\alpha}{\partial \dot{x}_k \partial x_j}\frac{\partial L_0}{\partial \dot{x}_\alpha}\dot{x}_j\right) + \sum_{\alpha=1}^{s_1}\frac{\partial^2 L_\alpha}{\partial \dot{x}_k \partial t}\frac{\partial L_0}{\partial \dot{x}_\alpha},
\end{aligned}$$

for $k = 1, \ldots, N$, and $s_1 = M$. Consequently such a force field does not exist in Newtonian mechanics if the constraints are nonlinear in the velocity.

II) *Principle of determinacy in* MVM. Equations (7.26) can be rewritten in the form

$$G\ddot{\mathbf{x}} + \mathbf{f}(t, \mathbf{x}, \dot{\mathbf{x}}) = 0, \qquad (7.77)$$

where $G = G(t, \mathbf{x}, \dot{\mathbf{x}})$ is the matrix $(G_{j,k})$ given by

$$G_{jk} = \frac{\partial^2 L_0}{\partial \dot{x}_j \partial \dot{x}_k} - \sum_{n=1}^{N} \frac{\partial A_{nk}}{\partial \ddot{x}_j} \frac{\partial L_0}{\partial \dot{x}_n}, \quad j, k = 1, \ldots, N,$$

and $\mathbf{f}(t, \mathbf{x}, \dot{\mathbf{x}})$ is a suitable vector function. If $\det G \neq 0$, then equation (7.77) can be solved with respect to $\ddot{\mathbf{x}}$. This implies, in particular, that the motion of the mechanical system at time $\bar{t} \in [t_0, t_1]$ is uniquely determined, i.e., the *principle of determinacy* (see for instance [4]) holds for the mechanical systems with equation of motion given in (7.26).

In particular, for the Appel–Hamel constrained Lagrangian systems we have (see formula (7.40)) that

$$\mathbf{x} = (x, y, z)^T, \quad \mathbf{f} = \left(\frac{a\dot{x}}{\sqrt{\dot{x}^2 + \dot{y}^2}} \lambda, \; \frac{a\dot{y}}{\sqrt{\dot{x}^2 + \dot{y}^2}} \lambda, \; g - \dot{\lambda} \right)^T$$

$$G = \begin{pmatrix} 1 + \dfrac{a^2 \dot{y}^2}{\dot{x}^2 + \dot{y}^2} & -\dfrac{a^2 \dot{x}\dot{y}}{\dot{x}^2 + \dot{y}^2} & 0 \\ -\dfrac{a^2 \dot{x}\dot{y}}{\dot{x}^2 + \dot{y}^2} & 1 + \dfrac{a^2 \dot{x}^2}{\dot{x}^2 + \dot{y}^2} & 0 \\ 0 & 0 & 1 \end{pmatrix}, \quad |G| = 1 + a^2.$$

So for the Appel–Hamel system the principle of determinacy holds.

III) *New point of view on the transpositional relations.* The next result is the *third point of view on the transpositional relations.*

Corollary 7.13.2. *For the constrained mechanical systems, virtual variations can produce zero or non-zero transpositional relations. For the unconstrained mechanical systems, virtual variations always produce zero transpositional relations.*

Proof. From Theorems 7.7.2 and 7.7.3 (see formulas (7.27) and (7.31)) and from all examples which we gave in this text demonstrate that there are systems with zero transpositional relations and systems for which all transpositional relations are not zero. By contrasting the MVM with the Lagrangian mechanics, we see that for the unconstrained Lagrangian systems the transpositional relations are always zero. This establishes the corollary. □

Bibliography

[1] M. ABRAMOWITZ AND I. STEGUN, Handbook of Mathematical Functions with Formulas, Graphs, and Mathematical Tables, Dover, New York, 1965.

[2] V.M. ALEKSEEV, V.M. TIXOMIROV AND S.V. FOMIN Optimal Control, *Springer-Verlag*, 1987.

[3] P. APPELL, Exemple de mouvement d'un point assujeti à une liaison exprimée par une relation non linéaire entre les composantes de la vitesse, *Rend. Circ. Mat. Palermo* **32** (1911), 48–50

[4] V.I. ARNOLD, Dynamical Systems III, *Springer-Verlag*, 1996.

[5] V.I. ARNOLD, V.V. KOZLOV, AND A.I. NEISHTADT, Mathematical Aspects of Classical Mechanics, Dynamical systems III, *Springer*, Berlin 1998.

[6] J.C. ARTÉS, B. GRÜNBAUM, AND J. LLIBRE, On the number of invariant straight lines for polynomial differential systems, *Pacific J. of Math.* **184** (1998), 207–230.

[7] J.A. AZCÁRRAGA AND J.M. IZQUIERDO, n–ary algebras: a review with applications, *J. Phys. A: Math. Theor.* **43** (2010), 293001.

[8] A.B. BASSET, The Lemniscate of Gerono, An elementary treatise on cubic and quartic curves, *Deighton Bell* (1901), 171–172.

[9] N.N. BAUTIN AND V.A. LEONTOVICH Metody i Priemy Kachestvennogo Issledovaniya Dinamicheskija System na Ploskosti, *Ed. Nauka*, Moscow 1979.

[10] M.I. BERTRAND, Sur la possibilité de déduire d'une seule de lois de Kepler le principe de l'attraction, *Comptes Rendues* **9** (1877).

[11] G.D. BIRKHOFF, Dynamical Systems, *Amer. Math. Soc.*, New York, 1927.

[12] A.M. BLOCH, Nonholonomic Mechanics and Control, *Springer*, Berlin, 2003.

[13] A.M. BLOCH, P.S. KRISHNAPRASAD AND R.M. MURRAY, Nonholonomic mechanical systems with symmetry, *Arch. Rational Mech. Anal.* **136** (1996), 21–99.

[14] A.M. BLOCH, J.E. MARSDEN, AND D.V. ZENKOV, Nonholonomic dynamics, *Notices Amer. Math. Soc.* **52** (2005), 324–333.

[15] A.V. BORISOV AND I.S. MAMAEV, Dynamics of a Rigid Body, *Institut Kompyuternykh Issledvanyi*, 2005 (in Russian).

[16] A.V. BORISOV AND I.S. MAMAEV, Conservation laws, hierarchy of dynam
ics and explicit integration of nonholonomic systems, *Regul. Chaotic Dyn.*
13 (2008), 443–490.

[17] L. CAIRÓ, M. R. FEIX, AND J. LLIBRE, Integrability and algebraic solution
for planar polynomial differential systems with emphasis on the quadratic
systems, Resenhas da Universidade de São Paulo **4** (1999), 127–161.

[18] A. CAMPILO AND M.M. CARNICER, Proximity inequalities and bounds fo
the degree of invariant curves by foliation of $\mathbb{P}^2_{\mathbb{C}}$, *Trans. Amer. Math. Soc*
349 (1997), 2221–2228.

[19] F. CARDIN AND M. FAVRETTI, On nonholonomic and vakonomic dynamics
of mechanical systems with nonintegrable constraints, *J. Geom. Phys.* **18**
(1996), 295–325.

[20] M.M. CARNICER, The Poincaré problem in the nondicritical case, Annals
of Math. **140** (1994), 289–294.

[21] D. CERVEAU AND A. LINS NETO, Holomorphic foliations in $\mathbb{C} \times \mathbb{P}^2$ having
an invariant algebraic curve, Ann. Inst. Fourier (Grenoble) **41** (1991), 883
903.

[22] S.A. CHAPLYGIN, On the theory of motion of nonholonomic systems. The
orems on the reducing multiplier, *Mat. Sb.* **28** (1911), 303–314 (in Russian)
see Regul. Chaotic Dyn. **13** (2008), 269–376.

[23] C.L. CHARLIER, Celestial Mechanics (Die Mechanik des Himmels, Leipzig
Veit, 1902–1907), *Nauka*, 1966 (in Russian);

[24] J. CHAVARRIGA AND M. GRAU, A family of non-Darboux-integrable quad
ratic polynomial differential systems with algebraic solutions of arbitrarily
high degree, Appl. Math. Lett. **16** (2003), 833–837.

[25] J. CHAVARRIGA AND J. LLIBRE, Invariant algebraic curves and rational firs
integrals for planar polynomial vector fields, J. Differential Equations **16**
(2001), 1–16.

[26] J. CHAVARRIGA, J. LLIBRE AND J.M. OLLAGNIER, On a result of Darboux
LMS J. Comput. Math. **4** (2001), 197–210 (electronic).

[27] N.G. CHETAEV, Modification of the Gauss principle, *Prikl. Mat. i Mekh.* **5**
(1941), 11–12 (in Russian).

[28] C. CHRISTOPHER, Polynomial systems with invariant algebraic curves
preprint, Univ. of Wales, Aberystwyth, 1991.

[29] C. CHRISTOPHER, Invariant algebraic curves and conditions for a center
Proc. Roy. Soc. Edinburgh **124A** (1994), 1209–1229.

[30] C. CHRISTOPHER Liouvillian first integrals of second order polynomial dif
ferential equations. *Electron. J. Differential Equations* **7** (1999).

[31] C. CHRISTOPHER, Polynomial vector fields with prescribed algebraic limi
cycles, *Geom. Dedicata* **88** (2001), 255–258.

[32] C. CHRISTOPHER, J. LLIBRE, Integrability via invariant algebraic curves for planar polynomial differential system, *Ann. Differential Equations* **16** (2000), 5–19.

[33] C. CHRISTOPHER AND J. LLIBRE, A family of quadratic differential systems with invariant algebraic curves of arbitrarily higt degree without rational first integrals, *Proc. Amer. Math. Soc.* **130** (2001), 2025–2030.

[34] C. CHRISTOPHER,D J. LLIBRE, C. PANTAZI, AND X. ZHANG, Darboux integrability and invariant algebraic curves for planar polynomial systems, *J. Physics A: Math. and Gen.* **35** (2002), 2457–2476.

[35] C. CHRISTOPHER, J. LLIBRE, C. PANTAZI, AND S. WALCHER, Inverse problems for multiple invariant curves, *Proc. Roy. Soc. Edinburgh* **137** (2007), 1197–1226.

[36] C. CHRISTOPHER, J. LLIBRE, C. PANTAZI, AND S. WALCHER, Inverse problems for invariant algebraic curves: explicit computations, *Proc. Roy. Soc. Edinburgh* **139** (2009) 287–302.

[37] C. CHRISTOPHER, J. LLIBRE, C. PANTAZI, AND S. WALCHER, Darboux integrating factors: Inverse problems, *J. Differential Equations* **250** (2010), 1–25.

[38] C. CHRISTOPHER, J. LLIBRE, C. PANTAZI, AND S. WALCHER, Inverse problems in Darboux' theory of integrability, *Acta Appl. Math.* **120** (2012), 101–126.

[39] C. CHRISTOPHER, J. LLIBRE, AND J. PEREIRA, Multiplicity of invariant algebraic curves in polynomial vector fields, *Pacific J. Math.* **229** (2007), 63–117.

[40] T. COURANT AND A. WEINSTEIN, Beyond Poisson structures, in: *Action Hamiltoniennes de groupes. Troisième théorème de Lie* (Lyon, 1986), 39–49. Travaux en cours, **27**, Hermann, Paris, 1998.

[41] J.L. COOLIDGE, A Treatise on Algebraic Plane Curves, *Dover*, New York, 1959.

[42] U. DAINELLI, Sul movimento per una linea qualunque, *Giorn. Mat.* **18** (1880).

[43] G. DARBOUX, Memoire sur les equations differentielles alebriques du premier ordre et du premier degré, *Bulletin Sci. Math.*, 2éme série **2** (1878), 151–200.

[44] P.A.M. DIRAC, Generalized Hamiltonian dynamics, *Canadian J. Math.* **2** (1950), 129–148.

[45] P.A.M. DIRAC, Lectures on Quantum Mechanics, Belfer Graduate School of Science, N.Y., 1964; Academic Press Inc., New York, 1967.

[46] M.V. DOLOV AND P.B. KUZMIN, On limit cycles of system with a given partial integral, *Different. Uravneniya*, **30** (1994), 1125–1132.

[47] V. DRAGOVIĆ , B. GAJIĆ, AND B. JOVANOVIĆ, Generalizations of classical integrable nonholonomic rigid body systems, *J. Phys. A: Math. Gen.* **31** (1998), 9861–9869.

[48] G.N. DUBOSHIN, Nebesnaya Mehanika, *"Nauka"*, Moscow, 1968 (in Russian).

[49] F. DUMORTIER, J. LLIBRE AND J.C. ARTÉS, Qualitative Theory of Planar Differential Systems, Universitext, Springer, Berlin, 2006.

[50] L.P. EISENHART, A Treatise on the Differential Geometry of Curves and Surfaces, Dover, New York, 1909.

[51] V.P. ERMAKOV, Determination of the potential function from given partial integrals, *Math. Sbornik* **15**, serie 4 (1881) (in Russian).

[52] N.P. ERUGIN, Construction of the whole set of Systems of differential equations having a given integral curve, *Akad. Nauk SSSR. Prikl. Mat. Meh.* **16** (1952), 659–670 (in Russian).

[53] M. FAVRETTI, Equivalence of dynamics for nonholonomic systems with transverse constraints, *J. Dynam. Differential Equations* **10** (1998), 511–536.

[54] Y.N. FEDOROV AND B. JOVANOVIC, Nonholonomic LR systems as generalized Chaplygin systems with an invariant measure and flows on homogeneous space, *J. Nonlinear Sci.* **14**,(2004) 341–381.

[55] Y.N. FEDOROV, A. MACIEJEWSKI, AND M. PRZYBYLSKA, The Poisson equations in the nonholonomic Suslov problem: integrability, meromorphic and hypergeometric solutions, *J. Nonlinear Sci.* **22**, (2009) 2231–2259.

[56] N.M. FERRERS, Extension of Lagrange's equations. *Quart. J. Pure Appl. Math.* **12** (1872), 1–5.

[57] V.T. FILIPPOV, On the *n*-Lie algebra of Jacobians, *Sibirsk. Mat. Zh.*, **39** (1998), 660–669.

[58] W. FULTON, Algebraic Curves. An Introduction to Algebraic Geometry W.A. Benjamin Inc., New York, 1969.

[59] W. FULTON, Introduction to toric varieties, Annals of Mathematics Studies **131**, Princeton University Press, Princeton, NJ, 1993.

[60] A.S. GALIULLIN, Inverse Problems of Dynamics, *Mir Publishers*, Moscow 1984.

[61] A.S. GALIULLIN, I.A. MUKHAMETZYANOV, AND R.G. MUKHARLYAMOV Investigations into the analytical design of the programmed-motion systems *Vestn. Ross. Univ. Druzh. Nar. Ser. Prikl. Mat. Inf.*, **98** No. 1 (1994), 5–21

[62] F.R. GANTMACHER, Lektsii po Analitisheskoi Mekhanike, *Ed. Nauka* Moscow, 1966 (in Russian).

[63] F.R. GANTMACHER, Lectures in Analytical Machanics, *"MIR"*, Moscow 1975.

[64] A. GASULL, SHENG LI AND, J. LLIBRE, Chordal quadratic systems, Rocky Mountain J. Math. **16** (1986), 751–782.

[65] I.M. GELFAND AND S.V. FOMIN, Calculus of Variations, *Prentice-Hall, Englewood Cliffs, Inc., N.J.*, 1963.

[66] H. GIACOMINI, J. LLIBRE, AND M. VIANO, On the nonexistence, existence and uniqueness of limit cycles, Nonlinearity **9** (1996), 501–516.

[67] C. GODBILLON, Géometrie Différentielle et Mécanique Analytique, Collection Méthodes, Hermann, Paris, 1969.

[68] A. GORIELY, Integrability and Nonintegrability of Dynamical Systems, *Advanced Series in Nonlinear Dynamics* **19**, World Scientific Publishing Co., Inc., River Edge, NJ, 2001.

[69] X. GRACIA, J. MARIN–SOLANO, M. MUÑOZ–LECANDA, Some geometric aspects of variational calculus in constrained systems, Rep. Math. Phys. **51** (2003), 127–148.

[70] P.A. GRIFFITHS, Exterior Differential Systems and the Calculus of Variations, *Birkhäuser, Boston MA*, 1983.

[71] G. HAMEL, Teoretische Mechanik, Grundlagen der Mathematischen Wissenschaften, 57, *Springer-Verlag, Berlin–New York*, 1978.

[72] H. HERTZ, Die Prinzipien der Mechanik in neue Zusammenhänge dargestellt, *Ges. Werke, Leipzig*, Barth, 1910.

[73] D. HILBERT, Mathematische Probleme, Lecture, Second Internat. Congr. Math. Paris (1900), Nachr. Ges. Wiss. Göttingen Math. Phys. KL. (1900), 253–297; English transl. *Bull. Amer. Math. Soc.* **8** (1902), 437–479.

[74] O. HÖLDER, Ueber die Prinzipien von Hamilton und Maupertius, *Nachr. Königl. Ges. Wiss. Göttingen Math.–Phys. Kl.* (1896), 122–157.

[75] H. HOSHIMURA AND J. MARSDEN, Dirac strutures in Lagrangian mechanics. I. Implicit Lagrangian systems, *J. Geom. Phys.* **57** (2006), 133–156.

[76] H. HOSHIMURA AND J.MARSDEN, Dirac strutures in Lagrangian mechanics, II. Variational structures, *J. Geom. Phys.* **57** (2006), 209–250.

[77] J.P. JOANOLOU, *Equations de Pfaff Algebraiques*, Lectures Notes in Mathematics, **708**, Springer-Verlag, Berlin, 1979.

[78] N.E. JOUKOVSKY, Construction of the potential functions from a given family of trajectories (in Russian), Gostekhizdat, 1948.

[79] P.V. KHARLAMOV, A critique of some mathematical models of mechanical systems with differential constraints, *J. Appl. Math. Mech.* **56** (1992), 584–594 (in Russian).

[80] E.I. KHARLAMOVA-ZABELINA, Rapid rotation of a rigid body about a fixed point under the presence of a nonholonomic constraints, *Vestnik Moskovsk. Univ., Ser. Math. Mekh., Astron., Fiz. Khim.* **6** (1957), 25 (in Russian).

[81] V.I. KIRGETOV, Transpositional relations in mechanics, *J. Appl.Math* *Mech.* **22** (1958), 490–498.

[82] R. KOOIJ AND C. CHRISTOPHER, Algebraic invariant curves and the inte grability of polynomial systems, Appl. Math. Lett. **6** (1993), 51–53.

[83] D.J. KORTEWEG, Über eine ziemlich verbreitete unrichtige Behandlungs weise eines Problemes der rollenden Bewegung, *Nieuw Archiv voor Wiskund* **4** (1899), 130–155.

[84] V.V. KOZLOV, Theory of integration of equations of nonholonomic mecha nics, *Uspekhi mekh.* **8** (1985), 85–101.

[85] V.V. KOZLOV, Realization of nonintegrable constraints in classical mechan ics, *Dokl. Akad. Nauk SSSR* **272** (1983), 550–554 (in Russian).

[86] V.V. KOZLOV, Gauss principle and realization of the constraints, *Reg Chaotic Dyn.* **13** (2008), 431–434.

[87] V.V. KOZLOV, Dynamics of systems with non-integrable constraints I *Vestn. Mosk. Univ., Ser. I Mat. Mekh.* **3** (1982), 92–100 (in Russian).

[88] V.V. KOZLOV, Dynamics of systems with non-integrable constraints II *Vestn. Mosk. Univ., Ser. I Mat. Mekh.* **4** (1982), 70–76 (in Russian).

[89] V.V. KOZLOV, Dynamics of systems with non-integrable constraints III *Vestn. Mosk. Univ., Ser. 3 Mat. Mekh.* **3** (1983), 102–111 (in Russian).

[90] V.V. KOZLOV, Symmetries, Topology and Resonances in Hamiltonian Me chanics, *Springer-Verlag*, Berlin, 1995.

[91] V.V. KOZLOV, On the integration theory of the equations of nonholonomi mechanics, *Regul. Chaotic Dyn.* **7** (2) (2002), 161–172.

[92] V.V. KOZLOV, Dynamical Systems X, General Theory of Vortices, *Spriger Verlag*, 2003.

[93] I. KUPKA AND W.M. OLIVA The nonholonomic mechanics, *J. Differentie Equations* **169** (2001), 169–189.

[94] R. IBAÑEZ, M. DE LEÓN, J. MARREROS, AND E. PADRÓN, Leibniz alge broid associated with a Nambu–Poisson structure, *J. Phys. A: Math. Ger* **32** (1999), 8129–8144.

[95] A.D. LEWIS AND R.M. MURRAY, Variational principle for constrained me chanical systems: Theory and experiments, *Internat. J. Nonlinear Mech.* **3** (1995), 793–815.

[96] F. CANTRIJN, J. CORTÉS, AND M. DE LEON, On the geometry of gener alized Chaplygin systems, *Math. Proc. Cambridge Philos. Soc.* **132** (2002 323–351.

[97] W. LI, J. LLIBRE, X. ZHANG, Planar vector fields with generalized rationa first integrals, Bull. Sci. Math. **125** (2001), 341–361.

[98] J. LLIBRE, Integrability of polynomial differential systems, in: Handbook c Differential Equations, Ordinary Differential Equations, *Eds. A. Cañada, I Drabek and A. Fonda, Elsevier* (2004), pp. 437–533.

[99] J. LLIBRE, Open problems on the algebraic limit cycles of planar polynomial vector fields, *Bul. Acad. Stiinte Repub. Mol. Mat.* **1** (56) (2008), 19–26.

[100] J. LLIBRE, On the integrability of the differential systems in dimension two and of the polynomial differential systems in arbitrary dimension, *J. Appl. Anal. Comp.* **1** (2011), 33–52.

[101] J. LLIBRE AND C. PANTAZI, Polynomial differential systems having a given Darbouxian first integral, *Bull. Sci. Math.* **128** (2004), 775–788.

[102] J. LLIBRE, C. PANTAZI, AND S. WALCHER, First integrals of local analytic differential systems, *Bull. Sci. Math.* **136** (2012), 342–359.

[103] J. LLIBRE, J.S. PÉREZ DEL RÍO, AND J.A. RODRÍGUEZ, Phase portraits of a new class of integrable quadratic vector fields, *Dynam. Contin. Discrete Impuls. Systems* **7** (2000), 595–616.

[104] J. LLIBRE, R. RAMÍREZ AND N. SADOVSKAIA, On the 16th Hilbert problem for algebraic limit cycles, *J. Differential Equations* **248** (2010), 1401–1409.

[105] J. LLIBRE, R. RAMÍREZ, AND N. SADOVSKAIA, On the 16th Hilbert problem for limit cycles on non-singular algebric curves, *J. Differential Equations* **250** (2011), 983–999.

[106] J. LLIBRE, R. RAMÍREZ, AND N. SADOVSKAIA, Integrability of the constrained rigid body, *Nonlinear Dynam.*, **73** (2013), 2273–2290.

[107] J. LLIBRE, R. RAMÍREZ, AND N. SADOVSKAIA, Inverse approach in ordinary differential equations: applications to Langrangian and Hamiltonian mechanics, *J. Dynam. Differential Equations* **26** (2014), 529–581.

[108] J. LLIBRE, R. RAMÍREZ, AND N. SADOVSKAIA, Planar vector fields with a given set of orbits, *J. Dynam. Differential Equations* **23** (2011), 885–902.

[109] J. LLIBRE AND G. RODRÍGUEZ, Configurations of limit cycles and planar polynomial vector fields, J. Differential Equations **198** (2004), 374–380.

[110] J. LLIBRE, S. WALCHER AND X. ZHANG, Local Darboux first integrals of analytic differential systems, to appear *Bull. Sci. Math.* **38** (2014), 71–88.

[111] J. LLIBRE AND Y. ZHAO, Algebraic limit cycles in polynomial systems of differential equations, *J. Phys. A: Math. Theor.* **40** (2007), 14207–14222.

[112] J. LLIBRE AND X. ZHANG, Darboux theory of integrability in $\mathbb{C}^n$ taking into account the multiplicity, *J. Differential Equations* **246** (2009), 541–551.

[113] A.I. LURIE, Analytical Mechanics, Translated from the Russian, Springer-Verlag, Berlin, 2002.

[114] A.J. MACIEJEWSKI AND M. PRZYBYLSKA, Non-integrability of the Suslov problem, *Regul. Chaotic Dyn.* **7** (1) (2002), 73–80.

[115] C.M. MARLE, Various approaches to conservative and nonconservative nonholonomic systems, *Rep. Math. Physics* **42** (1998), 211–229.

[116] J.M. MARUSKIN, A.M. BLOCH, J.E. MARSDEN, AND D.V. ZENKOV, A fiber bundle approach to the transpositional relations in nonholonomic mechanics, *J. Nonlinear Sci.* **22** (2012), 431–461.

[117] J. MOSER, Various aspects of integrable Hamiltonian systems, in: Dynami-
       cal Systems, C.I.M.E. Lectures, Bressanone 1978, Birkhäuser, Boston, 1980
       233–290.

[118] J. MOULIN OLLAGNIER, About a conjecture on quadratic vector fields, Jour
       Pure and Appl. Algebra **165** (2001), 227–234.

[119] Y. NAMBU, Generalized Hamiltonian dynamics, *Phys. Rev. D* **7** (1973)
       2405–2412.

[120] JU.I. NEIMARK AND N.A. FUFAEV, Dynamics of Nonholonomic Systems
       Translations of Math. Monographs, *American Mathematical Society, Rhode
       Island*, 1972.

[121] N.N. NEKHOROSHEV, Variables "action-angle" and their generalizations
       Trudy Mosk. Mat. Obshch. **26** (1972), 181–198 (in Russian).

[122] I. NEWTON, Philosophie Naturalis Principia Mathematica, London, 1687.

[123] V.S. NOVOSELOV, Example of a nonlinear nonholonomic constraints that
       is not of the type of N.G. Chetaev, *Vestnik Leningrad Univ.*, **12** (1957) (in
       Russian).

[124] G.G. OKUNEVA, Motion of a rigid body with a fixed point under the ac
       tion of a nonholomorphic constraint in a Newtonian force field, *Mekhanika
       Tverdogo Tela* **18** (1986), 40 (in Russian).

[125] W. MUNIZ OLIVA, Geometric Mechanics, Lecture Notes in Math. 1798
       Springer-Verlag, Berlin 2002.

[126] L.A. PARS, A Treatise on Analytical Dynamics, *John Wiley & Sons, New
       York*, 1965.

[127] L.S. POLAK, Variational Principles of Mechanics and Their Development
       and Applications to Physics, *Ed. Fisico-matematicheskoi literature*, 1960 (in
       Russian).

[128] H. POINCARÉ, Sur l'intégration des équations différentielles du premier ordre
       et du premier degré I and II, *Rendiconti del Circolo Matematico di Palermo*
       5 (1891), 161–191; **11** (1897), 193–239.

[129] H. POINCARÉ, Hertz's ideas in mechanics, in addition to H. Hertz, Die
       Prizipien der Mechanik in neuem Zusammemhange dargestellt, 1894.

[130] V.V. PRASOLOV AND O.V. SHVARTSMAN, The Alphabet of Riemann Sur-
       faces, *Fazis* Moscow, 1999 (in Russian).

[131] R. RAMÍREZ, Dynamics of nonholonomic systems, *Publisher VINITI* **387**
       (1985) (in Russian).

[132] R. RAMÍREZ AND N. SADOVSKAIA, Inverse problems in celestial mechanics
       Atti. Sem. Mat. Fis. Univ. Modena Reggio Emilia **52** (2004), 47–68.

[133] R. RAMIREZ AND N. SADOVSKAIA, On the dynamics of nonholonomic sys
       tems, *Rep. Math. Phys.* **60** (2007), 427–451.

134] R. Ramírez and N. Sadovskaia, Inverse approach into the study of ordinary differential equations, preprint Universitat Rovira i Virgili (2008), 1–49.

135] R. Ramirez and N. Sadovskaia, Cartesian approach for constrained mechanical systems, *Arxiv*:1011.3251v1 [math.DS], (2010).

136] R. Ramirez, R. and N. Sadovskaia, Differential equations having a complete set of independent first integrals, *Arxiv*, (2011).

137] V.N. Rubanovskii and V.A. Samsonov, Stability of Steady Motions, in Examples and Problems, *M.: Nauka* 1998 (in Russian).

138] V.V. Rumiansev, On Hamilton's principle for nonholomorphic systems, *J. Appl. Math. Mech.* **42** (1978), 387–399 (in Russian).

139] N. Sadovskaia, Inverse problems in the theory of ordinary differential equations, Ph. D. Thesis, Univ. Politécnica de Cataluña, 2002 (in Spanish).

140] N. Sadovskaia and R. Ramírez, Inverse approach to study the planar polynomial vector field with algebraic solutions, *J. Phys.* A **37** (2004), 3847–3868.

141] N. Sadovskaia and R. Ramírez, Polynomial planar vector field with two algebraic invariant curves, Preprint MA II-IR-00013 Universitat Politecnica de Catalunya (2004) (in Spanish).

142] N. Sadovskaia and R. Ramírez, Differential equations of first order with given set of partial integrals, Technical Report MA II-IR-99-00015 Universitat Politecnica de Catalunya, (1999) (in Spanish).

143] A.J. Van der Schaft Equations of motion for Hamiltonian systems with constraints, *J. Phys. A.: Math. Gen.*, **20**, (1987) 3271–3277.

144] M.F. Singer, Liouvillian first integrals of differential equations, Trans. Amer. Math. Soc. **333** (1992), 673–688.

145] S. Smale, Mathematical problems for the next century, Math. Intelligencer **20** (1998), 7–15.

146] A.S. Sumbatov, Nonholonomic systems, *Reg. Chaotic Dyn.* **7** (2002), 221–238.

147] G.K. Suslov, On a particular variant of d'Alembert principle, *Math. Sb.* **22** (1901), 687–691 (in Russian).

148] K. Sundermeyer, Constrained Dynamics, Lecture Notes in Physics 169, *Spriger-Verlag*, N.Y., 1982.

149] G.K. Suslov, Determination of the force function from given partial integrals, 1890, Doctoral Thesis, Kiev (in Russian).

150] G.K Suslov, Theoretical Mechanics, Gostekhizdat, Moscow–Leningrad 1946.

151] J.L. Synge, On the geometry of dynamics, Phil. Trans. Roy. Soc. London A **226** (1926), 31–106.

[152] V. SZEBEHELY, Open problems on the eve of the next millenium, *Celestia Mech.* **65** Dynam. Astronom. (1996/97), 205–211.

[153] L. TAKHTAJAN, On foundation of the generalized Nambu mechanics, *Comm Math. Phys.* **160** (1994), 295–315.

[154] A.M. VERSHIK AND L.D. FADDEEV, Differential geometry and Lagrangia mechanics with constraints, *Soviet Physics–Doklady* **17** (1972) (in Russian)

[155] L.E. VESELOVA, New cases of integrability of the equations of motion o a rigid body in the presence of a nonholomorphic constraint, in: Geometry differential equations and mechanics (Moscow, 1985), 64–68 Moskow Gor Univ., Moscow, 1986.

[156] A.P. VESELOV AND L.E. VESELOVA, Flows on Lie groups with nonholo nomic constraint and integrable non-Hamiltonian systems. *Funct. Anal Appl.* **20** (1986), 308–309.

[157] A. VIERKANDT, Über gleitende und rollende Bewegung, *Monatshefte de Math. Phys.* 3 (1982), 31–54.

[158] O. VIRO, From the sixteenth Hilbert problem to tropical geometry, Japa J. Math. **3** (2008), 185–214.

[159] P. VORONETS, On the equations of motion for nonholonomic systems, *Math Sb.* **22** (1901), 659–686 (in Russian).

[160] G. ZAMPIERI, Nonholonomic versus vakonomic dynamics, *J. Differentia Equations* **163** (2000), 335–347.

[161] D.V. ZENKOV, A.M. BLOCH, AND J.E. MARSDEN The energy-momentur method for the stability of non-holonomic systems, *Dynam. Stability System* **13** (1998) 123–165.

[162] X. ZHANG, The 16th Hilbert problem on algebraic limit cycles, *J. Differen tial Equations* **251** (2011), 1778–1789.

[163] H. ŻOŁADEK, The solution of the center-focus problem, preprint, Univ. o Warsaw, 1992.

[164] H. ŻOŁADEK, On algebraic solutions of algebraic Pfaff equations, *Studia Math.* **114** (1995), 117–126.

[165] G. WILSON, Hilbert's sixteenth problem, Topology **17** (1978), 53–74.

[166] E.T. WHITTAKER, A Treatise on the Analytic Dynamics of Particles and Rigid Bodies, *Cambridge University Press*, Cambridge, 1959.

Index

Printed in the United States
By Bookmasters